Urban Biodiversity and Equity

Urban Biodiversity and Equity

Justice-centered conservation in cities

EDITED BY

Max R. Lambert

&

Christopher J. Schell

Environmental Science, Policy, and Management, UC Berkeley, USA

OXFORD
UNIVERSITY PRESS

Great Clarendon Street, Oxford, OX2 6DP,
United Kingdom

Oxford University Press is a department of the University of Oxford.
It furthers the University's objective of excellence in research, scholarship,
and education by publishing worldwide. Oxford is a registered trade mark of
Oxford University Press in the UK and in certain other countries

Published in the United States of America by Oxford University Press
198 Madison Avenue, New York, NY 10016, United States of America

British Library Cataloguing in Publication Data
Data available

Library of Congress Control Number: 2023938516

ISBN 978–0–19–887727–1

ISBN 978–0–19–887728–8 (pbk.)

DOI: 10.1093/oso/9780198877271.001.0001

Printed and bound by
CPI Group (UK) Ltd, Croydon, CR0 4YY

'Tipping the Scales' by artist, Sonny 'Sundancer' Behan Painted in
2022 for Sea Walls: Artists for Oceans in Emeryville, CA

Note about the Cover Image

Sonny 'Sundancer' Behan is an acclaimed urban contemporary artist most renowned for his large-scale wildlife murals and intricately detailed oil paintings. Sonny's passion for using his creativity to highlight the need to protect and preserve the natural world has earned him the reputation of being an artist who is deeply engaged and driven to make an impact. This mural of a San Francisco garter snake [an endangered species] was painted in 2022 as part of the Emeryville edition of Sea Walls: Artists for Oceans, PangeaSeed Foundation's public art program that brings the oceans into streets around the world by creating large-scale public murals that address pressing environmental issues the oceans are facing.

Contents

Preface

This book arose from our understanding of scientific advances in urban ecology and conservation science, with an immediate acknowledgement that society's broad conservation deliverables were unobtainable without cities. Conservation in and with cities uniquely leverages society as place-based experts willing to engage nature in their backyards and understand the broader implications of doing so; but without centering justice and equity to include all peoples in that narrative, this newly emerging field would always be grossly incomplete. This book is therefore an attempt to genuinely embed environmental justice and equity both as outcomes from and the pathway towards biodiversity conservation.

The origins of this work in earnest began in the Fall of 2018. We began discussing the numerous ways that social inequities—particularly structural racism and classism—shaped urban development, nature, and biodiversity. What grew out of those conversations was a realization that it is impossible to comprehensively understand and manage urban biodiversity without first grappling with how social inequities and environmental injustices shape the urban landscapes that the species we care about inhabit. These discussions led to a synthesis we published in 2020 in the journal *Science* on "The ecological and evolutionary consequences of systemic racism in urban environments" with phenomenal colleagues Drs. Karen Dyson, Tracy Fuentes, Simone Des Roches, Nyeema C. Harris, Danica Miller, and Cleo Woelfle-Erskine (now Woelfle-Hazard).

We began organizing this book project later that year in Fall of 2020, less than a year into the Covid-19 pandemic that gripped the world. One outcome of the pandemic was that huge numbers of people around the world flocked to nature, not only in rural hinterlands but also in their backyards, in urban parks, and on city streets. Scientists were also quick to demonstrate that changes in human behaviors associated with the pandemic resulted in concomitant changes in the behavior of urban wildlife. As vehicle and pedestrian traffic, industry, and city noise ground to a halt, animals responded and people noticed. Society globally recognized that our cities and suburbs were flush with biodiversity to appreciate. Many people also recognized the complex roles that people played in shaping nature.

People across the globe but specifically in North America were also compelled to legitimately confront the role of White supremacy and structural racism in every facet of society. With respect to the pandemic itself, widespread anti-Asian hate, limited protective equipment, and delayed vaccine shipments to developing countries (especially in the Global South) further widened global health inequity gaps. The world also reeled from the murder of George Floyd, an unarmed Black man, who was killed by a White police officer in 2020. George Floyd's murder—and the unjust murder of other Black Americans, including Ahmaud Arbery and Breonna Taylor—sparked global protests in support of Black Lives and against systemic oppression. It appeared as though society was finally acknowledging the existence, terror, and pervasiveness of anti-Black violence.

On the same exact day of George Floyd's murder—May 25, 2020—Black birder Christian Cooper was watching birds in New York City's Central Park when a White woman called the police after he asked her to leash her dog. The woman had her dog unleashed in an area of the park that required dogs to be leashed for the safety of urban wildlife. This incident made headlines globally. Although the pandemic caused society to retreat into nature and appreciate biodiversity, it was clear that people of color do not share the same safe access to nature.

It became apparent that there was a need to comprehensively describe the remarkably tight associations between social justice, urban biodiversity, and conservation. To our knowledge, no such text exists.

There do exist outstanding texts that begin addressing the how and why of urban conservation (1,2). Moreover, there are international organizations and clearinghouses like the Nature of Cities and Urban Biodiversity Hub that have virtually published editorials and essays addressing justice and biodiversity themes. Even so, and as we discuss in the first chapter, conserving urban biodiversity has yet to fully blossom into fields of research or practice. Previous texts have also discussed the linkages among society, urban biodiversity, and design, occasionally addressing environmental justice or equity. However, it was not until the Fall of 2020 that discourses interrogating urban nature as governed by societal inequities became mainstream in research disciplines. We hope that our prior publication that year in *Science* contributed positive inertia to an emerging movement.

What our book therefore aims to accomplish is to update the philosophy and practice of conservation to better manage urban biodiversity in a just and equitable manner. Indeed, conservation in cities—and everywhere—cannot succeed without addressing justice and equity for society.

What this book is not is an urban ecology text per se. There are a host of urban ecological texts out there that cover urban ecosystems in general (3) or the urban ecology of certain taxa like bats (4) or birds (5). And this book is not a "how to" for setting camera trap surveys, conducting point counts, or dip netting for tadpoles in cities, although we highlight insights from these approaches. Most of the chapters, of course, draw on and synthesize information from the rich urban biodiversity research that has developed in recent decades and discuss examples that employ various ecological methods.

What this book is, however, is our best attempt to provide a roadmap for helping our cities and suburbs support thriving communities of people and biodiversity. We brought together a team of practitioners and researchers—in diverse biological, physical, and social science disciplines—to present the extraordinary breadth of science on urban biodiversity conservation and subsequently present rich examples for actually doing this work.

Conservation is for plants, animals, and all wild things—but it's fundamentally about people. Nowhere is that truer than in cities. Our own stories shaped how we came to understand biodiversity conservation in cities.

For instance, Max grew up in Phoenix, AZ and its surrounding suburbs. He watched throughout his childhood as desert and agricultural fields were converted into sprawling neighborhoods, strip malls, and industrial areas. Development is intense and expansive in the Phoenix Metro area. Yet there was a diversity of mammals, birds, reptiles, amphibians, and insects all over urban Phoenix. He just didn't recognize these species as "wildlife" because, at that time, dominant culture only told stories about wildlife in remote wild areas. To be considered wildlife meant an animal was removed from people. Interestingly, unbeknownst to him, during his childhood in 1997 Phoenix was founded as the Central Arizona–Phoenix Long-Term Ecological Research (CAP LTER) site, one of the founding programs on Earth for comprehensively studying urban ecology.[1] He received a classic education in wildlife conservation biology. But never in that schooling were urban areas taught as places to do conservation, just places to protect species from. As he developed as a researcher studying the impacts of urbanization on biodiversity, it became clear that cities could harbor an astonishing biodiversity, more than his education had taught him. At the same time, the social intersections with urban conservation were also becoming apparent. As an early example, he was invited to spend a few days with a local majority-minority and economically disadvantaged high school in New Haven, teaching about environmental science and showing students wildlife. As he and students were catching turtles

[1] CAP LTER was founded in 1997 to study the ecology and ecosystem processes of the expanding city of Phoenix, AZ and its suburbs (https://sustainability-innovation.asu.edu/caplter/about/).

near the school, he asked the students what wild animals they had seen. Not a single student raised their hand, none said they had ever seen wild animals. There certainly was not a lot of biodiversity to see at that school, but there was some. He asked them about the turtles they could see basking or the flock of feral monk parakeets flying around and squawking. Like his feelings growing up, many of these students expressed that these animals were not wildlife because they were in the city. Unlike the privilege Max had growing up to explore the hinterlands and see "nature," most of these students said they had never been outside of their city, especially not to look at wildlife. Even when cities do have biodiversity, we do not all have equal access to enjoy it.

Chris grew up in Pasadena, CA in the general Los Angeles area. He often walked the Angeles Crest trail in the National Forest hills north of his neighborhood, and was a junior herpetologist when corralling western fence lizards at his grandparents' home in Altadena, CA. His grandparents' home was nestled in the Altadena foothills, towering over Los Angeles County—which provided a tremendous view clear to downtown LA. However, that view in the 1980s–2000s was often obscured by the increasingly conspicuous layer of smog LA became so infamous for. Concurrently, many of the natural and concrete-laced rivers went from being annual streams to bone-dry neighborhood legends; and it just kept getting warmer. Fast-forward to his graduate work, and he was astonished that a coyote—those same species he saw on occasion back at home—walked into a Quizno's restaurant in downtown Chicago. Why did that animal choose to go into a populated restaurant? What urban features molded this coyote's behavior into securing sandwich-level glory? And how is it navigating all *this*? His fascination with behavioral ecology, human–wildlife interactions, and their far-reaching implications for how cities are structured/who makes those decisions, cracked open a panacea of abstract connections that were more credible than make-believe. Conserving water, remediating air pollution, and conserving species like these coyotes were more connected than we previously realized. Now as a father of two young children and partner to a public health official and cultural anthropologist, he is reminded that love of urban biodiversity is emblematic of a deep calling to protect it for family and future Black scientists.

We also recognize our positionality and how our privileges limit this work's reach. Because this is an edited volume, we necessarily invited contributors for each chapter from within the networks of people we knew. And because a throughline across chapters was the interplay of environmental equity and biodiversity, this further limited the network of practitioners and researchers we knew who were doing this work. Accordingly, this book is dominated by individuals who currently reside in North America. The lessons within each chapter should be malleable to cities throughout the world but the perspectives will be biased towards North American insights. We have challenged the authors of each chapter to try to draw on global examples but also acknowledge the constraints of their conclusions.

In the first chapter, we (Lambert and Schell) make the case for urban conservation's role in advancing the intertwined goals of our broader conservation and environmental justice disciplines. From there, this book is broadly organized into three sections focused on (1) urban nature's social fabric, (2) the innovative approaches to understanding and prioritizing equitable urban conservation, and (3) emergent urban planning and management frameworks for addressing societal and conservation goals. In the first section, Hoover and Scarlett lay out the history and hopeful future of urban environmental justice, Larson and Brown unravel the complex layers of human motivations that constrain or facilitate urban biodiversity, and Pejchar and Reed extend urban conservation from city boundaries to suburbs and exurbs, as well as illustrate approaches that center housing justice and biodiversity as priorities. This section ends with a rich case study by Guderyahn and Logalbo documenting how conservation organizations in Portland, OR are grappling with and trying to address inequities in their practices. In the second section, Locke et al. make an accounting of urban forests as habitats for diverse species but also as critical venues for how we understand and practice urban environmental justice. Perkins et al. synthesize rich information on and best practices for participatory science methods, their increasing use for urban nature, and how these can be tools to improve or perpetuate social inequities. This second section also includes work from Magle et al., Avilés-Rodríguez et al., Gupta et al., and Stanton et al. who discuss the ways that multicity researcher/practitioner networks help us develop generalizable urban conservation approaches, how cutting-edge genetic tools allow us to study and prioritize biodiversity conservation at the scale of individual buildings or neighborhood blocks, ways for using proxy or indicator species for urban conservation, and how the interplay between wildlife and human behaviors provides tools for promoting coexistence between people and biodiversity. In the third section, Byers et al. detail a One Health approach for human and wildlife health in cities, Spotswood et al. help us begin building a new toolbox for urban conservation, and Stanford et al. synthesize the key principles of urban landscape ecology that allow urban planning to incorporate biodiversity. We end with a chapter by us and colleagues (Schell et al.) on what current and future conservationists will need to consider when developing an ecologically resilient, just, and biodiverse city.

In the time it has taken to finish this project, our worlds and the whole world have transformed dramatically. When we started, Max was a Postdoctoral Fellow based out of UC Berkeley and Chris was an Assistant Professor at the University of Washington in Tacoma. Partway into the project, Max took a position as a senior research scientist with the Washington Department of Fish and Wildlife and Chris became faculty at UC Berkeley—we literally drove past each other on I-5 as Max moved from Berkeley to Olympia and Chris from Tacoma to Berkeley. The pandemic continues but society in many ways has shifted into a new normal. As we stopped typing, representatives from 188 governments met in Montreal at the United Nations Biodiversity Conference (COP 15) in December 2022, adopting the Kunming-Montreal Global Biodiversity Framework to tackle the biodiversity crisis, enhance global ecosystem restoration, and protect Indigenous rights. The goal is to protect 30% of the planet and 30% of degraded land by 2030 to halt and reverse biodiversity loss. How equitable urban conservation fits into this ambitious initiative is yet to be seen.

We hope that this book provides a starting place to reimagine a just and equitable conservation.

References

1. Adams CE Lindsey KJ. *Urban Wildlife Management*. Boca Raton, FL: CRC Press; 2009. 432 p.
2. McCleery RA Moorman CE Peterson MN. *Urban Wildlife Conservation: theory and practice*. New York: Springer, 2014. 417 p.
3. Niemala J Breuste JH Elmquiest T Gunternspergen G James P McIntyre NE. *Urban Ecology: patterns, processes, and applications*. Oxford: Oxford University Press, 2011. 392 p.
4. Moretto L Coleman JL Davy CM Fenton MB Korine C Patriquin KJ. *Urban Bats: biology, ecology, and human dimensions*. Cham, Switzerland: Springer, 2022. 190 p.
5. Lepczyk CA Warren PS. *Urban Bird Ecology and Conservation*. Berkeley: University of California Press, 2012. 344 p.
6. Boal CW Dykstra CR. *Urban Raptors: ecology and conservation of birds of prey in cities*. Washington, DC: Island Press, 2018. 320 p

List of Contributors

Lisa Angeloni Colorado State University, USA
Myla F. J. Aronson Rutgers University, USA
Kevin Avilés-Rodríguez Louis Calder Biological Field Station, Fordham University, USA
Rajeev Bacchu Urban Slender Loris Project, India
Micaela Bazo Second Nature Ecology and Design, USA
Erin Beller Google Inc., USA
Matthew Benjamin City of San Jose, USA
Sarah Benson-Amram University of British Columbia, Canada
Kesang Bhutia Ashoka Trust for Research in Ecology and Environment, India
Jeffrey A. G. Brown Natural Areas Conservancy, 1234 Fifth Avenue, New York, USA
Kaylee A. Byers Simon Fraser University, Canada
Simone Des Roches University of Washington, USA
Robert R. Dunn North Carolina State University, USA
Alexander Felson University of Melbourne, Australia
Mason Fidino Urban Wildlife Institute, Lincoln Park Zoo, USA
Travis Gallo University of Maryland, USA
J. Letitia Grenier San Francisco Estuary Institute, USA
Peter Groffman City University of New York and Cary Institute of Ecosystem Studies, USA
Robin Grossinger Second Nature Ecology and Design, USA
J. Morgan Grove Northern Research Station, USDA Forest Service, USA
Laura Guderyahn City of Portland Parks & Recreation, USA
Kaberi Kar Gupta Urban Slender Loris Project, India
Nyeema C. Harris Applied Wildlife Ecology (AWE) lab, Yale University, USA
Nicole Heller Carnegie Museum of Natural History, USA
Fushcia-Ann Hoover University of North Carolina, Charlotte, USA
Kimberly Hughes Louis Calder Biological Field Station, Fordham University, USA
Kelly Ikyanan San Francisco Estuary Institute, USA
Madhusudan Katti North Carolina State University, USA
Soumya Kori Ashoka Trust for Research in Ecology and Environment, India
Vidisha Kulkarni Center for Ecological Science, Indian Institute of Science, India
Harshitha C. Kumar Urban Slender Loris Project, India
Max R. Lambert Environmental Science, Policy, and Management, UC Berkeley, USA
Kelli L. Larson School of Geographical Sciences and Urban Planning and School of Sustainability, Arizona State University, USA
Elizabeth W. Lehrer Urban Wildlife Institute, Lincoln Park Zoo, USA
Dexter H. Locke Northern Research Station, USDA Forest Service, USA
Mary Logalbo West Multnomah Soil & Water Conservation District, USA

Seth Magle Urban Wildlife Institute, Lincoln Park Zoo, USA

Tobin Magle Research Computing Services, Northwestern University, USA

Jason Munshi-South Louis Calder Biological Field Station, Fordham University, USA

Maureen H. Murray Lincoln Park Zoo, USA

Joanne E. Nelson University of British Columbia, Canada

Lauren M. Nichols North Carolina State University, USA

Selena Pang San Francisco Estuary Institute, USA

Liba Pejchar Colorado State University, USA

Deja J. Perkins North Carolina State University, USA

Steward T. A. Pickett Cary Institute of Ecosystem Studies, USA

Hari Prakash J. Ramesh Urban Slender Loris Project, India

Sarah E. Reed Robert and Patricia Switzer Foundation, USA

Jonathan L. Richardson University of Richmond, USA

Rachel D. Scarlett Georgia State University, USA

Christopher J. Schell Environmental Science, Policy, and Management, UC Berkeley, USA

Erica Spotswood San Francisco Estuary Institute and Second Nature Ecology and Design, USA

Bronwen Stanford San Francisco Estuary Institute, USA

Lauren A. Stanton Environmental Science, Policy, and Management, UC Berkeley, USA

Lewis Stringer The Presidio Trust, USA

Arun P. Visweswaran Urban Slender Loris Project, India

Christine E. Wilkinson Environmental Science, Policy, and Management, UC Berkeley, USA

Jonathan Young The Presidio Trust, USA

Julie K. Young Utah State University, USA

CHAPTER 1

Cities as the Solution to the Biodiversity Crisis

Max R. Lambert and Christopher J. Schell[‡]

Introduction

Biodiversity is most certainly in crisis. In this new era aptly titled the Anthropocene, Earth has witnessed losses of nature and biodiversity at unprecedented rates—only rivaled by the previous five mass extinctions that have ravaged life on this planet. The 21st century's unsustainable use of natural capital and ecosystems by humans is dismantling biodiverse ecosystems en masse (1). This, while the Intergovernmental Panel for Climate Change (IPCC) produced its Sixth Assessment Report in 2022 warning that if we do not institute drastic changes to reduce humanity's carbon footprint, global climate will surpass the ominous 1.5°C threshold, leading to numerous catastrophic impacts for both nature and society (2,3). The frequency of severe weather events (e.g., heat waves, storms, drought) is accelerating, jeopardizing our efforts to conserve biodiversity. These dual crises of biodiversity and the climate are serving as the one-two punch to global ecosystem stability, that, if unresolved, may lead to the catastrophic destabilization of the very ecosystems that support and maintain humanity's health and well-being (4).

It is no surprise, then, that this mass loss of biodiversity has garnered considerable global attention in recent years. In the spring of 2019, the Intergovernmental Science-Policy Platform on Biodiversity and Ecosystem Services (IPBES) issued an comprehensive report suggesting that nearly one million plants and animals are threatened with extinction, with many likely to go extinct within a matter of decades (5). The proverbial alarm bells would inspire critical discourses on how to build comprehensive strategies to halt global biodiversity loss, culminating in the 15th meeting of the Conference of the Parties to the Convention on Biological Diversity (CBD COP 15, referred to hereon as COP 15). Held in 2021 and 2022 in Kunming, China and Montreal, Canada, respectively, COP 15 served as a proving ground to reimagine our relationships with nature and biodiversity. In December 2022, a summative report—now referred to as the Kunming-Montreal Global Biodiversity Framework (GBF)—saw signatories from over 190 member nations agree to 23 ambitious goals, some of which included conserving at least 30% of lands and water by 2030 (known colloquially as the 30 by 30 initiative), amplifying Indigenous knowledge and peoples in conservation discourses, and developing sustainable solutions to future development (6).

During the closing ceremony of COP 15, the Executive Director at the UN Environment Programme stated that these historic and bold recommendations represented "a first step in resetting our relationship with the natural world."[1] Likewise, the World Wildlife Federation's (WWF's) International chief Marc Lambertini made parallels to the 2015 Paris

[‡] Environmental Science Policy and Management, UC Berkeley, USA

[1] Quote from Inger Anderson at the closing plenary of COP 15, December 2022 (https://news.un.org/en/story/2022/12/1131837).

Max R. Lambert and Christopher J. Schell, *Cities as the Solution to the Biodiversity Crisis*. In: *Urban Biodiversity and Equity*. Edited by: Max R. Lambert & Christopher J. Schell, Oxford University Press. © Oxford University Press (2023). DOI: 10.1093/oso/9780198877271.003.0001

Agreement, stating that "Halting and reversing biodiversity loss by 2030 is the equivalent of [the global warming limit of] 1.5C."[2] However, the clock is ticking: as of the publication of this writing in 2023, we have less than seven years to achieve the stated deliverables before reaching the 2030 deadline. If past is prologue, then concern should be considerably high, as the globe fell short of the Aichi Biodiversity Targets previously established in 2010 (7). Researchers suggest we failed to meet the Aichi goals because they lacked a mechanism for accountability, allowing companies to agree to the targets in principle without mandating that those companies monitor or disclose their impacts on biodiversity (7). The lack of effective "teeth" to Aichi (and arguably, the current Kunming-Montreal directives) serves as a stark reminder that future efforts will require unconventional, malleable, and transformative strategies that avoid replicating the same futile outcomes.

The Kunming-Montreal GBF almost buried the lede on that unconventional solution: that *cities* are the key to achieving our biodiversity goals. Emerging urban growth projections underscore this point. Currently more than half (55% and climbing) of the 8 billion people residing on this planet live in cities.[3] That number is projected to increase to nearly 70% by 2050, with new urban land-cover projected to increase by 0.82–1.53 million km^2 (8). In the US alone, dense urban area coverage increased nearly 500% between 1950 and 2000, covering roughly 2% of the contiguous US. Global modeling efforts suggest that urban land cover may increase by up to 500% by 2100, covering nearly 2.4% of the Earth's land surface (9). Moreover, unmitigated urban expansion is projected to directly endanger 855 species, with critical habitat loss projected to impact over 30,000 species (8). These projections are concerning, as future development is expected to occur in key biodiversity hotspots (10). Coincidentally, the same conditions that enticed people to settle an area and develop cities are often also the geographical conditions that promote biodiversity, such as along rivers, deltas, and estuaries or adjacent to mountains (11,12). It is not hyperbolic to suggest that cities are situated at the literal and figurative frontlines of biodiversity conservation. Thus, neglecting biodiversity conservation in cities necessarily means torpedoing our chances of meeting recently established conservation directives (13).

Despite their pivotal role on this global conservation stage, implementing biodiversity conservation policies in cities will not be without significant challenges. Urban ecosystems have a stubborn—but unwarranted—reputation as inhospitable wastelands riddled with environmental disturbances and toxicants that significantly compromise conservation efforts (14). People living in cities may also become increasingly disconnected from the natural world (i.e., the extinction of experience) (15). The disconnect from nature is exacerbated by the inequitable distribution of environmental amenities and access to urban green spaces that persistently obstructs the participation of the most minoritized and disenfranchised peoples in cities (16). Inequities perpetuated by industry and economic incentives neglect impacted residents' basic right to a healthy environment, worsening the extinction of experience. It may seem, then, that urban biodiversity conservation is in an unenviable quagmire: successful conservation action will require support from those very impacted residents that may be progressively separated from nature (i.e., the pigeon paradox) (17), while conservationists must concurrently interrogate the social and economic injustices that aggravate environmental harms and sabotage collective progress on conserving species (18).

The traditional narratives around cities, however, are a misdirection. While urban development has inarguably transformed natural ecosystems worldwide, cities are neither uninhabitable nor void of conservation value (19). The untold subtext in these depictions is that most species residing in and near human settlements are a lost cause, and, by extension, so too are urban residents. However, recent reckonings across the broader conservation discipline have opened the door for critical discourses on how we perform conservation science in our own backyards (20). To that end, communities, grassroots organizations, and some local city governments have *already* been doing urban conservation (Figure 1.1), including by working on their own biodiversity conservation plans (19,21,22).

[2] Marc Lambertini's final press briefing during COP 15, December 2022 (https://www.carbonbrief.org/explainer-can-the-world-halt-and-reverse-biodiversity-loss-by-2030/).

[3] World Economic Forum report on global human population and consumption (https://www.weforum.org/agenda/2022/04/global-urbanization-material-consumption/).

Figure 1.1 Coyotes like these seen here in San Francisco are increasingly using cities across North America as habitat, living near places where people live, work, and play. Instead of removing coyotes, cities like San Francisco are working to coexist with biodiversity and use signage to promote awareness and limit human–wildlife conflict.
Photos courtesy of Stephen Riffle.

City officials have established biodiversity-centered strategies that utilize various conservation tools to tackle highly dynamic, social-ecological issues (21,22). Residents cultivate backyards and community gardens that both support habitat connectivity and provide space for social gathering and connection with urban nature (23). Further, a growing number of local governments have cocreated community programs that stress the need for balancing conservation action and societal needs (24,25). But these are not enough on their own without deep, meaningful, and coordinated investment in conservation within and across cities (13).

There is tremendous opportunity to radically transform conventional conservation methods to build a wholly novel agenda—one that seamlessly incorporates the needs of biodiversity and people. However, our traditional approaches are ill-equipped for meeting these goals, and decades of biodiversity research and policy interventions have not alleviated our current biodiversity crisis (26,27). If we are to rightfully acknowledge that cities and the people that reside in them hold the key to conservation success, then we must be willing to interrogate and unpack the societal inequities—from local to global scales—that disrupt our ability to conserve biodiversity.[4] Treating social inequity as an ecological issue unlocks a pathway for reconciliation and restorative justice that builds a better conservation practice. Such a process also shines a bright light on how existing programs are woefully under-resourced and infrequently recognized. Hence, we poits that the skeleton key to mobilize cities as biodiversity stewards will ultimately rest in our ability to embed environmental justice and social equity as the core mission of an adaptable conservation agenda (28,29).

Our objective is to highlight the dynamism and transdisciplinary nature of a growing discipline: urban biodiversity conservation. We first briefly outline the recent history of urban ecology and biodiversity research. We then detail urban biodiversity patterns and habitats, subsequently leading to a discourse on how blended conservation strategies can promote biodiversity in cities. Successively, we propose a multifunctional, social-ecological framework for urban conservation that represents an extension of the agenda articulated by urban ecologists on the "ecology *in*," "ecology *of*," and "ecology *for*" paradigm (30–32). Our philosophy emphasizes how humanity and biodiversity are inextricably linked, especially in urban areas. As such, policy interventions are compelled to consider solutions simultaneously for both people and biodiversity (26,33). In sum, biodiversity conservation more broadly cannot achieve its fullest potential without intentional and deep investment in urban conservation, and that work in cities is impossible without authentic commitment to equity and justice.

A (very) brief history of urban ecology

Urban ecology is an interdisciplinary field largely dedicated to understanding how ecological relationships across scales (i.e., organismal-, population-, community-, and ecosystem-level ecology) are both implicitly and explicitly governed by people (34). This is a strong deviation from ecological research and thought in the 20th century, as most biological scientists were convinced that legitimate ecological research could only be conducted in spaces where humans were nearly absent. Indeed, the late blossoming of this field can be attributed to the reluctance of ecologists to consider studying species where people were omnipresent (35,36). However, startling evidence in the 1960s of rising global temperatures due to human-released carbon emissions sounded the figurative alarm bells for scientists worldwide (35). It quickly became evident that our reach as a species was global, and we needed more sustained research on human–environment interactions to determine how to mitigate and halt future emissions.

In response to emerging climate science, the United Nations Educational, Scientific and Cultural Organization (UNESCO) established the Man and the Biosphere (MAB) program in the 1970s to specifically investigate how humans influenced the ecosystem-level processes of cities. Results from the

[4] Guderyahn's and Logalbo's chapter details how Portland's conservation groups are reckoning with the city's historical and contemporary environmental injustices.

study languished, leading to brief spurts of interest in the proceeding decades until the end of the 1990s (37). Then in the late 1990s, the US National Science Foundation (NSF) funded two interdisciplinary research groups as the first two long-term ecological research (LTER) programs principally based in urban ecosystems: the Baltimore Ecosystem Study (BES) and the Central Arizona-Phoenix (CAP) research program (starting in 2021, there is a third: Minneapolis-St. Paul, or MSP LTER). Both programs were unique in their approach to blending social and ecological data to understand the biology of cities, and subsequently produced foundational studies on urban–rural gradients, nutrient cycling, and energy flows (34). Those works accelerated the methodologies and advances that now constitute the field's foundation.

Urban ecology has grown rapidly over the past several decades from a discipline that largely applied standard ecological methods to cities (the "ecology *in* cities" approach) to one that recognized cities as functioning ecosystems where human and "natural" processes were deeply intertwined as quintessential social-ecological systems (the "ecology *of* cities" approach) (30,31). This recognition underscored humans as ecosystem engineers, and consequently elevated the importance of understanding human motivations and constraints surrounding urban biodiversity.[5] Notably, biologists were relatively late to this party: since the 1960s, social scientists have produced thorough and meticulous works that detailed the complexity of human–environment interactions in cities (35). By the time the NSF awarded the two urban LTER grants, sociologists, anthropologists, and political ecologists had already articulated the intricate and nuanced social-ecological processes that persist in cities (38). Now, those early principles are being resurrected to propel transdisciplinary research examining the biology of cities beyond the simplistic urban–rural gradient and coarse urban versus nonurban approaches.

The guiding paradigm of urban ecology would soon emanate from the established urban LTER hubs. Most notably, the ecology "*in*," "*of*," and "*for*" cities would serve as a foundational framework to establish the scale and target of urban ecological science. The "ecology *in*" paradigm primarily focused on using traditional ecological principles to study the habitats and organisms within cities. Comparatively, the expanded "ecology *of*" paradigm treated entire urban environments as the amalgamation of social, ecological, and infrastructural components to identify how society and ecology coalesce. Both the "ecology *in*" and "ecology *of*" paradigms, though unique in their approach to investigating the social-ecological dynamics of cities, were not equipped for application. Rather, the third paradigmatic expansion, "ecology *for*" cities, is meant to combine urban design and ecological sciences to advance urban sustainability and climate resiliency goals which inherently link social, environmental, and economic integrity. These paradigms now serve to scaffold approaches in the urban ecological discipline (30,31).

In parallel to these ecological movements, biologists started to gather evidence suggesting that species were evolving in and because of cities (39–41). The comingling of social, ecological, and evolutionary research in cities has produced new paradigms around urban evolutionary ecology and "socio-eco-evolutionary dynamics" that are beginning to untangle how human society structures the interplay between ecological and evolutionary processes in cities and how those feedbacks reciprocally influence human society and culture (42,43).

This brief accounting of urban ecology's development as a discipline represents only a fraction of the developmental narrative and we encourage urban researchers and practitioners to explore the remarkable and rapid transformations in this field.

The rise of urban biodiversity conservation

To understand the contemporary form of urban biodiversity conservation, it is necessary to articulate how conservation science recognized urban environments. In chronicling the history of conservation, Dr. Dorcetta Taylor highlights the irony that cities were both the catalyst for the conservation movement and also the venues for some of the first

[5] Larson's and Brown's chapter delves into the complexity of human decisions for urban nature and how these motivations guide urban conservation.

conservation actions. Yet the conservation movement quickly abandoned urban areas as lost causes in favor of "remote" nature (20). Undergirding the rise of conservation by urbanites was rampant social inequality that was intertwined with degraded urban environmental conditions. Early conservation advocates focused their efforts away from cities to remove themselves and "nature" from the social and environmental conditions they despised (20). Conservation itself was thus conceived from privilege and a philosophy of exclusionary practices. Dominant conservation approaches and philosophy still emphasize places where people are not—particularly nonurban areas—and seldom consider social needs as priorities (26,44).

In 1991, proceedings of nearly 300 pages from the "Wildlife Conservation in Metropolitan Environments" symposium were published (45). Much of this symposium was dedicated to "basic" urban ecology. However, these proceedings contain rich data and conversations about urban planning for biodiversity, urban refuges and national parks, community inclusion in managing biodiversity, and the interplay between urban conservation and policy, education, and economics. Discussion in this volume even addressed the tremendous value that urban conservation would provide for advancing social justice goals and providing access to nature for marginalized communities. What is particularly profound about this symposium is that the Deputy Director of the U.S. Fish and Wildlife Service—Bruce Blanchard—highlighted the interplay between urban conservation and advancing equity, stressing the essential importance of coordination between federal, state, and local governments, conservation groups, and others (45). Shortly after in 1994, *The Ecological City* was published as an edited volume that largely covered the social, legal, educational, and political imperatives of urban conservation (46). These two volumes represent a proactive and arguably prescient agenda for advancing urban conservation.

The three decades to follow would see an explosion of research documenting biodiversity patterns, assessing how urban infrastructural components shape those patterns, and positing management strategies for conserving more native wildlife. Two additional volumes—*Urban Wildlife Conservation* and *Urban Biodiversity: from research to practice*—would be published in 2014 and 2018, respectively, both of which sought to provide basic and applied recommendations for biodiversity management in cities. These volumes were complemented by a healthy and growing literature of empirical and conceptual works that interrogated the social-ecological factors impacting biodiversity. Many of these publications were squarely focused on the drivers of biodiversity patterns in cities (47–50), or reviews of conservation policies across cities (22), though rarely did those interventions directly address actions that enhanced urban biodiversity.

The academic zeitgeist in urban biodiversity conservation subsequently propelled parallel interest across other sectors. Organizations like The Nature Conservancy[6] and the World Economic Forum[7] have recently amplified the importance of cities to the biodiversity crisis, broadening the conversation to communities outside academia. Collective working groups like Cities for Biodiversity (C4B)[8] and the NATURA[9] project have convened researchers, practitioners, and urban planners to develop conservation roadmaps for synthesizing scientific evidence and applying it to local policy interventions. Despite these advancements in discourse, however, academic recommendations have seldom been put into practice. Translating this newer research from an academic exercise to actionable policy levers still requires substantial development that will need to be accelerated to meet the tremendous demands for urban biodiversity conservation (27). Moreover, there is still a critical need to build the justice and equity components of these emerging urban conservation programs (44).

[6] Editorial by The Nature Conservancy on cities as necessary for conservation (https://www.nature.org/en-us/newsroom/urban-expansion-impacts-for-biodiversity-planning-yale/).

[7] Story by the World Economic Forum on cities as hubs for biodiversity conservation (https://www.weforum.org/agenda/2021/06/cities-ecosystems-biodiversity-climate-change/).

[8] Global Platform for Sustainable Cities plan to conserve biodiversity (https://www.thegpsc.org/cities-biodiversity-c4b).

[9] The Global Roadmap for Urban Nature-Based Solutions (https://natura-net.org/nbs-global-roadmap).

Equitable urban conservation advances all conservation

At the heart of conservation discourses rests a quintessential, resonant truth: without a solid foundation built on environmental justice and societal equity, current and future conservation policy interventions are doomed to fail. This truth has become abundantly clear in recent decades, as both the Aichi and Kunming-Montreal proposals consistently stressed that biodiversity loss will have the most severe impacts on the world's poorest countries (7). These losses extend far beyond gross domestic product (GDP) and economic growth, infringing on the livelihoods, health, and well-being of the most marginalized and often neglected communities (51,52). Hence, despite counternarratives that attempt to separate conservation science from global equity, the two are inseparable entities that will dictate our success in halting biodiversity loss.

Adjusting our vantage point to illustrate the importance of biodiversity to society starts to unpack the need for merging conservation and justice movements. Biodiversity serves as the connective tissue that sustains ecosystem function and overall health, providing an extraordinary array of supportive (e.g., primary productivity and nutrient recycling), regulating (e.g., climate regulation and disease control), provisioning (e.g., energy and food resources), and cultural (e.g., recreation, education, and spiritual connection) ecosystem services that greatly benefit society (53). Species consequently serve as nodes in a vast biological network that, when intact, sustain life. The eradication of nodes (i.e., species extinctions) therefore compromises network integrity, weakening both environmental and societal resilience toward emerging crises. When the equity lens is applied, the inequitable calculus of global biodiversity loss becomes apparent: the poorest, most marginalized communities across neighborhoods, regions, and countries are systemically being denied access to sustaining ecosystem services (54).

Because cities currently contain more than 50% of the global human population and climbing, they serve as the nexus points for addressing biodiversity loss and social injustices. It is in these urban areas that grassroots movements for environmental and social justice became forces to restore access to nature and a healthy environment for minoritized and neglected communities. From protests in 1982 against the unjust deposition of toxic waste in a predominantly Black county of North Carolina, to the formation of the 17 foundational principles of the environmental justice (EJ) discipline in 1991, Indigenous and Black activists across the US recurrently decried how racism and classism underpinned numerous offenses of environmental degradation and dispossession (55,56). Activism and scholarship in the 1980s and 1990s documented how spatial segregation, unjust policies and zoning laws, and disinvestment in predominantly poor neighborhoods and communities of color impacted access to nature for the most minoritized, violating the overarching tenet of EJ: access to a healthy and safe environment is a fundamental right (55).

Fast-forward three decades later, and persistent evidence across cities underscores that injustices are built into the concrete and social fabric of cities, fundamentally controlling urban nature (57–59). In cities across the globe, socioeconomic wealth positively predicts green space size and health, as well as overall species diversity (60). This luxury effect emerges because individuals, households, and neighborhoods with more monetary (and often political) capital have a greater capacity to steward habitat and biodiversity, whereas poorer communities are denied investments that would support greater local biodiversity (60,61). Historical patterns of residential and spatial segregation also dictate biodiversity patterns, especially of primary producers that structure ecological communities within cities (62–64). Policies like redlining—the American government-supported Homeowners' Loan Corporation—mapped investment "riskiness" from 1933 to 1968 across more than 230 cities based on neighborhood race characteristics. Neighborhoods outlined in red (thus red-lined) were predominantly Black and perceived as the riskiest (59). Racial segregation policies informed the spatial distribution of (dis)amenities, directly associated with the reduced distribution, diversity, and size of urban trees[10] in

[10] Locke et al.'s chapter synthesizes the history of identifying inequities in the urban forest and how practitioners are addressing these injustices.

formerly red-lined neighborhoods (65–68). Echoes of these legacy effects continue to compromise environmental and human health today, as those living in previously redlined areas are more likely to develop cancer (69), have more frequent emergency room visits (70), and experience more air pollutants (71), intensified urban heat (72), and amplified densities of oil and gas wells (73). Consequently, if known disturbances like habitat degradation, pollution, heat, and infectious disease all cause compounding harms to biodiversity, then it becomes glaringly apparent that historical and contemporary injustices must be resolved to begin the arduous task of saving urban biodiversity (58).

The next steps are therefore abundantly clear: if we hope to conserve species in cities and beyond, reconciliation of past ills and intentional development of equitable policy interventions need to be the norm. Achieving that goal will require a transformational approach not just in how cities are planned and built, but also in conservationists' willingness to operate as scientist activists. Traditional conservationists and practitioners have been reluctant to infuse antiracist and decolonized principles into urban biodiversity conservation (28). Such a reluctance is not unexpected, as the Western (specifically North American) model of conservation is inherently exclusionary and exploitative (74–78). This model, founded on settler colonialist ideals, actively strips Indigenous persons and communities of their ancestral ties to the land while only allowing the privileged few to enjoy wilderness (77,78). This model also perpetuates harms on communities of color, denying various cultural experiences with nature because they do not conform to the White supremacist ideals that such practices were born from (76,79). The ghosts of these ideals are often echoed in urban planning, in which officials sponsor action that positively impacts tree planting and greening efforts, yet leads to increased property and rent values, displacing low-income residents and communities of color (16,29). Certainly, the next evolution of urban biodiversity conservation must be an active movement toward a liberated and decolonized approach that decenters myopic views of conservation and leverages cocreation and coproduction to legitimately achieve social equity (28).

Despite the grim past and uncertain future, there is reason for optimism. We are learning more about how beneficial urban biodiversity is, with a remarkable body of research emphasizing that access to green spaces and biodiversity is beneficial for human health and well-being, thus enhancing the restorative and psychological health benefits we receive (80–84). This means that people can perceive different amounts of biodiversity and those places with more biodiversity have increased benefits to human well-being. This also means that despite evidence identifying green spaces as a societal good, green spaces with minimal biodiverse flora and fauna are not nearly as impactful in benefiting human well-being. Harnessing these data can be a launchpad for conversations about access, biodiversity, and health,[11] spotlighting environmental equity policies as unified with conservation efforts. Such efforts also situate urbanites as local experts with experiential knowledge on how to address EJ issues and conservation.

In her *Scientific American* article "Cities Build Better Biologists,"[12] Yale mammal ecologist Dr. Nyeema C. Harris reflects on her experiences growing up in Philadelphia:

> Nature is no longer only pristine wilderness; it includes sounds of human laughter, trash trucks, and sirens. We urbanites are gritty, resourceful, and imaginative. We need more capacity, more participation, more energy, and more innovation in science to create solutions to combat environmental degradation and halt biodiversity loss. Identifying this talent across cities presents an easy remedy.

Harris articulates how biologists who grow up in cities have unique experiences and ways of understanding nature that differ from and complement biologists who grew up in nonurban communities. These unique experiences can translate into innovative approaches to conserving biodiversity. Accordingly, an equity-centered approach to urban conservation provides direct benefits to

[11] Byers et al.'s chapter takes a One Health approach to understand the interconnectedness of human, wildlife, and environmental health and how to address these challenges.

[12] Article by Nyeema C. Harris in *Scientific American* on being a biologist who grew up in Philadelphia (https://www.scientificamerican.com/article/cities-build-better-biologists/).

biodiversity in cities and to surrounding nonurban regions, generates equitable benefits across diverse communities, and can also develop a larger, more diverse group of conservationists who will advance new ideas and approaches to conserve biodiversity everywhere. Although there is an urgent need for equitable urban conservation, marginalized communities throughout the US and across the world have long been at the forefront of urban environmental stewardship.[13]

Biodiversity patterns and habitat considerations in the city

Urbanization often uniquely favors species guilds or traits that result in a distinctive assemblage of native, imperiled, and introduced species, referred to as no-analog communities (39).[14] These no-analog communities represent assemblages of species that have never existed before in space or time. As such, urbanization produces novel communities of species that require careful and, often, different management considerations. The value of conceptualizing urban biodiversity as no-analog communities is to reorient our misconceived perceptions of cities as biologically impoverished to a refined and nuanced framework that values the remarkable uniqueness of urban biodiversity. Urban conservation is intrinsically different from traditional conservation approaches in that the goal is to maintain a diversity of species in an arguably different habitat context than its prior nonurban situation. Consequently, the overarching goal of urban conservation is not to conserve to a prior baseline condition but, rather, to work towards a new equilibrium. From this vantage point, it becomes evident that conserving urban biodiversity requires a diverse multitude of perspectives, experiences, and approaches to authentically articulate the beautiful complexity of the urban environment.

[13] Hoover's and Scarlett's chapter traces the history of environmental justice and urban conservation work, underscoring that marginalized communities are consistently leading this work.

[14] Magle et al.'s chapter illustrates how standardized comparisons across multiple cities can teach us what aspects of urbanization favor certain species in general and how an individual city's conditions may favor one species better than another city does.

Though it is true that some species will simply never make it in cities, urban areas are far from devoid of biodiversity. In fact, cities and suburbs[15] play a pivotal role in sustaining a diverse array of threatened and endangered species (14,19,50,85). One global study found that nearly one-third of cities contained bird species considered to be threatened with extinction on the International Union for Conservation of Nature's (IUCN's) Red List and nearly 10% of cities contained threatened plants (50). However, these are almost certainly underestimates given the IUCN often does not fully account for species on national or regional lists. Case in point: nearly one-quarter of all endangered plants occur within the 40 largest cities (12).

Of course, urban areas also harbor a variety of introduced species including human commensals (e.g., Norway rats, feral pigeons, and house sparrows), non-native, and invasive species. Many plants in urban areas are cultivated species that are purposefully planted or invasive species that proliferate on their own, although the extent of these and other non-native species varies geographically (86–88). Human commensals are found in cities throughout the world, often restricted to areas with dense human development where they depend on human-constructed resource subsidies (e.g., habitat conditions and human-provisioned foods) (89,90). Cities can also be vectors of transport and dispersal for non-native species, frequently serving as the first arrival points for new invasive species that then spread into nonurban landscapes from cities (91,92). In a few cases, non-native species that are threatened or endangered in their native ranges are thriving and doing well in cities where they have been introduced outside their native ranges (11). Examples include the green and gold bellfrog (*Litoria aurea*) and the red-crowned amazon (*Amazona viridigenalis*). These examples notwithstanding, urban biodiversity is overwhelmingly comprised of native species—including threatened and endangered species—and generally reflects the regional pool of native biodiversity (50).

[15] Pejchar et al.'s chapter discusses the importance of suburban ecosystems in sustaining biodiversity and advancing equity.

What constitutes an urban habitat is also subverted in cities, as traditional frameworks for the conditions that support wildlife are far more malleable than how we conceive of them in nonurban systems. Certainly, remnant or relictual patches of native or restored habitat provide invaluable ecological niches that scaffold species persistence in cities (24). However, viable urban habitat is not simply a conglomerate patchwork of residual "natural" habitat and impenetrable gray "other." Rather, habitats can also include vacant lots, backyards, cemeteries, golf courses, street trees, city parks, green roofs, and an array of other human-modified spaces that also are home to nonhuman species. The diversity and configuration of these kinds of green spaces are what allow for different types of species to survive in cities (24,93). Although some of these green spaces are somewhat designed to both serve society and cultivate habitat for nonhuman organisms (e.g., green roofs, urban gardens, cultivated backyards), many others are not. Nevertheless, a cadre of species utilize these nontraditional habitat types regardless of whether habitat was an intended management goal (48). To illustrate the point, rooftop gardens in Singapore were shown to host over 100 butterfly and bird species, including 24 uncommon or rare species, as were narrow green spaces like roadside medians (94). And road medians and cemeteries that were not intended as habitat act as conduits for dispersal and gene flow among larger habitat patches in New York City (95).

Viable habitat also extends beyond the urban green spaces that exist within a city. Infrastructure, too, can often be incorporated by urban wildlife into their expanding habitat niche in cities. Sometimes this includes forms of "green" or "blue" infrastructure which are a mixture of natural or created vegetated and watered features that can be linked across the urban landscape to buffer the impacts of development (96). Such blue-green infrastructure provides habitat, facilitates dispersal, and supports habitat linkages within urban landscapes, as well as among urban and nonurban areas (97,98). Paradoxically, gray infrastructure can also serve as habitat. Inarguably, gray infrastructure like roads and buildings cause direct mortality to wildlife or limit dispersal. However, diverse bird and bat species often use buildings, bridges, light and electricity poles, and other built structures as primary foraging and roosting sites (99,100). As a result, managing biodiversity in cities means recognizing that urban infrastructure is not simply an impediment to biodiversity but also constitutes habitat, identifying those elements as targets for conservation action.

Strategies for conserving urban biodiversity

For some urban conservation issues, we may be able to successfully translate interventions from nonurban conservation frameworks into cities.[16] For instance, modern molecular tools allow us to understand population biology and connectivity at scales that are finer than human neighborhoods, providing extraordinary precision in where we prioritize stewardship, restoration, and species conservation efforts.[17] Further, there are tremendous opportunities to establish multicity networks that collectively study urban biodiversity to identify which patterns are generalizable across cities and which are unique to specific cities, thus requiring tailored conservation actions.[18]

How biodiversity responds to traditional conservation actions in cities, however, may deviate significantly from our expectations. We can often plant desired species in new urban habitats, but some areas might remain species depauperate. Low success rates of restoring species, even when urban habitats are enhanced or modified for species colonization and persistence, could have multiple explanations. For instance, there is serious concern of "extinction debts" occurring as a function of urbanization, where the loss of biodiversity is not immediate but rather manifests over time as populations die out (101,102). Newer or enhanced urban habitats may also experience a "colonization debt," where it may be challenging for species to access new or recently enhanced urban habitats. Highly mobile

[16] Spotswood et al.'s chapter discusses which standard conservation tools can be applied to cities and which approaches need modifying to fit the unique situation of cities.

[17] Avilés-Rodriguez et al.'s chapter illustrates the diverse ways that modern genetic approaches help us not only better understand but better manage urban biodiversity.

[18] Magle et al.'s chapter explores insights from multiple multicity biodiversity projects.

(e.g., squirrels), volant (e.g., bats), or cognitively advanced (e.g., crows) species may be more likely to colonize new and enhanced urban habitats. However, other species that are more human-avoidant or movement-constrained may have a difficult time navigating the urban matrix to find these new habitats. Such species therefore require additional management actions like translocations coupled with urban habitat enhancement,[19] or conflict mitigation interventions that correct problem behaviors and promote human–wildlife coexistence.[20]

The creativity needed for urban conservation relies on identifying conservation opportunities that might be unique to cities. A widespread opportunity could lie in embracing conservation opportunities associated with urban infrastructure. For example, North American peregrine falcons declined in or were extirpated from much of their range by the mid-1900s, primarily due to reproductive failure from pesticides like DDT. In 1970 the U.S. Fish and Wildlife Service listed this raptor as endangered but subsequently removed them from the endangered species list in 1999 (103). Removal from the list was facilitated by the banning of pesticides like DDT. However, raptor experts believe that urbanization expedited the recovery of this once near-extinct species (104). This is because tall buildings in urban areas have provided numerous breeding opportunities for peregrine falcons which naturally nest on very tall cliffs. At present, peregrine falcons have become so common that they are iconic across North American cities, nesting on skyscrapers and bridges in New York City, the Campanile at UC Berkeley, and on the Washington State Capitol Building in Olympia (Figure 1.2). Practitioners now sometimes capitalize on buildings by providing additional structure to encourage falcon nesting.

There are also ways to integrate communities into managing infrastructure for habitat. Stormwater ponds—a form of gray/green infrastructure—are used to minimize flooding and damage to streams during rainstorms. These ponds also act as wetland habitat that was previously lost to urban development and host diverse species. Although this infrastructure is typically not designed with habitat in mind, frogs and salamanders thrive in stormwater ponds, including species that are otherwise sensitive to urban development (105,106). There are also opportunities to enhance habitat conditions in built stormwater ponds where amphibians do not exist yet. For instance, the City of Gresham in the broader Portland, Oregon Metro Area organizes community science groups to monitor urban stormwater ponds for amphibians. City planners and engineers use data from community scientists to modify stormwater pond design and management to improve conditions for biodiversity. This includes modifying how long stormwater ponds retain water or altering the timing or extent of maintenance activities like dredging sediment (Figure 1.3). Such work underscores the potential for community scientists to advance urban conservation, especially as participatory science methods continue expanding in cities.[21]

The charismatic nature of species like urban peregrine falcons and salamanders underscores a fruitful approach—designating "proxy" or "surrogate" species. Conservation biology has a rich history of using proxy species as (1) umbrellas whose protection will subsequently protect other species, (2) indicators of environmental health, (3) keystone species whose elimination would destabilize ecosystems, and/or (4) flagship species whose charisma stimulates support for conservation organizations (107). For example, consider how the World Wildlife Fund uses the charismatic panda to bring in support for global conservation endeavors or how nonprofit organizations like Ducks Unlimited leverage support for duck hunting to protect, restore, and create expansive wetlands for waterfowl and diverse other co-occurring species. Proxy species can also work in cities.[22] For instance, upon the death of

[19] Spotswood et al.'s chapter discusses a diversity of tools for urban conservation, including various forms of translocation coupled with habitat restoration.

[20] Stanton et al.'s chapter discusses how animal behavior can teach us about human–wildlife conflict in urban areas but also whether conservation actions are successful.

[21] Perkins et al.'s chapter reviews the various forms of participatory, community, and citizen science methods and how these tools can be used to guide urban conservation efforts.

[22] Kar Gupta et al.'s chapter introduces the idea of using proxy and surrogate species for urban conservation and provides a detailed example of imperiled lorises in Bengaluru, India.

Figure 1.2 Peregrine falcons—once endangered—like these on the state capitol building in Olympia, Washington commonly use buildings and other tall infrastructure across North America for perching, hunting, mating, and nesting. Tall urban infrastructure likely contributed to the relatively rapid population rebound of this species after the pesticide DDT was banned.

Photos courtesy of Nora Hawkins.

Figure 1.3 Dr. Katie Holzer, an environmental specialist with the City of Gresham, Oregon, trains community members to survey constructed stormwater ponds (a) for sensitive amphibian species and their egg masses like those of northwestern salamanders (b). This community science helps identify previously unknown urban wildlife populations (c) and provides data for city engineers and planners to modify constructed ponds that are choked with urban sediment and invasive plants (d) so that they can manage them (e), and provide better open water habitat for biodiversity (f). Urban stormwater ponds are typically not designed as habitat but to manage flooding damage during storms, yet creative work like this in Gresham allows constructed urban infrastructure to better serve habitat functions as well.
Photos courtesy of Katie Holzer.

the mountain lion P-22, "the Brad Pitt of mountain lions," California's Governor Gavin Newsom, multiple agencies, and myriad conservation groups reflected on the role this individual played in passing legislation and acquiring funding for the Wallis Annenberg Wildlife Crossing. This wildlife crossing will traverse 10 lanes of the U.S. 101 freeway in Los Angeles, making it the world's largest wildlife crossing when it is completed in 2025.[23] P-22 garnered tremendous public support for urban wildlife and wildlife crossings more broadly because of his captivating appearances throughout multiple Los

[23] KQED story about the legacy of the mountain lion P-22 by Rachel Treisman, December 2022 (https://www.npr.org/2022/12/21/1144627754/p22-mountain-lion-wildlife-crossing-los-angeles).

Angeles neighborhoods, an iconic image beneath the famous Hollywood sign, and by remarkably surviving multiple freeway crossings.

Which habitats are prioritized for conservation and management action also needs to be reevaluated in urban settings. Recent research has demonstrated that small habitat patches play disproportionately important roles for maintaining biodiversity including imperiled species (108). As such, conservation in cities cannot only focus on large patches of habitat but will have to inherently include assembling an array of small patches for management. This means that conservation entities will likely have to become comfortable with doing conservation on lands that they do not own (19). Urban conservation planning that includes smaller parcels with diverse ownership will need to leverage emerging frameworks and new planning paradigms that account for the outsized roles that small landscape features play for urban conservation.[24] This is typified by recent work in Los Angeles that emphasizes the need to work across spatial scales and identify networks of smaller parcels that together create habitat corridor networks across large urban areas (109).

Conservation *in*, *of*, *for*, and *with* cities

Enacting any urban biodiversity conservation strategy is predicated on effective collaboration across diverse sectors, communities, and perspectives. Part of that process is understanding the skill sets and expertise possessed by each member of a working coalition. Success is also predicated on knowing how to mesh those complementary pieces together in an inclusive way that forms a collective agreement and pathway for moving forward. One potential way to facilitate this convening process may be to create a conceptual architecture that provides guidelines that recognize the various forms of expertise and their relationship(s) to each other. Hence, we propose a framework that partitions the diversity of urban conservation research and practice into its component parts as a means of making the enormity of the discipline more manageable and less overwhelming (Figure 1.4). This framework concatenates urban conservation into four primary arcs: conservation *in* cities, conservation *of* cities, conservation *for* cities, and conservation *with* cities.

Conservation *in* cities focuses on protecting and enhancing genetic and phenotypic diversity, populations, and biotic communities to maintain healthy ecosystem function in urbanized and urbanizing landscapes. Conservation in cities is heavily influenced by conservation biology principles and adapted to work in urban environments. Research that focuses on maintaining the genetic connectivity of species across urban metapopulations or manages the proportion of native to non-native species fits squarely in the conservation *in* paradigm, using contemporary conservation biology tools locally adapted for cities.

Conservation *of* cities incorporates the biological, social, and built components of urban ecosystems to manage how ecosystem processes like hydrological regimes, nutrient flows, fluxes, and exchanges are informed by social-ecological processes. Conservation of cities seeks to influence urban heterogeneities (e.g., green space distribution, socioeconomics, legacy effects, gentrification) that can enhance ecosystem conditions for species and society by leveraging urban process-based approaches. Investigations into the social and ecological drivers of biodiversity, as well as the societal forces that govern biotic and abiotic processes in cities, employ tools from the natural and social sciences to assess the urban ecosystem.

Conservation *for* cities operationalizes lessons from the *"in"* and the *"of"* components to establish a blueprint for action. Conservation *for* provides a justice-informed, solutions-based framework that emphasizes a collective, unified approach that simultaneously applies unconventional conservation praxis to tackle nuanced social-ecological issues and embeds environmental justice principles into the fabric of this praxis. This includes justice-centered nature-based solutions, sustainable urban planning, equitable urban greening, and biodiversity management that guards against displacement. This paradigm extends beyond the academic realm,

[24] Stanford et al.'s chapter draws together landscape ecological principles to provide a tangible framework for urban planning that promotes biodiversity conservation at multiple scales in cities.

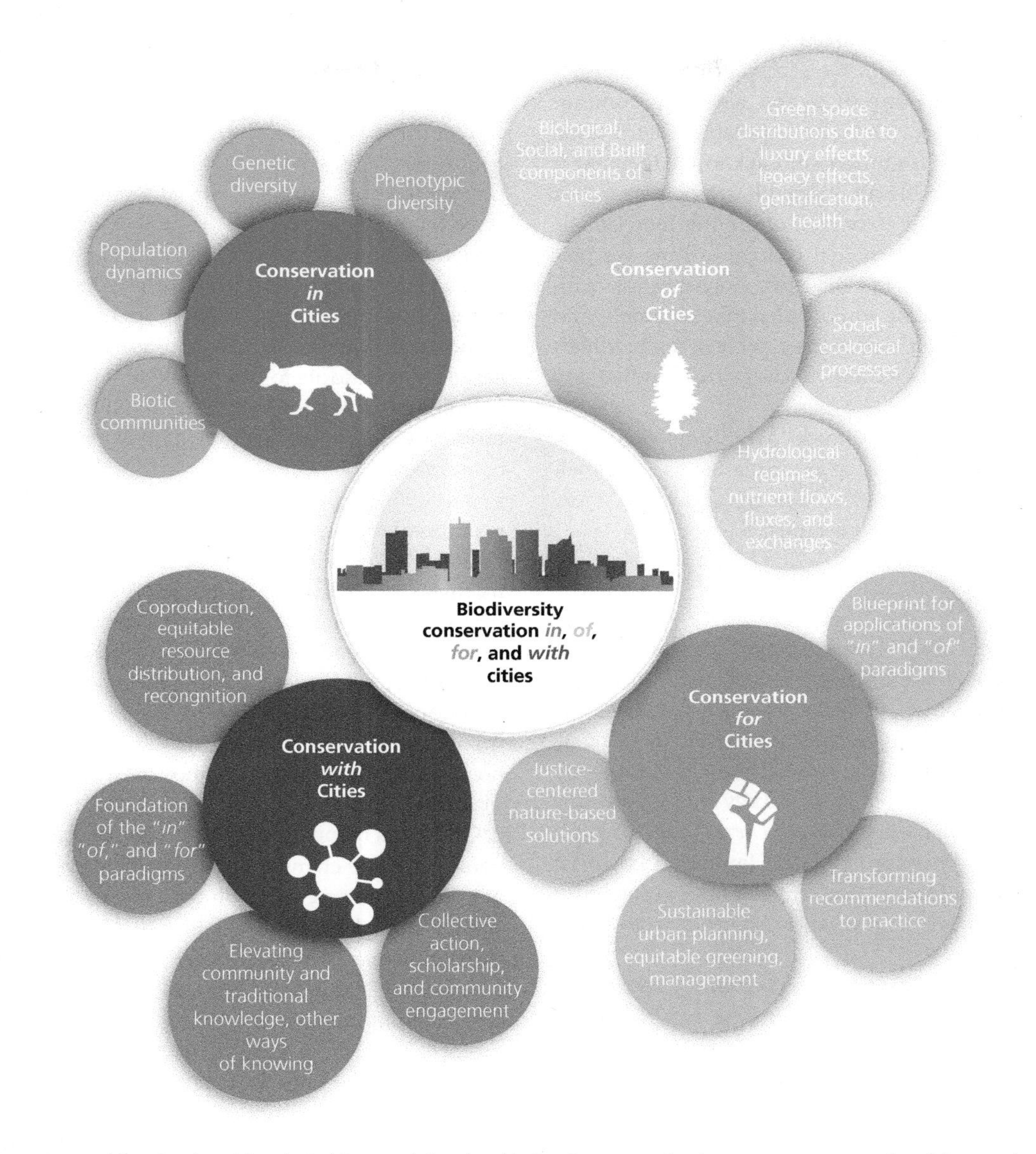

Figure 1.4 A multifunctional, social-ecological framework for urban biodiversity conservation that represents an extension of the agenda articulated by urban ecologists. The conservation *in*, *of*, *for*, and *with* cities explicate the numerous ways in which researchers, practitioners, managers, community members, policymakers, and others conduct biodiversity conservation across human-dominated landscapes, while simultaneously providing guiding principles to promote transdisciplinary integration across paradigms.

providing a roadmap for application that directly puts recommendations into practice.

Finally, conservation *with* cities encompasses a holistic approach that moves beyond understanding urbanization only as a threat to biodiversity, to understanding it also as an essential tool to counteract social injustices and biodiversity loss within urban areas and beyond. Conservation with cities heavily relies on the collective action, scholarship, and ways of knowing that emanate from communities of practice. Conservation with cities must serve as the initial foundation for conservation *in*, *of*, and *for* to be realized, and doing so requires active coproduction, the decentering of normative research perspectives and elevating community knowledge, and the equitable distribution of resources, recognition, and acknowledgement.[25] Conservation with cities provides the blueprint for a decolonized tapestry that offers the mechanistic pathway for fundamental change in how we perform our science and practice. Conservation with cities embraces the philosophy that equitable and just urbanization is necessary to conserve biodiversity in urban areas and address the biodiversity crisis globally.

It is our hope that these paradigms provide a roadmap for integrating multiple forms of knowledge into how we plan, design, and build more equitable and biodiverse cities.

Conclusion

In the 1993 film "Jurassic Park," capitalist naturalist John Hammond (played by Richard Attenborough, brother of famed real-life naturalist Sir David Attenborough) uses genetic engineering to create a zoo-like theme park filled with real-life "dinosaurs" ranging from turkey-sized *Compsagnathus* to iconic *Tyrannosaurus rex*. After giving a tour of the dinosaur exhibits, Hammond is aghast when lawyer Donald Genarro gleefully quips about charging tourists thousands of dollars to enter Jurassic Park.

Hammond: "This park was not built to cater only for the super-rich. Everyone in the world has the right to enjoy these animals."

Lawyer: "Sure. They will. We'll have a coupon day or something."

In many ways, dominant conservation approaches emulate Hammond's vision by focusing on captivating species in faraway places. Conservation also simultaneously models the lawyer Genarro's ethos—even if inadvertently—by centering biodiversity management that is only accessible to a relatively small portion of humanity who has the privilege to access it. Various regional and national parks may have discounted days (e.g., the U.S. National Park Service currently has a few free entrance days each year) that are analogous to Jurassic Park's "coupon days." However, these discounted days still require that would-be visitors have the relative capital (e.g., finances, transportation, time off work) and privileges to experience the biodiversity that exists within the confines of those parks. If we authentically and thoughtfully embed justice and equity into urban conservation, it can serve as an antidote to this problem.

Doing conservation in cities and suburbs can promote environmental justice and advance broader conservation objectives at larger scales beyond a given city's footprint. Perhaps more important is the promise of robust synergies and feedbacks between these two outcomes. Equitable practices that bring biodiversity to broader and more diverse communities in our cities are poised to usher in a new generation of conservationists. Beyond broadening support for conservation and enlarging the global coalition of conservationists, engaging diverse urbanites with biodiversity in their own neighborhoods offers the promise of innovative conservation practices.

Our era is being defined by compounding biodiversity, climate change, public health, and environmental justice crises. Urban conservation certainly cannot solve these issues on its own. Yet addressing these challenges cannot be done without conserving biodiversity in cities and suburbs. Neglecting urban conservation means abandoning the large swathes of Earth with the greatest potential for a net gain or lift in biodiversity. But urban conservation still requires centering equity, and equity relies on justice. A central focus only on backyard or urban spaces that are already large and green is almost guaranteed to perpetuate classist and racist constraints in who has biodiversity experiences. Equity-centered urban conservation can bring biodiversity to everyone's front door and bolster biodiversity globally.

Are we suggesting that we release dinosaurs into our cities? Well, yes, maybe we are. Perhaps not in the traumatizing way that the sequel "The Lost World: Jurassic Park" showed when a *T. rex* is unleashed into San Diego. But birds are dinosaurs

[25] Schell et al.'s chapter discusses the importance of integrating multiple forms of knowledge, amplifying educational engagement, and composing transdisciplinary coalitions across sectors to build just futures.

after all (110). Urban conservation presents the prospect of enriching our cities with a diversity of fauna and flora. We can build better cities that can serve as a refuge for humans and nonhuman organisms alike. Doing so takes an enormous step toward reaching a just ethos where, as Hammond articulates, "everyone in the world has the right to enjoy these animals."

References

1. Bellard C, Bertelsmeier C, Leadley P, Thuiller W, Courchamp F. Impacts of climate change on the future of biodiversity. Ecology letters. 2012;15(4):365–77.
2. IPCC. Climate Change 2022: Mitigation of Climate Change. Contribution of Working Group III to the Sixth Assessment Report of the Intergovernmental Panel on Climate Change. Cambridge, UK and New York, NY, USA: Cambridge University Press; 2022.
3. IPCC. Climate Change 2022: Impacts, Adaptation and Vulnerability. Working Group II Contribution to the Sixth Assessment Report of the Intergovernmental Panel on Climate Change. Cambridge, UK, and New York, NY, USA: Cambridge University Press; 2022. 3056 p.
4. Turney C, Ausseil A-G, Broadhurst L. Urgent need for an integrated policy framework for biodiversity loss and climate change. Nature Ecology and Evolution. 2020;4(8):996.
5. IPBES. Summary for Policymakers of the Global Assessment Report on Biodiversity and Ecosystem Services of the Intergovernmental Science-Policy Platform on Biodiversity and Ecosystem Services. Bonn, Germany: IPBES Secretariat; 2019. 56 p.
6. Obura D. The Kunming-Montreal Global Biodiversity Framework: business as usual or a turning point? One Earth. 2023;6(2):77–80.
7. Gilbert N. Nations forge historic deal to save species: what's in it and what's missing. Nature. 2022. https://doi.org/10.1038/d41586-022-04503-9
8. Simkin RD, Seto KC, McDonald RI, Jetz W. Biodiversity impacts and conservation implications of urban land expansion projected to 2050. Proceedings of the National Academy of Sciences of the United States of America. 2022;119(12):e2117297119.
9. Gao J, O'Neill BC. Mapping global urban land for the 21st century with data-driven simulations and shared socioeconomic pathways. Nature Communications. 2020;11(1):2302.
10. Seto KC, Güneralp B, Hutyra LR. Global forecasts of urban expansion to 2030 and direct impacts on biodiversity and carbon pools. Proceedings of the National Academy of Sciences of the United States of America. 2012;109(40):16083–8.
11. Shaffer HB. Urban biodiversity arks. Nature Sustainability. 2018;1(12):725–7.
12. Schwartz MW, Jurjavci NL, O'Brien JM. Conservation's disenfranchised urban poor. Bioscience. 2002;52(7):601–6.
13. Oke C, Bekessy SA, Frantzeskaki N, Bush J, Fitzsimons JA, Garrard GE, et al. Cities should respond to the biodiversity extinction crisis. npj Urban Sustainability. 2021;1(1):9–12.
14. Spotswood EN, Beller EE, Grossinger R, Grenier JL, Heller NE, Aronson MFJ. The biological deserts fallacy: cities in their landscapes contribute more than we think to regional biodiversity. Bioscience. 2021;71(2):148–60.
15. Miller JR. Biodiversity conservation and the extinction of experience. Trends in Ecology & Evolution. 2005;20(8):430–4.
16. Wolch JR, Byrne J, Newell JP. Urban green space, public health, and environmental justice: the challenge of making cities "just green enough." Landscape and Urban Planning. 2014;125:234–44.
17. Dunn RR, Gavin MC, Sanchez MC, Solomon JN. The pigeon paradox: dependence of global conservation on urban nature. Biological Conservation. 2006;20(6):1814–16.
18. Martin A, McGuire S, Sullivan S. Global environmental justice and biodiversity conservation. Journal of Geographic Information System. 2013;179(2): 122–31.
19. Soanes K, Sievers M, Chee YE, Williams NSG, Bhardwaj M, Marshall AJ, et al. Correcting common misconceptions to inspire conservation action in urban environments. Biological Conservation. 2019;33(2):300–6.
20. Taylor DE. The Rise of the American Conservation Movement: power, privilege, and environmental protection. Durham, NC: Duke University Press; 2016. 496 p.
21. Puppim de Oliveira JA, Balaban O, Doll CNH, Moreno-Peñaranda R, Gasparatos A, Iossifova D, et al. Cities and biodiversity: perspectives and governance challenges for implementing the convention on biological diversity (CBD) at the city level. Biological Conservation. 2011;144(5):1302–13.
22. Nilon CH, Aronson MFJ, Cilliers SS, Dobbs C, Frazee LJ, Goddard MA, et al. Planning for the future of urban biodiversity: a global review of city-scale initiatives. Bioscience. 2017;67(4):332–42.
23. Goddard MA, Dougill AJ, Benton TG. Scaling up from gardens: biodiversity conservation in urban

environments. Trends in Ecology & Evolution. 2010;25(2):90–8.

24. Aronson MFJ, Lepczyk CA, Evans KL, Goddard MA, Lerman SB, MacIvor JS, et al. Biodiversity in the city: key challenges for urban green space management. Frontiers in Ecology and the Environment. 2017;15(4):189–96.
25. Turo KJ, Gardiner MM. The balancing act of urban conservation. Nature Communications. 2020;11(1):3773.
26. Wyborn C, Montana J, Kalas N, Clement S, Davila F, Knowles N, et al. An agenda for research and action toward diverse and just futures for life on Earth. Biological Conservation. 2021;35(4):1086–97.
27. Soanes K, Taylor L, Ramalho CE, Maller C, Parris K, Bush J, et al. Conserving urban biodiversity: current practice, barriers, and enablers. Conservation Letters. 2023:e12946.
28. Mullenbach LE, Breyer B, Cutts BB, Rivers L, Larson LR. An antiracist, anticolonial agenda for urban greening and conservation. Conservation Letters. 2022;15(4):e12889.
29. Pineda-Pinto M, Frantzeskaki N, Nygaard CA. The potential of nature-based solutions to deliver ecologically just cities: lessons for research and urban planning from a systematic literature review. Ambio. 2022;51(1):167–82.
30. Pickett STA, Cadenasso ML, Childers DL, McDonnell MJ, Zhou W. Evolution and future of urban ecological science: ecology in, of, and for the city. Ecosystem Health and Sustainability. 2016;2(7):e01229.
31. Schwarz K, Herrmann DL. The subtle, yet radical, shift to ecology for cities. Frontiers in Ecology and the Environment. 2016;14(6):296–7.
32. McPhearson T, Pickett STA, Grimm NB, Niemelä J, Alberti M, Elmqvist T, et al. Advancing urban ecology toward a science of cities. Bioscience. 2016;66(3):198–212.
33. Nilon CH. Urban biodiversity and the importance of management and conservation. Landscape and Ecological Engineering. 2011;7(1):45–52.
34. Pataki DE. Grand challenges in urban ecology Frontiers in Ecology and the Environment. 2015;3:57.
35. McDonnell M. The history of urban ecology: an ecologist's perspective. In: Niemelä J (ed.) Urban Ecology: patterns, processes and applications. Oxford: Oxford University Press; 2011. p. 5–13.
36. Grimm NB, Grove JM, Pickett STA, Redman CL. Integrated approaches to long-term studies of urban ecological systems: urban ecological systems present multiple challenges to ecologists—pervasive human impact and extreme heterogeneity of cities, and the need to integrate social and ecological approaches, concepts, and theory. Bioscience. 2000;50(7): 571–84.
37. McDonnell MJ, MacGregor-Fors I. The ecological future of cities. Science (80-.). 2016;352(6288):936–8.
38. Young RF. Interdisciplinary foundations of urban ecology. Urban Ecosyst. 2009;12(3):311–31.
39. Donihue CM, Lambert MR. Adaptive evolution in urban ecosystems. Ambio. 2014;44(3):194–203.
40. Johnson MTJ, Munshi-South J. Evolution of life in urban environments. Science (80-). 2017;358(6363):eaam8327.
41. Szulkin M, Munshi-South J, Charmantier A. Urban Evolutionary Biology. New York: Oxford University Press; 2020.
42. Des Roches S, Brans KI, Lambert MR, Rivkin LR, Savage AM, Schell CJ, et al. Socio-eco-evolutionary dynamics in cities. Evolutionary Applications. 2021;14(1):248–67.
43. Alberti M, Palkovacs EP, Des Roches S, De Meester L, Brans KI, Govaert L, et al. The complexity of urban eco-evolutionary dynamics. Bioscience. 2020;70(9):772–93.
44. Montambault JR, Dormer M, Campbell J, Rana N, Gottlieb S, Legge J, et al. Social equity and urban nature conservation. Conservation Letters. 2018;11(3):e12423.
45. Adams LW, Leedy DL (eds.) Wildlife Conservation in Metropolitan Environments. Columbia, MD: National Institute for Urban Wildlife; 1991. National Symposium on Urban Wildlife, theme "Meeting Public and Resource Needs," Nov 11–14, 1990, Cedar Rapids, IA.
46. Platt RH, Rowntree RA, Muick PC. The Ecological City: preserving and restoring urban biodiversity. Amherst, MA: The University of Massachusetts Press; 1994. 291 p.
47. Beninde J, Veith M, Hochkirch A. Biodiversity in cities needs space: a meta-analysis of factors determining intra-urban biodiversity variation. Ecology Letters. 2015;18(6):581–92.
48. Faeth SH, Bang C, Saari S. Urban biodiversity: patterns and mechanisms. Annals of the New York Academy of Sciences. 2011;1223(1):69–81.
49. Uchida K, Blakey RV, Burger JR, Cooper DS, Niesner CA, Blumstein DT. Urban biodiversity and the importance of scale. Trends in Ecology and Evolution. 2021;36(2):123–31.
50. Aronson MFJ, La Sorte FA, Nilon CH, Katti M, Goddard MA, Lepczyk CA, et al. A global analysis of the impacts of urbanization on bird and plant diversity reveals key anthropogenic drivers. Proceedings of the Royal Society B: Biological Sciences. 2014;281(1780):20133330.

51. Moranta J, Torres C, Murray I, Hidalgo M, Hinz H, Gouraguine A. Transcending capitalism growth strategies for biodiversity conservation. Biological Conservation. 2022;36(2):e13821.
52. Agrawal A, Redford K. Conservation and displacement: an overview. Conservation and Society. 2009;7(1):1.
53. Mace GM, Norris K, Fitter AH. Biodiversity and ecosystem services: a multilayered relationship. Trends in Ecology and Evolution. 2012;27(1):19–26.
54. Vallet A, Locatelli B, Levrel H, Dendoncker N, Barnaud C, Conde YQ. Linking equity, power, and stakeholders' roles in relation to ecosystem services. Ecology and Society. 2019;24(2):30.
55. Mohai P, Pellow D, Roberts JT. Environmental justice. Annual Review of Environment and Resources. 2009;34:405–30.
56. Murray MH, Buckley J, Byers KA, Fake K, Lehrer EW, Magle SB, et al. One Health for all: advancing human and ecosystem health in cities by integrating an environmental justice lens. Annual Review of Ecology, Evolution, and Systematics. 2022;53:403–26.
57. Pickett STA, Boone CG, Cadenasso ML. Ecology and environmental justice: understanding disturbance using ecological theory. In: Boone CG, Fragkias M (eds.) Urbanization and Sustainability: linking urban ecology, environmental justice and global environmental change. Dordrecht: Springer Netherlands; 2013. p. 27–47.
58. Schell CJ, Dyson K, Fuentes TL, Des Roches S, Harris NC, Miller DS, et al. The ecological and evolutionary consequences of systemic racism in urban environments. Science (80-.). 2020;369(6509):eaay4497.
59. Rothstein R. The Color of Law: a forgotten history of how our government segregated America. New York: W.W. Norton & Company; 2017. 342 p.
60. Leong M, Dunn RR, Trautwein MD. Biodiversity and socioeconomics in the city: a review of the luxury effect. Biology letters. 2018;14(5):20180082.
61. Hope D, Gries C, Zhu W, Fagan WF, Redman CL, Grimm NB, et al. Socioeconomics drive urban plant diversity. Proceedings of the National Academy of Sciences of the United States of America. 2003;100(15):8788–92.
62. Roman LA, Pearsall H, Eisenman TS, Conway TM, Fahey RT, Landry S, et al. Human and biophysical legacies shape contemporary urban forests: a literature synthesis. Urban Forestry and Urban Greening. 2018;31:157–68.
63. Warren PS, Harlan SL, Boone C, Lerman SB, Shochat E, Kinzig AP. Urban ecology and human social organisation. In: Gaston KJ (ed.) Urban Ecology. Cambridge: Cambridge University Press; 2013. p. 172–201.
64. Grove M, Ogden L, Pickett S, Boone C, Buckley G, Locke DH, et al. The legacy effect: understanding how segregation and environmental injustice unfold over time in Baltimore. Annals of the Association of American Geographers. 2018;108(2):524–37.
65. Burghardt KT, Avolio ML, Locke DH, Grove JM, Sonti NF, Swan CM. Current street tree communities reflect race-based housing policy and modern attempts to remedy environmental injustice. Ecology. 2023;104(2):e3881.
66. Locke DH, Hall B, Grove JM, Pickett STA, Ogden LA, Aoki C, et al. Residential housing segregation and urban tree canopy in 37 US cities. npj Urban Sustain. 2021;1(1):15.
67. Anderson EC, Locke DH, Pickett STA, LaDeau SL. Just street trees? Street trees increase local biodiversity and biomass in higher income, denser neighborhoods. Ecosphere. 2023;14(2):e4389.
68. Nowak DJ, Ellis A, Greenfield EJ. The disparity in tree cover and ecosystem service values among redlining classes in the United States. Landscape and Urban Planning. 2022;221(Jan):104370.
69. Nardone A, Chiang J, Corburn J. Historic redlining and urban health today in U.S. cities. Environmental Justice. 2020;13(4):109–19.
70. Nardone A, Casey JA, Morello-Frosch R, Mujahid M, Balmes JR, Thakur N. Associations between historical residential redlining and current age-adjusted rates of emergency department visits due to asthma across eight cities in California: an ecological study. Lancet Planet. Health. 2020;4(1):e24–31.
71. Lane HM, Morello-Frosch R, Marshall JD, Apte JS. Historical redlining is associated with present-day air pollution disparities in U.S. cities. Environmental Science & Technology Letters. 2022;9(4):345–50.
72. Hoffman JS, Shandas V, Pendleton N. The effects of historical housing policies on resident exposure to intra-urban heat: a study of 108 US urban areas. Climate. 2020;8(1):12.
73. Gonzalez DJX, Nardone A, Nguyen AV, Morello-Frosch R, Casey JA. Historic redlining and the siting of oil and gas wells in the United States. Journal of Exposure Science & Environmental Epidemiology. 2023;33(1):76–83.
74. Peterson MN, Nelson MP. Why the North American model of wildlife conservation is problematic for modern wildlife management. Human Dimensions of Wildlife. 2017;22(1):43–54.
75. Eichler L, Baumeister D. Hunting for justice. Environment Society. 2018;9(1):75–90.

76. Morales N, Lee J, Newberry M, Bailey K. Redefining American conservation for equitable and inclusive social-environmental management. Ecological Applications. 2023;33(1):e2749.
77. Feldpausch-Parker AM, Parker ID, Vidon ES. Privileging consumptive use: a critique of ideology, power, and discourse in the North American model of wildlife conservation. Conservation & Society. 2017;15(1):33–40.
78. Hessami MA, Bowles E, Popp JN, Ford AT. Indigenizing the North American model of wildlife conservation. FACETS. 2021;6:1285–306.
79. Bailey K, Morales N, Newberry M. Inclusive conservation requires amplifying experiences of diverse scientists. Nature Ecology and Evolution. 2020;4(10):1294–5.
80. Fuller RA, Irvine KN, Devine-Wright P, Warren PH, Gaston KJ. Psychological benefits of greenspace increase with biodiversity. Biology letters. 2007;3(4):390–4.
81. Wood E, Harsant A, Dallimer M, de Chavez AC, McEachan RRC, Hassall C. Not all green space is created equal: biodiversity predicts psychological restorative benefits from urban green space. Frontiers in Psychology. 2018;9:2320.
82. Schebella M, Weber D, Schultz L, Weinstein P. The wellbeing benefits associated with perceived and measured biodiversity in Australian urban green spaces. Sustainability. 2019;11(3):802.
83. Methorst J, Rehdanz K, Mueller T, Hansjürgens B, Bonn A, Böhning-Gaese K. The importance of species diversity for human well-being in Europe. Ecological Economics. 2021;181:106917.
84. Callaghan A, McCombe G, Harrold A, McMeel C, Mills G, Moore-Cherry N, et al. The impact of green spaces on mental health in urban settings: a scoping review. Journal of Mental Health Counseling. 2021;30(2):179–93.
85. Ives CD, Lentini PE, Threlfall CG, Ikin K, Shanahan DF, Garrard GE, et al. Cities are hotspots for threatened species. Global Ecology and Biogeography. 2016;25(1):117–26.
86. Padullés Cubino J, Cavender-Bares J, Hobbie SE, Hall SJ, Trammell TLE, Neill C, et al. Contribution of non-native plants to the phylogenetic homogenization of U.S. yard floras. Ecosphere. 2019;10(3):e02638.
87. Padullés Cubino J, Avolio ML, Wheeler MM, Larson KL, Hobbie SE, Cavender-Bares J, et al. Linking yard plant diversity to homeowners' landscaping priorities across the U.S. Landscape and Urban Planning. 2020;196:103730.
88. Avolio M, Blanchette A, Sonti NF, Locke DH. Time is not money: income is more important than lifestage for explaining patterns of residential yard plant community structure and diversity in Baltimore. Frontiers in Ecology and Evolution. 2020;8:85.
89. Ravinet M, Elgvin TO, Trier C, Aliabadian M, Gavrilov A, Sætre GP. Signatures of human-commensalism in the house sparrow genome. Proceedings of the Royal Society B: Biological Sciences. 2018;285(1884):20181246.
90. Carlen E, Munshi-South J. Widespread genetic connectivity of feral pigeons across the Northeastern megacity. Evolutionary Applications. 2021;14(1):150–62.
91. Reed EMX, Serr ME, Maurer AS, Burford Reiskind MO. Gridlock and beltways: the genetic context of urban invasions. Oecologia. 2020;192(3):615–28.
92. Shochat E, Lerman SB, Anderies JM, Warren PS, Faeth SH, Nilon CH. Invasion, competition, and biodiversity loss in urban ecosystems. Bioscience. 2010;60(3):199–208.
93. Lepczyk CA, Aronson MFJ, Evans KL, Goddard MA, Lerman SB, MacIvor JS. Biodiversity in the city: fundamental questions for understanding the ecology of urban green spaces for biodiversity conservation. Bioscience. 2017;67(9):799–807.
94. Wang JW, Poh CH, Tan CYT, Lee VN, Jain A, Webb EL. Building biodiversity: drivers of bird and butterfly diversity on tropical urban roof gardens. Ecosphere. 2017;8(9):e01905.
95. Munshi-South J. Urban landscape genetics: canopy cover predicts gene flow between white-footed mouse (*Peromyscus leucopus*) populations in New York City. Molecular Ecology. 2012;21(6):1360–78.
96. Donati GFA, Bolliger J, Psomas A, Maurer M, Bach PM. Reconciling cities with nature: identifying local blue-green infrastructure interventions for regional biodiversity enhancement. Journal of Environmental Management. 2022;316:115254.
97. Furberg D, Ban Y, Mörtberg U. Monitoring urban green infrastructure changes and impact on habitat connectivity using high-resolution satellite data. Remote Sensing. 2020;12(18):3072.
98. Nguyen TT, Meurk C, Benavidez R, Jackson B, Pahlow M. The effect of blue-green infrastructure on habitat connectivity and biodiversity: a case study in the Ōtākaro/Avon River catchment in Christchurch, New Zealand. Sustainability. 2021;13(12):6732.
99. MacGregor-Fors I, Schondube JE. Gray vs. green urbanization: relative importance of urban features for urban bird communities. Basic and Applied Ecology. 2011;12(4):372–81.
100. Russo D, Ancillotto L. Sensitivity of bats to urbanization: a review. Mammalian Biology. 2015;80(3):205–12.

101. Soga M, Koike S. Mapping the potential extinction debt of butterflies in a modern city: implications for conservation priorities in urban landscapes. Animal Conservation. 2013;16(1):1–11.
102. Hahs AK, McDonnell MJ. Extinction debt of cities and ways to minimise their realisation: a focus on Melbourne. Ecological Management & Restoration. 2014;15(2):102–10.
103. Padayachee K, Reynolds C, Mateo R, Amar A. A global review of the temporal and spatial patterns of DDT and dieldrin monitoring in raptors. Science of the Total Environment. 2023;858:159734.
104. McCabe JD, Yin H, Cruz J, Radeloff V, Pidgeon A, Bonter DN, et al. Prey abundance and urbanization influence the establishment of avian predators in a metropolitan landscape. Proceedings of the Royal Society B: Biological Sciences. 2018;285(1890):20182120.
105. Guderyahn LB, Smithers AP, Mims MC. Assessing habitat requirements of pond-breeding amphibians in a highly urbanized landscape: implications for management. Urban Ecosystem. 2016;19(4): 1801–21.
106. Holzer KA. Amphibian use of constructed and remnant wetlands in an urban landscape. Urban Ecosystem. 2014;17(4):955–68.
107. Caro T. Conservation by Proxy. Washington, DC: Island Press; 2010. 400 p.
108. Wintle BA, Kujala H, Whitehead A, Cameron A, Veloz S, Kukkala A, et al. Global synthesis of conservation studies reveals the importance of small habitat patches for biodiversity. Proceedings of the National Academy of Sciences of the United States of America. 2019;116(3):909–14.
109. Zellmer AJ, Goto BS. Urban wildlife corridors: building bridges for wildlife and people. Frontiers in Sustainable Cities. 2022;4:954089.
110. Gauthier J. Saurischian monophyly and the origin of birds. Memoirs of the California Academy of Sciences. 1986;8:1–55.

SECTION 1

Urban Nature's Social Fabric

The decisions and actions emanating from human society—whether intentional or not with respect to nature—fundamentally shape where and how species exist and thrive within and across cities. Human dimensions and societal drivers are therefore essential for understanding why urban biodiversity seems to flourish or fail sporadically. This section explores the intimate links between society's decisions and urban nature.

CHAPTER 2

Escaping the Practice of Exclusion

Conservation, Green Space, and Urban Planning in the Age of Environmental Justice

Fushcia-Ann Hoover[‡] and Rachel D. Scarlett[§]

Introduction

Environmental exclusion

Environmental racism is often discussed as a problem of the past or the outcome of individual decisions (1) but environmental racism is supported by policies at the federal, state, and local levels making it an institutionalized process (2,3). Environmental racism, defined as "any policy, practice, or directive that differentially affects or disadvantages (whether intended or unintended) individuals, groups, or communities based on race or color" (4), concentrates environmental benefits and amenities in predominantly white spaces, while communities of color are burdened with the costs. Ultimately, all planning decisions continue to be influenced by racism, discrimination, and the devaluation of Black spaces (5) driving the uneven spatial patterns we observe across our ecosystems, including those in urban conservation.

Following the public murder of George Floyd by three Saint Paul, MN police officers in 2020, a sudden awareness and interest in the impacts of environmental racism, environmental justice (EJ), and racism in urban ecology, conservation, and planning was observed across media, job postings, and academic journals.[1] Unfortunately, this engagement has yet to produce meaningful change in outcomes or conservation practices. While awareness and interest are crucial, efforts to change these systemic processes remain insufficient. To understand why and how to overcome this, ecologists, planners, and others must first understand then confront the racialized processes undergirding urban conservation and planning practices. We explore how historical and contemporary inequalities influence the spatial distribution and abundance of biodiversity in cities, by examining environmental racism, urban planning, and conservation. Using three case cities, we illustrate how racism and injustice drive ecological outcomes, the exclusion of minoritized communities from decision-making processes, and produce "race-neutral" biodiversity research outcomes. Finally, we present examples that center minoritized people *and* nature through EJ, collective action, and the integration of multiple research approaches. We define EJ as a set of values and practices that prioritize equity, accountability, transparency, and democratic processes; and collective action as community-led organizing and

‡ University of North Carolina, Charlotte, USA
§ Georgia State University, USA

[1] Journals across the social and physical sciences, and humanities issued calls for submissions to special issues focused on environmental justice and injustice. Similarly, a vast number of tenure track job calls had a sudden focus on Black studies, EJ, and equity research. For example, *The American Studies Journal*. "2021 AMSJ Special Issue call for papers—Our Shared Planet: The Environment Issue." July 21, 2021: https://amsj.blog/calls-for-papers/; Stanford cluster hire on impact of race in STEM. Job post date: January 11, 2021: http://elementsmagazine.org/doc/Stanford_2021.pdf

Fushcia-Ann Hoover and Rachel D. Scarlett, *Escaping the Practice of Exclusion.* In: *Urban Biodiversity and Equity.* Edited by: Max R. Lambert & Christopher J. Schell, Oxford University Press. © Oxford University Press (2023). DOI: 10.1093/oso/9780198877271.003.0002 self-determination that promote shifts in power and decision-making.

Identifying structural inequalities in conservation

Historically, conservation ideologies (e.g., biophilia versus stewardship) and biodiversity protections have come at the expense of Black (and)[2] Indigenous practices and caretaking, leading to our exclusion and erasure from nature altogether (9). The conservation movement exists on a "backdrop of racism, sexism, class conflicts, and nativism..." (10). In fact, many of conservation's "founding fathers" (e.g., John Muir, Henry David Thoreau) were explicitly racist and articulated green space exclusively for whites, advocating for the exclusion of people of color from nature and recreation (10,11). Additionally, most urban planning deprioritized nature particularly as highway development expanded. Already racist and exclusionary (e.g., housing segregation, urban renewal), the emergence of environmental planning management decisions rarely considered, consulted, or included people of color, compounding existing inequalities and entrenching environmental racism.

Despite legal mechanisms like the National Environmental Protection Act (NEPA)[3] and the Civil Rights Act,[4] environmental racism persists because planners, engineers, and conservationists continue to use the same metrics and tools (e.g., benefit-costs analysis, reliance on neighborhood/organization, community-science tools like iNaturalist[5]) to assess the value of neighborhoods, species, and neighborhood biodiversity without a critical examination of biases. Community-science and homeowner personal preference, like the likeability of a species, can play key roles in improving conservation efforts (13,14). However, when community-science is compounded by racialized understandings of space and place (5,15), communities of color and their biodiversity are undervalued, under-measured, or ignored. Despite the exclusion and underrepresentation in mainstream conservation management, communities of color have engaged in novel practices of mutual aid, and stewardship and conservation practices (16–18), including resistance based on ancestral knowledge and a necessity to reshape and create alternative landscapes (8), and ecologies of resistance[6] (e.g., 20). They have created stewardship practices tailored to community and environmental needs.

Environmental racism, environmental justice

The EJ movement emerged from three pivotal cases starting with a series of lawsuits from 1978–1979 filed against Olin, a DDT manufacturing facility, by Black residents and government agencies in Triana, Alabama. This was followed, most notably, by lawsuits filed against the State of North Carolina, Robert Burns, and the Ward Transformer Company, over the siting of a polychlorinated biphenyl (PCB) landfill outside Afton, Warren County, NC. While unsuccessful in preventing the siting of the facility, these lawsuits[7] resulted in national attention, US Environmental Protection Agency (EPA)-commissioned impact studies, and eventual federally led lawsuits and criminal charges. Finally, residents in Alsen, LA, in the region known as Cancer Alley,[8] filed suit in 1999 against Condea

[2] We place "and" in parenthesis to illuminate Black Indigenous identities, both in and outside of the US. To be Black and to be Indigenous are not mutually exclusive (6). We also want to honor the Black stewardship practices and relations established by our enslaved ancestors; through a forced settler/non-settler identity (7), Black people have cultivated healing relationships with, for, and to the land, creating strong emotional bonds to our geographies of place (8).

[3] National Environmental Policy Act of 1969 § 83, 42 U.S.C. § 4321 et seq. (1970).

[4] Civil Rights Act of 1964 § 7, 42 U.S.C. § 2000 et seq. (1964).

[5] iNaturalist is a nature app that helps the user identify plants and animals through photo and location data. It is sponsored by National Geographic and the California Academy of Sciences. More about the app can be found at https://www.inaturalist.org/. Because the database is built from individual observations, many of the observations made are in white neighborhoods, compounding the exclusion of neighborhoods of color from biodiversity data analysis, grant funding, and urban conservation conversations. See also Perkins et al.'s chapter and (12).

[6] For additional works, see the nine articles in the recently published Global Black Ecologies series in the *Journal of Environment and Society*, Volume 13, Issue 1, edited by Justin Hosbey, Hilda Lloréns, and J. T. Roane (19).

[7] Filed by the National Association for the Advancement of Colored People (NAACP), landowners, and Warren County Board of Commissioners (21).

[8] A stretch of land from Baton Rouge to New Orleans where the second highest cancer rate among men and the fourth highest among women in the US occur. This area coincides with heavy oil and gas extraction and processing plants (21).

Vista (Conoco) for groundwater contamination; and a community buyout was negotiated by residents in Diamond, LA against Shell Oil Company after going to trial in 1997 over exposure, decreased property values, etc. In all cases, the majority of residents living in these communities were poor, Black, and ignored for the "common good" of a state or company needs (21–24).

As it grew into a national movement, 17 principles of EJ were formed at the First National People of Color Environmental Leadership Summit (*Principles of Environmental Justice*, 1996: https://www.ejnet.org/ej/principles.html), following which researchers theorized three distinct facets of EJ: distributional justice, recognitional justice, and procedural justice. Distributional justice examines the spatial relationship of the siting and location of environmental hazards (and amenities) with race. It is foundational work that established a relationship between majority-Black zip codes and prevalence of environmental hazards (23,25). Recognitional justice focuses on how minoritized individuals and communities are recognized and given voice in environmental planning processes (26). Procedural justice engages and questions the process for participation, including how meaningful that participation is (25,27). Put in terms of the proverbial *seat at the table*, these theories ask where is the table, when is the meeting, who is at the table and how were they invited, are they empowered to speak, is their voice valued and listened to, and what happens when the meeting ends?

Many researchers have focused most heavily on distributional justice. There are many reasons but the desire for generalization and scaling findings is a significant one. Government-level questions around resources often rely on systematic citing opposed to individual experiences within a place. While it is possible to generalize from qualitative work, critical engagement with geographies of place would conclude that generalization is antithetical to sense of place, resulting in an "orientation to generalization rooted in the same system that generated the inequalities in the first place" (in conversation with Dr. Candace Miller,[9] October 11, 2022). This is bolstered by the EPA's definition of EJ as, "the fair treatment and meaningful involvement of all people *regardless* of race, color, national origin, or income, with respect to the development, implementation, and enforcement of environmental laws, regulations, and policies" (28, emphasis added). The EPA's definition has proved limiting for researchers pursuing more holistic methodologies as the principles of EJ *do* engage race. The absence of procedural justice is iniquitous given that many federal and nonprofit conservation efforts have perpetuated the exclusion and deprioritization of the communities most heavily impacted by environmental hazards (e.g., 21). Thus, erasing the influence of racism from systems where race was foundational to resource allocation is injudicious. Further egregious is that many conservation organizations and federal agencies were founded on conservationist ideologies of human and nature separation, Indigenous (and) Black land dispossession, and racialized access to these natural spaces (10,11,29).[10]

Our understanding of urban conservation and biodiversity is a direct extension of these ideologies. Planning and conservation decisions traverse such social and physical boundaries, yet conservationists often fail to recognize that they are working in racialized ecologies. Urban conservation compounds conservation's racist frameworks through urban planning, mainstream environmental movements, and white elite activism (e.g., not-in-my-backyard (NIMBY)). Such negligence can lead to repetition and reinforcement of environmental racism and failing to meet biodiversity conservation goals.

Segregation and urban biodiversity

Environmental racism shapes urban habitat and biodiversity in ways that researchers are beginning to unravel (30). In US cities, racist policies

[9] Dr. Miller is a mixed-methods urban sociologist and assistant professor in the Department of Sociology and Organizational Science at the University of North Carolina, Charlotte. She explores racial inequalities embedded in and stemming from contemporary urban political economy, and is a brilliant sociologist.

[10] Dr. Taylor does an excellent job detailing communications and rationales between administrators at the Department of the Interior and the National Park Service director to preserve segregated park facilities for White and Black visitors. See Chapter 12 in (10).

and practices were used to segregate neighborhoods whereby people of different races are relegated to differential spaces (15). The legacy of these practices not only resulted in the present-day segregation of urban populations based on race and class (31), but also in lasting economic disinvestment from Black communities, and more recently uncovered biophysical patterns in the abundance and biodiversity of plants and animals (30,32,33). These patterns are consequences of explicit racist practices, policies, and discourse which have the potential to drive patterns in biodiversity within cities and disproportionate access to biodiverse landscapes by race and social status.

Residential segregation not only served to separate people based on race but also contributed to the racial wealth gap[11] and the structural and service disparities impacting racially marginalized people and the ecosystems they live in. Zoning regulations were designed to keep unwanted land uses out of residential neighborhoods; however, in practice, they served as the foundation for the ongoing concentration of industrial pollution and toxic waste in (mostly but not only) Black and Latine neighborhoods (23,39,40). One of the first comprehensive zoning regulations was established in New York City. On the one hand, the zoning regulations served wealthy and white residents well in that their property was protected from unwanted land uses, and the regulations even included requirements for open public space. On the other hand, non-white and working-class communities were designated as manufacturing districts which allowed for industrial buildup alongside residential neighborhoods (40). Black, Latine, and Asian/Pacific Islander overrepresentation in industrial zones is a clear pattern across US metropolitan areas (4), and these patterns continue today (41).

Polluting industries have local impacts on people and biodiversity, and can contribute to unique urban species assemblages by intensifying urban heat island effects and pollutant exposure which both have the potential to drive species evolution (30).[12] For example, pollutant exposure has the potential to increase heritable genetic mutations in rodents and birds (42).

Overtly racist segregation practices followed exclusionary zoning and led to disproportionate investment in Black communities across the US. In cities, the Home Owner's Loan Corporation (HOLC), reinforced through the Federal Housing Authority (FHA), sanctioned segregation by creating security grading maps—the system that gave way to redlining. Throughout the 1940s and 1950s, predominately Black neighborhoods were assessed as high risk or "Hazardous" for defaulting on home loans and given a D rating or the color red. Meanwhile, predominately white, Christian neighborhoods were rated as "Best" or given A grades (noted in green). "Still Desirable" or "Declining" neighborhoods received a B or C grade, respectively (43). This practice was used to justify the allocation of federally insured loans to segregated white neighborhoods and deny federally insured home loans to Black communities. Redlining and other racialized zoning policies denied a critical source of generational wealth—via subsidized home ownership and affordable rental rates—to the Black community. Neighborhood income is positively associated with vegetation cover, plant diversity, bird diversity, and mammal diversity,[13] such as squirrels, coyotes, and raccoons (e.g., 44–48). This phenomenon is known as the luxury effect, and researchers hypothesize that it is predicated on the fact that those with higher capital (and thus resource supply) exert stronger influence on plant assemblages (47); however, neighborhood wealth is limited in its explanation of patterns in urban biodiversity, and other structural inequities are important (30,49).

Beyond individual wealth, disparities in public investment and services across segregated

[11] The racial wealth gap represents the extreme wealth inequality between white families and families of color. In 2016, the median net worth of a white family was on average ten times greater than that of a Black family ($171,000 compared to $17,100, respectively) (34). For further reading on how public policies, including residential and educational segregation, played a critical role in creating the wealth gap see (35–38).

[12] Byers et al.'s chapter details the complex impacts of urban heat and pollution on people and biodiversity.

[13] Magle et al. (48) found that species richness was associated with income in 9 out of 20 observed cities.

neighborhoods can lead to differing habitat and species assemblages within cities. Redlined neighborhoods were further denied green space, parks, and infrastructural investments which structurally shaped canopy cover in many US cities. Green space in Black neighborhoods was seen as a wasted investment, and a hindrance to police surveillance (50). Such lack of investment is evidenced by lower tree canopy cover in redlined neighborhoods in cities across the US (33) (Figure 2.1). Neighborhoods graded in "D" in St. Louis not only were denied federally insured home loans, but also denied basic sanitation services that they paid into as taxpayers (51). Lack of basic

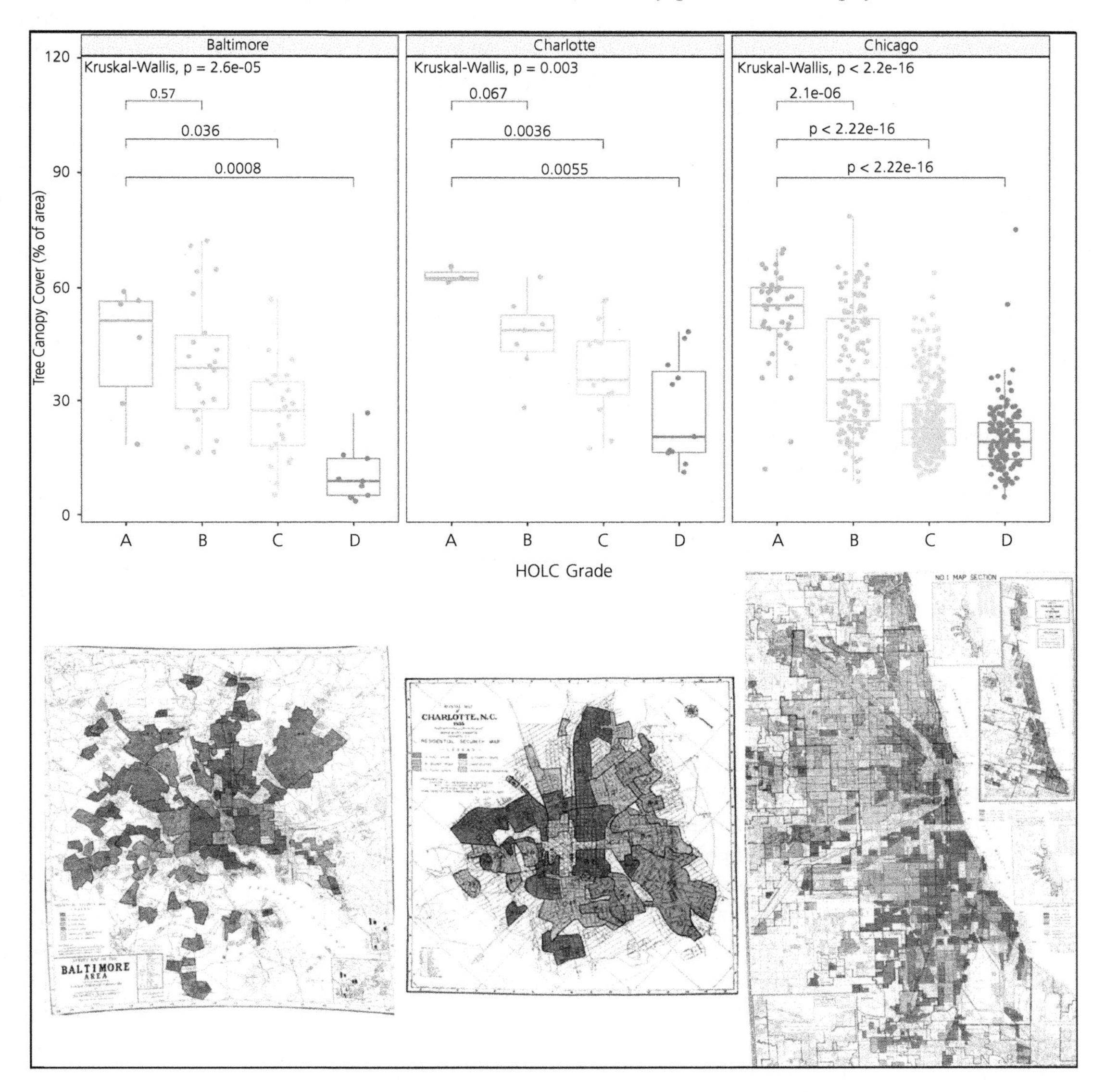

Figure 2.1 Top row: comparison of the historic neighborhood [Home Owner's Loan Corporation (HOLC)] grading system and the percentage of tree canopy cover for (left to right) Baltimore, MD, Charlotte, NC, and Chicago, IL. Redlined areas or "D" graded neighborhoods shown in red have significantly less tree canopy than "A" graded neighborhoods shown in green (116). Bottom row: digitized maps demonstrating the distribution of historical HOLC graded neighborhoods for (left to right) Baltimore, MD, Charlotte, NC, and Chicago, IL (117).
Data courtesy of Dexter Locke.

sanitary services, like consistent trash pickup, can lead to disease vectors through the overabundance of generalist pest species like rats and mosquitoes who survive on human waste and standing water.[14]

While we know how exclusionary practices can shape the landscape and habitat quality, relatively little is known about the importance of these changes for biodiversity (30). One can infer that habitat changes will shape ecological community structure and function, such as limiting carrying capacity or genetic diversity (32). Still, emerging research also shows that in many instances, densely populated urban landscapes can support various species (52–54).

Exclusionary practices also importantly shape conservation discourse in historically excluded communities. In conservation science and practice, there is often a lack of recognition that harms to biodiversity uniquely impact marginalized communities with noncommodified ties to local ecologies (e.g., 8,55,56). Pollution in Black, Latine, and Indigenous communities is often correlated with dirty and uncivilized people rather than historical and contemporary practices of racism and colonization (57). These communities are also often overlooked as spaces of value to biodiversity and remain undersampled for biodiversity metrics (58).[15]

Landscape inequality shapes the presence and absence of urban biodiversity in several distinct yet important ways. First, as it pertains to cities where communities of color *have* green space, and second, where communities of color *lack* green space. Where communities of color have green space, institutional processes devalued the land itself, based on the real but mostly perceived land value that included physical (e.g., proximity to a floodplain), economic viability (e.g., racial capitalism), and social and environmental characteristics (e.g., racism, infrastructure disinvestment)—resulting in communities of color being ignored, environmental protections being ignored, and these areas being left out of biodiversity monitoring and other ecological studies. Ironically, these same areas may be more likely to contain higher levels of species richness or abundance (30) due to the disinvestment (e.g., vacant lots, rural or edge communities, and unincorporated townships (59,60)).

The result of these two patterns of spatial inequality is the creation of two potential urban worlds: one where green space and biodiversity thrive but the biodiversity and amenities are to the benefit and access of whites only; the other where green space is present but amenities and biodiversity are unknown or unmeasured and communities of color are ignored.

A tale of three cities

We present three examples to illustrate how the histories of conservation and ecology directly connect to contemporary approaches to urban conservation and biodiversity, with the following objectives:

(a) illustrate how historic environmental racism influences contemporary decision-making and drives practices of exclusion in urban biodiversity and species management;
(b) analyze why practices of exclusion persist even in cities where a goal of EJ or sustainability is present; and,
(c) explore the social inequalities and impacts on planning and decision-making for green space generally.

Portland

Portlandia is rooted in the containment, exploitation, and exclusion of people of color.

Walidah Imarisha[16]

Over the past 30 years, Portland has been branded as a sustainability leader (61–63). Leading the way in environmental policy and management, it is home to Forest Park, one of the largest urban forests in the nation, a statewide urban growth boundary that protects natural land and agriculture from urban expansion, and it is at the forefront of green

[14] Byers et al.'s chapter takes a One Health approach to understanding linkages between social inequality, wildlife pathogen vectors, and disease in both wildlife and people.

[15] Perkins et al.'s chapter reviews racial-spatial biases in biodiversity sampling, particularly from citizen science approaches, that skew urban conservation and environmental justice.

[16] "Why aren't there more Black people in Oregon? A hidden history," public lecture by Walidah Imarisha organized by the Jordan Schnitzer Museum of Art in 2014.

infrastructure initiatives (64). Underlying these successes, however, is a long history of racial exclusion that has led to Portland being the Whitest major city in the US. In this case study, we will explore how racial exclusion has enabled much of the environmental progress in Portland, leading to uneven urban habitat across the city.

Urban spatial inequality: whites-only sustainability

Portland has over 13 million large trees in the city; however, tree canopy cover is unevenly distributed (Figure 2.2) (64), with some communities having up to 82% tree coverage, living within walking distance of Forest Park, while others have 4% coverage (64). Unsurprisingly, lower-income communities of color disproportionately live in the neighborhoods with lower mixed and woody vegetation and higher impervious surface cover (65). However, the uneven distribution of trees and park access is not coincidental; we highlight a few key policies and practices that contributed to this disparity.

The early white settlers of Oregon criminalized Blackness through The Organic Law of 1844—giving free Black men and women two years to leave Oregon or risk public whippings if caught. In 1857, a racial exclusionary clause in the state constitution banned Black people from moving to Oregon, owning property, and using the legal system. Although unlawful, this clause was preserved in the constitution until 1926. Racism continued into the early to mid-20th century as the small Portland Black population was squeezed into Northeast Portland, the Albina District. Portland Realty Board's code of ethics, federally backed redlining, and housing covenants restricted home ownership for Black and Asian families to this

Figure 2.2 Street tree density and urban parks (orange polygons) in Portland. East Portland is outlined in grey on the far right/east of the city. From the map, the highest tree density is present in toward the center of the city, close to the Willamette River.

Figure modified from Shandas and Hellman (63) courtesy of Max Lambert using City of Portland open access data.

inner core Portland neighborhood (66–68). Moreover, exploitive lending practices kept residents from building wealth through home ownership in Northeast Portland up until the 1990s (69).

Like many other cities, Portland was also experiencing "white flight'[17] from the cities—an exodus of white middle-class residents from the inner core to unincorporated suburban towns. Driven by government-sponsored redlining and white families' refusal to integrate with Black families, white families moved to East Portland,[18] which quickly developed as a suburban haven (67). The flight of middle-class white families from the inner city also drained the city's tax base, leaving inner core neighborhoods, like Albina, lacking public funds for infrastructural updates and development (67). In short, this was part of a long process that marked Black neighborhoods for devaluation and disinvestment up until the 1990s.[19]

The devaluation of inner core Portland made the city ripe for reinvestment and gentrification because profit margins were high (66,67). Federally funded urban renewal projects revitalized Portland's inner core; however, these projects failed to protect the homes and cultural community assets of the Albina neighborhoods. These projects forcibly removed hundreds of Albina homes and pushed out Black families (66). In an effort to increase the city's tax base, Portland annexed East Portland unincorporated towns throughout the 1980s and 1990s. During this time, Portland aggressively branded itself as a sustainable and green city to attract affluent white suburbanites back to the inner core and justify the hike in rent and home prices (67).

Sustainable branding attracted capital to Portland's inner core and resulted in urban greening that coincided with development. Developers and city planners attracted affluent residents with plans for improved city livability like public transit, green roofs, and river restoration (67,72,73). For example, a city ordinance required green roofs on new building developments in downtown, which explains why green roofs are clustered in this area (64). Likewise, street tree planting correlates with an increase in housing prices in Portland (74). Along the Willamette River, which is a superfund site, early ecological restoration efforts were economically dependent on the development of elite mixed-use riverfront condos (72). Meanwhile, restoration projects did not serve the subsistence and noncommodified needs of Indigenous, Black, Latine, and houseless populations along the river (75). While case studies clearly show that sustainability rhetoric was used to attract affluent residents back to the city, it is unclear how much this green investment contributed to citywide patterns in green space (76). More apparent is how uneven development within the city led to poor infrastructure and green space in East Portland.

Urban spatial inequality: neglect and disinvestment in East Portland

Haphazard development of East Portland was likely a large contributor to disparities in green space across Portland. Today's East Portland is a legacy of the auto-centric suburban towns post-WWII. The towns were designed to house working-class white families who needed to travel to the city for work; they were built haphazardly with lax environmental regulation (67). The revitalization of the inner city brought East Portland families back into the city and resulted in a dramatic shift in inner core and suburban population demographics where inner core neighborhoods like Albina were upwards of 75% Black in 1970 and have declined to less than 10% percent Black in the 2010s (67). As affluent (mostly) white Portlanders moved back to the inner core, suburban East Portland was vacated, and people of color that were outpriced in the inner core relocated to East Portland. East Portland is now one of the most diverse neighborhoods in Oregon. However, money continues to be funneled into the city's inner core, which is now mostly occupied by white more affluent residents, and much-need infrastructural updates in East Portland have been ignored or underfunded for decades resulting in Black (and other communities of color) Portland communities'

[17] Pejchar's and Reed's chapter addresses racial segregation and contemporary demographic patterns in suburban areas and associations with biodiversity.

[18] The most affluent white families moved to the west of Portland, west of the Willamette River (67).

[19] For a detailed history of gentrification and devaluation of Black neighborhoods in Portland see (66,67,70). For ongoing displacement processes in Portland see (71).

continuous displacement to devalued urban spaces (67).[20]

Such haphazard development, auto-centric planning, and disinvestment in East Portland created the landscape that we see today. Auto-centric planning created a landscape full of impervious surfaces. Additionally, the transition from single family homes (a marker of white suburban lifestyle) to multifamily units, as people of color moved to East Portland, resulted in urban densification with no city-mandated provisions for green space (77). Not only is street tree canopy cover lower in East Portland (Figure 2.2), but also East Portland has a higher potential for urban heat island effects and urban flooding than more central areas of the city (78). These changes to the urban ecosystem can impact biodiversity in a myriad of ways that ecologists are still uncovering (30). Conservationists must recognize that nature in cities is intertwined with social and political processes, and conservation practices can reproduce or, with justice-centered practice, disrupt uneven urban ecosystems. Conservation organizations in Portland are beginning to grapple with social inequality but have a long way to go.[21] Conservation practice is always already part of structural racism in cities, and making EJ a priority requires understanding and disrupting the processes that have unevenly and inequitably shaped our urban landscape.

Baltimore

Conservation and management of urban biodiversity is about people.

Charles H. Nilon (79)

Located in the Mid-Atlantic of the US, Baltimore, MD is a city of 576,498 persons, plus an additional 272,818 in Baltimore County. While 62.3% and 29.7% of Baltimore City residents are Black or white, respectively, 31.3% and 58.8% of residents in Baltimore County identify as Black or white, respectively. Perhaps unsurprisingly, 20% of residents in the city live in poverty, but that number decreases to 8.9% when expanding to include the county. Median housing values and homeownership rates at the county level all exceed those of the city as well (80).

Situated in the Chesapeake Bay watershed and harbor, the region also plays a critical role as an important estuary for a variety of migratory birds, aquatic and terrestrial species, as well as sustaining oyster and crabbing livelihoods, a naval academy, and a prominent shipping harbor. As a result, there are key conservation considerations in addition to the spatial eco heterogeneity of green space across the city and county. These distinctions between city and county are in part due to the long history of racial segregation, white flight, and efforts to exclude Baltimore's Black residents from accessing various resources. In Baltimore, systemic biases shape environmental planning, resource management, and green space distribution. As mentioned, spatial inequality presents itself in two ways: one where green space and biodiversity thrive but the biodiversity and amenities are to the benefit and access of whites only and one where green space is present but amenities and biodiversity are unknown or unmeasured, and communities of color are ignored. The City of Baltimore provides a case where both urban worlds exist.

Urban spatial inequality: whites-only biodiversity

In addition to state- and municipal-based segregation tools, green space itself was used as a tool for reinforcing segregation and an indicator for whiteness through zoning, city ordinances, and more (81). Across the country, predominantly white neighborhoods were intentionally greened and planned with integrated green space along parkways, roads, and residential streets, while predominantly Black neighborhoods were not (82,83). Particularly as whites fled the urban core during the 1960s and 1970s in pursuit of spaces devoid of Black and poor residents through white flight, surrounding suburbs and incorporated townships were designed with abundant trees, greened boulevards, and expansive parks (84).

[20] Goodling et al. (67) provide a detailed account of lax enforcement of environmental regulation and lack of city investment in East Portland.

[21] Guderyahn's and Logalbo's chapter details how West Multnomah Soil & Water Conservation District committed fewer resources to racially diverse and redlined neighborhoods in Portland. Illuminating these discrepancies was a key step to understanding conservationists' role in environmental racism. The District and Portland Parks & Recreation are beginning to address environmental inequities and injustices.

Locke et al. (33) found that through the process of government-sanctioned "redlining," formerly "D" graded neighborhoods currently have 21% less tree canopy cover than "A" graded neighborhoods (Figure 2.1). Other studies have found similar results between tree canopy and ecosystem services (85), canopy distribution (86), and more, including variations in the quality and typology of tree canopy or green space along racialized neighborhoods.[22] For example, Boone et al. (87) examined park distribution, size, access (approximated by walking distance), and estimates of park use density. Combined, the city and surrounding counties boast a total of 22,823 hectares of parks, or 9.1 hectares per 1000 people, but that number ranges from 7.6 ha in-city to 46.3 ha across the counties. Additionally, parks adjacent to predominantly white neighborhoods were larger, with similar comparisons made among park access and size in cities like Los Angeles for white and Latinx communities (87). In summary, few Black Baltimoreans have access to large parks and, where access exists, they share less space with more people. Given the benefits to improving air quality that trees can provide, it is no surprise that environmental disamenities also map alongside racial segregation in Baltimore. For example, negative health outcomes associated with exposure to ozone, $PM_{2.5}$, and PM_{10} rate higher for "D" graded neighborhoods than "A" graded neighborhoods (88).

Urban spatial inequality: unmeasured biodiversity in Black neighborhoods

Urban conservation and biodiversity research has been increasing in focus and attention, particularly as urban ecological research centers and networks have expanded (e.g., Long-Term Ecological Research Project (LTER) urban sites in Baltimore, MD (established 1997), Phoenix, AZ (1997), and most recently Minneapolis, MN (2021); see Magle et al.[23] for more information on multicity research sites), including perceptions of urban areas as sites for great species richness and local to regional biodiversity. Even for these urban LTERs, studies that engage both conservation and race are limited. Using the LTER Network Bibliography database, a keyword search for "biodiversity" and "conservation" identified 12 and nine peer-reviewed studies (with overlap) out of 263 and 137 studies, respectively.[24] Of the 21 conservation/biodiversity studies, just four engaged race or ethnicity.

The oldest temporal study (in Phoenix) examined socioeconomic and cultural characteristics in efforts to explain biodiversity patterns; a counter approach to traditional biodiversity gradient methods[25] (89). They reported variations in species richness by residential areas based on socioeconomic status (SES) and cultural characteristics, but omitted detail on those characteristics or patterns, and relied on median household income (as a proxy for SES) for the final analysis. A study examining stormwater ponds in Columbia and Chesterfield, MD, cities just outside Baltimore County (90), recognized the importance of human interests in shaping management decisions and biodiversity outcomes, but did not engage with the community characteristics where ponds were surveyed, despite concluding that biodiversity patterns show divergent outcomes based on different management practices due to regional species pond constraints. One Baltimore, MD study (91) found "distinct patterns of tree structure and biodiversity across clusters that could clarify links between land use, community composition, and ecosystem function in the urban forest" (p. 1). These outcomes were organized into "clusters" for analysis and discussed based on landuse type (e.g., cluster 4 was predominantly residential); however, no connections between landuse

[22] Locke et al.'s chapter details the history of urban forests for understanding the connections between biodiversity and systemic racism in cities.

[23] Magle et al.'s chapter synthesizes how multicity comparisons have been powerful tools for understanding general patterns and city-specific patterns in biodiversity, including associations with inequities.

[24] The search was restricted to peer-reviewed journal articles that were sponsored by the LTER site. A total of 16 studies were identified for "conservation" for both the Baltimore and Central-Arizona Phoenix (CAP) urban sites out of the 137. The search for "biodiversity" yielded 18 studies across both urban sites. It is possible that more studies exist using published LTER data, and are just not captured in the database.

[25] The gradient approach referred to relies on landuse without considering how the characteristics of the people may influence or change biodiversity. Distance is the most commonly used metric, thus assuming that neighborhood social characteristics do not influence biodiversity.

types and sociodemographic or other characteristics were performed, despite one cluster's canopy cover decreasing by 21% (91). Finally, another Baltimore, MD study (92) examined biodiversity and residents' preferences of vacant lots aimed to maximize biodiversity goals and human values in Baltimore, MD. Their work included a random sample of 150 lots, electronic surveys targeting 20 community groups and park associations, site visits, and questions targeting participants' race, sex, and other socioeconomic information (SES). They reported that the location of a vacant lot may have stronger effects on the lot's vegetation versus other lot or neighborhood characteristics (e.g., landuse type, SES), and that most respondents preferred lots with evidenced management efforts and multipurpose uses. However, the authors did not discuss the participants and the organization makeup itself. Thus, despite location being a driving factor for lot vegetation, the authors did not provide even aggregated results of the SES data collected from participants.[26] This is an important detail in a city where a majority of vacant lots are in low-income neighborhoods, which would have limited resources to modify the lots compared to suburban lots. Management decisions could be based on these findings, but whom are the respondents representing? More importantly, the authors ignore what location as the determining factor for vacant lot vegetation means in a highly segregated city. Is it because vacant lots in Black communities are ignored and left to grow at will? Or is it because lots in suburban areas are planted with more diverse plant species?

With the exception of tree data studies, these studies all point to a repeating pattern where the importance of race and ethnicity is erased or pushed to the background of urban conservation and biodiversity work; a shocking finding given that Baltimore is a foundational city for informing our understanding of urban ecology, *and* where redlining was first implemented. The absence of race from these studies' results *and* discussion leaves the findings with minimal context, nuance, and, most importantly, solutions that might benefit conservation efforts and are based on residents' resources and the city's investment strategies.

Global perspectives: Cintruón Verde Metropolitano, Medellín, Colombia

Urban spatial inequality: green gentrification[27]

Conflicts and trade-offs in planning that prioritizes land or species conservation at the expense of vulnerable and minoritized communities are not unique to the US. In particular, conflicts between ecological preservation and land development come on the heels of urban population growth in the form of settlements, barrios, and other makeshift communities (93,94). In addition, many of these communities are individuals displaced due to ecological, economic, or social crises occurring in their home communities, regions, and countries (95). In Mexico City, MX for example, conflicts over population changes are creating pressures on an ecological preservation zone. Despite regulatory zoning laws intended to protect the more than 1800 species of plants and animals in the Suelo de Conservacion (SC), informal settlements are infringing on the SC, an area comprising forest, grasslands, agriculture, and Indigenous towns. At the time of the study, Mexico City registered 19.2 million residents with growth primarily occurring in subdistricts that comprise parts of the SC. Simultaneously, land for poor or low-income residents has not been made available (93).

Similar concerns between population growth, settlements, and conservation goals exist in the northern city of Medellín, Colombia, South America, and the capital of Antioquia province. As the second largest city in Colombia, Medellín boasted a population of 2.27 million in 2012, which has seen increased internally displaced persons (IDPs), with many settling in various informal settlements (96).

[26] Among ecologists' research, the use of SES data is often collapsed into "income" alone, and the other social indicators like race are left out of the analysis. This is a problem.

[27] This case study was adapted from the International Planning Case Studies Project, a library of case studies for international planning and development. They were developed by practitioners and researchers through a collaboration between Dr. Lesli Hoey of University of Michigan and Dr. Andrew Rumbach of Texas A&M University. The case study used in this chapter is cited as (97).

One such informal community, named Villa Hermosa or Comuna Ocho, is home to more than 155,000 residents based on a 2010 survey, with 40% of those being IDPs. Comuna Ocho also ranks as the most socioeconomically vulnerable community in Medellín; 80% of residents sit in the two lowest socioeconomic classes the country uses to categorize residents. As a country, Colombia ranks highest in the world for the number of IDPs, with estimates of 4.9–5.5 million at the time of the proposed plan (96).

Activism and outcomes

Around 2012, the city proposed a Greenbelt Plan, Cintruón Verde Metropolitano (CVM), outlining four zones intended to address the rapid population growth, degraded barrios, land squatting, and ecosystem loss as a long-term growth strategy. However, one of the four zones titled the "protection zone" would create an ecological conservation area that would include bicycle and hiking paths, and would pass through a portion of Comuna 8, requiring the displacement of over 6000 households, while ignoring needs identified by the community (96) (Figure 2.3). Fortunately, the Comuna 8 residents are well-organized and in 2007 formed the Comuna 8 Planning and Local Development Council. Additionally, Colombia's 1991 constitution established goals that uphold land as having both social and ecological functions, citizen opinions in informing urban policy, and that those trade-offs should be fairly distributed across stakeholders (96,97).

In response to the CVM, the community's Council published a Declaration of Wants and Needs outlining eight priorities, that included adequate, safe, and permanent housing; public facilities like schools and health clinics; space for increased urban agriculture; income generation; and meaningful participation. Additionally, they collaborated with several universities to draft alternatives to the proposed Greenbelt route, in addition to plans that integrated the Comuna 8 priorities. In 2014, a Local Land Use Plan outlining the creation of a Jardin Circvunvalar was signed into law in Medellín, Columbia. Since then, the Jardin Circvunvalar has

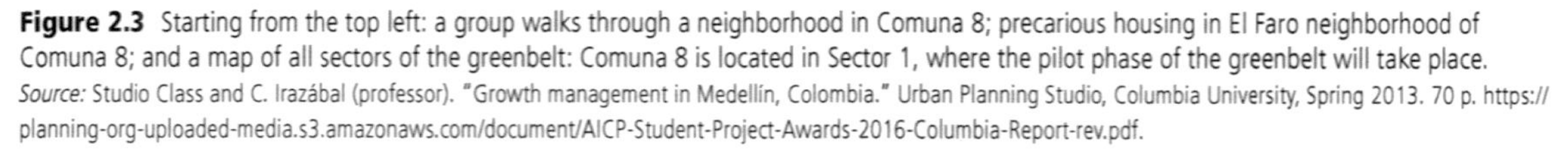

Figure 2.3 Starting from the top left: a group walks through a neighborhood in Comuna 8; precarious housing in El Faro neighborhood of Comuna 8; and a map of all sectors of the greenbelt: Comuna 8 is located in Sector 1, where the pilot phase of the greenbelt will take place.
Source: Studio Class and C. Irazábal (professor). "Growth management in Medellín, Colombia." Urban Planning Studio, Columbia University, Spring 2013. 70 p. https://planning-org-uploaded-media.s3.amazonaws.com/document/AICP-Student-Project-Awards-2016-Columbia-Report-rev.pdf.

increased accessibility and improved pedestrian pathways while providing stronger infrastructure to protect against landslides, flooding, and other environmental hazards. Unfortunately, other disamenities have taken hold that include nuanced displacement and dispossession of lands for the construction of projects without compensation or relocation provisions. Additionally, these outcomes are impacting lower-income families as higher-income communities within the region are being allowed to remain in place (95).

In this section, we illustrated how contemporary decision-making influenced by historic environmental racism drives exclusionary practices in conservation studies in Baltimore, MD. We discussed why practices of exclusion persist even in cities where a goal of sustainability is present in Portland, OR, and we explored the social inequalities and impacts on planning and decision-making for green space in Medellín, Colombia. Below, we present examples of Black, Indigenous, and Latinx/e stewardship, conservation, and ecologically focused practices.

Social-ecological applications: environmental justice

Biodiversity conservation and environmental goals are impossible to achieve without centering equity and justice, and mainstream environmentalists have failed to recognize much less address how the environmental movement is produced from the dispossession, appropriation, and containment of people of color (10); we also add the exclusion of people of color from conversation efforts to this list. Historically, conservation took a utilitarian approach to natural resources; preservation valued natural resources for their intrinsic value. As such, conservation "implied planning and efficiency" (10), that eventually included the management of water and land resources, going so far as to be characterized as development (10). This detail of conservation history is precisely why planning and conservation are tightly connected and why racialized planning processes greatly influence conservation practices. Which is why meaningful engagement without the reproduction of social inequities is often a challenge in mainstream conservation organizations; these organizations are overrepresented by white men and women who uphold policies and practices in support of racial and economic exclusion from ecosystem benefits (98,99). At the same time, people of color and low-income communities are not seen as conservationists by the public or mainstream environmentalism.[28] Globally, communities of color are deeply concerned about environmental issues, and often at the forefront of environmental organizing, implementing stewardship practices that simultaneously address social and ecological inequities, including the privatization of land, water resources (100), and rapid development.[29] Conservationists of color prioritize the needs of communities and empower those on the margins of society to lead conservation efforts while generating income and sustaining their livelihood (56).

We highlight conservation practitioners and Black stewards as evidence of communities of color that have long been doing the work of promoting and fighting for better environmental stewardship and conservation practices. These examples (1) provide a counternarrative to the racist narrative that urban populations of color are unconcerned or disconnected with their environment, (2) that people of color do not already practice environmental stewardship in cities, (3) that people of color are unaware of environmental concerns, and (4) highlight the paltry availability of research centering people of color and our stewardship practices. We will conclude with ways to move forward in

[28] For a wonderful overview from Dr. Dorceta Taylor on the ways the White conservation/environmental movement views people of color, and how that needs to change, see this interview with Diane Toomey from Yale 360: "How green groups became so white and what to do about it," June 21, 2018. https://e360.yale.edu/features/how-green-groups-became-so-white-and-what-to-do-about-it (accessed October 11, 2022).

[29] Despite legislation preserving all of Puerto Rico's beaches as public, the people are actively fighting against private developers, the government, and wealthy, mostly White US expatriates to keep the playas public and accessible. Puerto Rican rapper and activist Bad Bunny highlights the battle and activism in his visual documentary "El Apagón - Aquí Vive Gente," from his newest album release "Un Verano Sin Ti." https://www.youtube.com/watch?v=1TCX_Aqzoo4 (accessed October 11, 2022). We highly recommend watching it.

urban conservation and planning that align with EJ. Ultimately, if conservationists (researchers and practitioners) are serious about EJ and building inclusive conservation practices, then we have to recognize the hidden ways inequality and environmental racism are built into contemporary decision-making and land management practices. Conversely, if conservationists are serious about biodiversity conservation, then EJ must be a priority, otherwise environmental and conservation goals are not obtainable.

People for Community Recovery (PCR)

Hazel Johnson, the mother of the Environmental Justice Movement, founded People for Community Recovery (PCR) in 1979, one of the nation's first environmental organizations headquartered in public-housing, Altgeld Gardens in southeast Chicago, IL. Altgeld Gardens, one of the nation's first housing projects, was a flourishing Black community from the mid-1940s to late 1960s when the Chicago Housing Authority (CHA) placed an income cap on residents, forcing middle-class Black families out of the neighborhood.[30] Today, more than two-thirds of the adults are unemployed, and the 2014 median household income was $16,000 (101). Geographically situated in what Hazel Johnson coined a "toxic donut" (102), the community is surrounded by the Calumet Sewage Treatment Plant's sludge beds to the north, landfills to the east, an active steel plant to the west, and the Little Calumet River to the south, previously used as a dumping pipeline for the surrounding industries. Additionally, over 21 landfills occupy the larger southeast Chicago region (103). Meanwhile, the region contains critical habitat for endangered species and was historically marked by prairie wetlands, oak savannas and woodlands, and meandering rivers. Although industrialization destroyed many of these natural habitats, remnant wetlands, prairies, and savannas sustain breeding populations of up to 15 endangered or threatened bird species in Illinois. The Calumet region is now recognized as one of the highest priorities for biodiversity conservation in the state (104).

[30] Job loss resulting from deindustrialization was also detrimental to the Altgeld Gardens (101).

Activism and outcomes

After noticing the prevalence of cancer and respiratory diseases in Altgeld Gardens—her husband died of lung cancer—Johnson founded PCR to address human health and environmental impacts of industrialization in the region (102). Johnson was keen to the impacts of toxins on human health and wildlife:

> You know, a lot of times you do not hear how it affects human health. You do not hear how it affects the wildlife. But it does... If you do not think it is affecting you, then you are wrong. This could affect you twenty years from now or an unborn child. So this is a problem not just for people that live in these communities, it is a problem for everybody. (102)

Led by Johnson, PCR released one of the first US community health studies, collecting Altgeld Gardens community members' health symptoms in 1985. While the EPA did not recognize this study as scientific, it was one of the first community-engaged science health studies of its kind and aided in garnering media attention to the "toxic donut" (105). PCR protested the siting of a waste incinerator and landfills leading to a moratorium on new landfills in the area, and, in 1999, held the CHA accountable for cleaning up pollution in the community's soils.

While the long-term impacts of PCR's work on biodiversity have yet to be established, one of the largest impacts that PCR and Hazel Johnson had on the Calumet region is the pride and recognition of Altgeld Gardens as a community that is cared for and loved despite being a community that has been effectively abandoned by government (106). While Altgeld Gardens is still plagued with pollution, Frazier notes that Hazel Johnson and other Black women in the community developed an ethic of community care that arguably has a longer-lasting impact on building an environmental movement that places care for the community and ecosystem at the forefront. PCR's work illustrates how justice-centered ecological restoration can benefit both the community and wildlife.

Little Village Environmental Justice Organization

Little Village (LV) is a community in the South Lawndale neighborhood of Chicago, IL. One of the densest communities in Chicago, and 82% Latinx/e, primarily Mexican-American, this community has a strong history of political and environmental activism (107). Asthma, obesity, mental-health, and teen birth rates were some of the highest in Chicago, and despite their population, the community has the least amount of green space per capita among Chicago neighborhoods. The Little Village Environmental Justice Organization (LVEJO) was founded in 1994 by parents of Chicago Public School children after learning of the possible exposure their children were experiencing during building renovations. The parents organized to halt the work, and soon thereafter organized to develop a youth leadership program, and surveyed residents to learn about environmental problems facing the community, all the while lobbying to close and remediate various industrial sites. LVEJO's mission is to "organize with our community to accomplish environmental justice in Little Village and achieve the self-determination of immigrant, low-income, and working-class families," with an organizing model that: (1) sustains community self-determination through intergenerational leadership; (2) assumes that those directly affected have the solutions to solve their own problems; and (3) builds upon the existing assets and resources of the community for social change. Furthermore, they organize their work around the theory of social change to carry out those principles (108).

Activism and outcomes

LVEJO identified two coal power plants (Fisk and Crawford), the Celotex insulation company, and an absent 31st bus line as key EJ concerns (108). Joining with PERRO, another EJ group in predominantly Latine neighborhood Pilsen, LVEJO was successful in retiring both coal power plants in 2012, and in 2013 won their Celotex superfund campaign resulting in the first public park in Little Village in 75 years. They also reestablished the 31st bus line to the community. In 2013, in collaboration with Little Village businesses, schools, 325 residents, and over 40 organizations, they put together the Little Village Quality-of-Life Plan, outlining eight metrics and a vision of building a community that is "educated, peaceful, united, clean, and prosperous" (107). Then in 2014, LVEJO commissioned the aid of a consultant and, after a series of interviews, listening sessions, and engagement with LVEJO leadership, staff, Little Village residents, and youth, issued a 2015–2020 Strategic Plan. Connected to an EJ2020 Plan, they outline actions like brownfield and riverfront redevelopment, EJ education, and energy resilience, with an overarching goal of remediating the impacts of industrial waste (109). Their EJ2020 Plan on the other hand is intended to "promote equitable redevelopment practices that foster collective action and strengthen the self-determination of Little Village" (109).

LVEJO continues to serve as an advocate and liaison for the EJ needs of Little Village residents, and continues to provide guidance and input to future planning or development (110), EJ, and climate change (109).

Conclusion

Conservation is a racialized process rooted in planning and decision management (10), and the exclusion of communities of color has been to the detriment of achieving biodiversity and conservation goals. Throughout, we have detailed the way environmental racism and other systemic segregation policies and practices aided and drove the exclusion-based conservation practice we see today. We provided examples of communities of color and organizations that demonstrate how people of color not only are concerned and connected with their environment, but have been practicing environmental stewardship since time immemorial. A significant barrier to achieving urban conservation goals remains the lack of involvement of, focus on, and research centering people of color, communities of color, and the stewardship practices they employ.

Urban residents are more likely to be interested and engaged with urban conservation practices and biodiversity if the experiences they have in green and open spaces are positive. "The brownfield, remnant woodland, local park, and green space may be important for species conservation, but they are also

Case study The Green Belt Movement, Kenya

The Green Belt Movement (GBM) is one organization leading the way in Indigenous conservation efforts to protect biodiversity, community, and livelihoods. GBM used a tree planting campaign as an Indigenous initiative to avert desertification, increase biodiversity, improve food security and environmental consciousness, and empower African women in Kenya (Figure 2.4). The GBM was founded by Nobel Peace Prize winner, Dr. Wangari Maathai, in 1977 in response to the increasing food insecurity for rural families that was a direct result of the desertification of Kenya's forests. The GBM is notable for its blending of conservation, Indigenous knowledge, and women's empowerment.

Activism and outcomes

Dr. Maathai and her colleagues spent years identifying the needs of Kenyan communities, building conservation efforts to address those needs and conserve biodiversity (111). Dr. Maathai mobilized (mostly) rural Kenyan women to plant "green belts," a strategy of planting rows of at least 1000 tree seedlings (111). The women led and sustained this initiative, despite opposition from the government. As the cost of purchasing seedlings increased, Dr. Maathai mobilized the women to start nurseries, and professional foresters offered to train the women. However, forester teachings were often overly technical and difficult to understand, so instead the women relied on Indigenous knowledge and innovative techniques to teach themselves how to sustain tree nurseries. The GBM also focused on planting native trees and crops as an effort to both improve biodiversity and maintain Indigenous knowledge of plants, promoting local crops like arrowroots and yams, and native trees like the Nandi flame tree (111).

One notable victory in Nairobi, Kenya, was an organized movement to stop development in Uhuru Park. Uhuru Park is a central, large park and green space in Nairobi, and became the site of one of the first GBM green belts (56). In 1989, the Kenyan government and developers were developing a project that would place a 60-storey building in Uhuru Park. Dr. Maathai protested the skyscraper through written campaigns, and led protests against the development. Their direct action successfully blocked the development (56), and in 2019, when the Kenyan government again planned on encroaching on Uhuru Park to build a highway, the GBM and Kenyans again successfully halted this development (112). The movement continues to advocate for the protection and restoration of urban green space including Karura and Ngong Forests (113). Nationally, the GBM has planted over 53 million trees in the last 45 years and grown to be an international organization dedicated to tree planting and regreening communities (114,115).

Figure 2.4 Three members of the Kiharu Kahuro Women's Group plant seeds for future tree plantings.
Image Courtesy of the Green Belt Movement. https://www.greenbeltmovement.org/news-and-events/media-resources.

places where messages concerning local biodiversity projects can be shared and where the participation and inclusion of local residents' conservation and management projects can be encouraged" (79). Persistent challenges include moving away from a focus on rare species and engaging the people who live in cities; recognizing that goals for ecosystem function and biodiversity may differ, and residents often have broader and more expansive goals; moving away from the role as educator; and recognizing that residents are just as likely to serve as the educator (78). Cities are complex systems with many climate and ecosystem challenges, but addressing these challenges with an inclusive "one health" approach,[31] where communities of color lead, is the best pathway forward.

References

1. Pulido L. Rethinking environmental racism: White privilege and urban development in southern California. Annals of the American Association of Geographers. 2000;90(1):12–40.
2. Bullard R. Environmental racism. Environmental Protection. 1991;2:25–6.
3. Omi M, Winant H. Racial Formation in the United States. New York: Routledge; 2014.
4. Bullard RD, Mohai P, Saha R, Wright B. Toxic wastes and race at twenty: why race still matters after all of these years. Environmental Law. 2008;38(2): 371–411.
5. Rothwell J, Loh TH, Perry A. The Devaluation of Assets in Black Neighborhoods: the case of commercial property. Washington, DC: Brookings Institute; 2022.
6. Miles T. Ties that Bind: the story of an Afro-Cherokee family in slavery and freedom. Berkeley, CA: University of California Press; 2015.
7. Hernandez J. Fresh Banana Leaves: healing indigenous landscapes through indigenous science. Huichin, unceded Ohloe land, aka Berkeley, CA: North Atlantic Books; 2022. 260 p.
8. Roane JT. Plotting the Black Commons. Souls. 2018;20(3):239–66.
9. Murdock EG. Nature where you're not: rethinking environmental spaces and racism. In: Hosbey J, Lloréns H, Roane JT (eds.) The Routledge Handbook of Philosophy of the City [Internet]. 1st ed. Abingdon: Routledge; 2020. p. 13. Available from: https://www.routledge.com/The-Routledge-Handbook-of-Philosophy-of-the-City/Meagher-Noll-Biehl/p/book/9781138928787
10. Taylor DE. The Rise of the American Conservation Movement: power, privilege, and environmental protection. Durham, NC: Duke University Press; 2016. 486 p.
11. Murdock EG. Conserving dispossession? A genealogical account of the colonial roots of Western conservation. Ethics, Policy, and Environment. 2021;24(3):235–49.
12. Mahmoudi D, Hawn CL, Henry EH, Perkins DJ, Cooper CB, Wilson SM. Mapping for whom? Communities of color and the citizen science gap. ACME International Journal of Critical Geography. 2022;21(4):372–88.
13. Collins L, Paton GD, Gagné SA. Testing the likeable, therefore abundant hypothesis: bird species likeability by urban residents varies significantly with species traits. Land. 2021;10(5):487.
14. Jordan R, Gray S, Sorensen A, Newman G, Mellor D, Newman G, et al. Studying citizen science through adaptive management and learning feedbacks as mechanisms for improving conservation. Conservation Biology. 2016;30(3): 487–95.
15. Lipsitz G. The racialization of space and the spatialization of race: theorizing the hidden architecture of landscape. Landscape Journal. 2007;26(1): 10–23.
16. Finney C. Black Faces, White Spaces: African Americans and the great outdoors [Internet] [Ph.D. Thesis]. ProQuest Dissertations and Theses. [Ann Arbor, MI]: Clark University; 2006. Available from: https://www.proquest.com/dissertations-theses/black-faces-white-spaces-african-americans-great/docview/305357888/se-2
17. Voices from the Valley [Internet]. Voices from the Valley. [cited 2022 Jul 8]. Available from: http://www.voicesfromthevalley.org/
18. Democracy Green Homepage [Internet]. Democracy Green. [cited 2022 Oct 12]. Available from: https://democracygreen.wixsite.com/democracygreen
19. Hosbey J, Lloréns H, Roane JT (eds.) Environment and Society. 2022 [cited 2022 Oct 18];13(1). Available from: https://www.berghahnjournals.com/view/journals/environment-and-society/13/1/environment-and-society.13.issue-1.xml
20. Purifoy D, Wilson J. To live and thrive on new Earths. Southern Cultures. 2020;26(4):78–89.

[31] Byers et al.'s chapter details a One Health approach for cities.

21. Taylor DE. Toxic Communities: environmental racism, industrial pollution, and residential mobility. New York: New York University Press; 2014. 343 p.
22. Bullard R, Mohai P, Saha R, Wright B. Toxic Wastes and Race at Twenty 1987—2007. Cleveland, OH: United Church of Christ; 2007. 160 p.
23. Bullard RD. Dumping in Dixie: race, class, and environmental quality. 3rd ed. Boulder, CO: Westview Press; 2000.
24. Murdock EG. A history of environmental justice: foundations, narratives, and perspectives. In: Coolsaet B (ed.) Environmental Justice: key issues. Abingdon: Routledge; 2020. p. 6–17.
25. Schlosberg D. Defining Environmental Justice: theories, movements, and nature. Oxford: Oxford University Press; 2007.
26. Honneth A. Recognition and justice: outline of a plural theory of justice. Acta Sociologica. 2004;47(4): 351–64.
27. Young IM. Justice and the Politics of Difference. Princeton, NJ: Princeton University Press; 1990. 286 p.
28. US EPA. Environmental justice [Internet]. US Environmental Protection Agency; 2014 [cited 2022 Jul 8]. Available from: https://www.epa.gov/environmentaljustice
29. Spence MD. Dispossessing the Wilderness: the preservationist ideal, Indian removal, and national parks [Internet] [Ph.D. Thesis]. ProQuest Dissertations and Theses. [Ann Arbor, MI]: University of California, Los Angeles; 1996. Available from: https://www.proquest.com/dissertations-theses/dispossessing-wilderness-preservationist-ideal/docview/304228023/se-2
30. Schell CJ, Dyson K, Fuentes TL, Des Roches S, Harris NC, Miller DS, et al. The ecological and evolutionary consequences of systemic racism in urban environments. Science. 2020;369(6510):eaay4497.
31. Delmelle EC. The increasing sociospatial fragmentation of urban America. Urban Science. 2019; 3(1):9.
32. Schmidt C, Garroway C. Systemic racism alters wildlife genetic diversity [Internet]. EcoEvoRxiv; 2022 [cited 2022 Oct 7]. Available from: https://osf.io/wbq83
33. Locke DH, Hall B, Grove JM, Pickett STA, Ogden L, Aoki C, et al. Residential housing segregation and urban tree canopy in 37 US cities. Urban Sustainability. 2021;1(15):1–9.
34. McIntosh K, Moss E, Nunn R, Shambaugh J. Examining the Black-White Wealth Gap. Washington, DC: Brookings Institution; 2020.
35. Baradaran M. The Color of Money: Black banks and the racial wealth gap. Cambridge, MA: Harvard University Press; 2017.
36. Darity W, Hamilton D, Paul M, Aja A, Price A, Moore A, et al. What we Get Wrong about Closing the Racial Wealth Gap. Durham, NC: Samuel DuBois Cook Center on Social Equity and Insight Center for Community Economic Development; 2018. 67 p.
37. Shapiro TM. The Hidden Cost of Being African American: how wealth perpetuates inequality. New York: Oxford University Press; 2004.
38. Shapiro T, Meschede T, Osoro S. The Roots of the Widening Racial Wealth Gap: explaining the Black-White economic divide. Waltham, MA: Institute on Assets and Social Policy; 2013.
39. Wilson S, Hutson M, Mujahid M. How planning and zoning contribute to inequitable development, neighborhood health, and environmental injustice. Environmental Justice. 2008;1(4):211–16.
40. Sze J. Noxious New York: the racial politics of urban health and environmental justice. Cambridge, MA: MIT Press; 2007.
41. Kramar DE, Anderson A, Hilfer H, Branden K, Gutrich JJ. A spatially informed analysis of environmental justice: analyzing the effects of gerrymandering and the proximity of minority populations to U.S. superfund sites. Environmental Justice. 2018;11(1):29–39.
42. Somers CM, McCarry BE, Malek F, Quinn JS. Reduction of particulate air pollution lowers the risk of heritable mutations in mice. Science. 2004;304(5673):1008–10.
43. Rothstein R. The Color of Law: a forgotten history of how our government segregated America. New York: Liveright; 2017. 368 p.
44. Avolio M, Blanchette A, Sonti NF, Locke DH. Time is not money: income is more important than lifestage for explaining patterns of residential yard plant community structure and diversity in Baltimore. Frontiers in Ecology and Evolution. 2020;8:85.
45. Chamberlain D, Reynolds C, Amar A, Henry D, Caprio E, Batáry P. Wealth, water and wildlife: landscape aridity intensifies the urban luxury effect. Global Ecology and Biogeography. 2020;29(9): 1595–605.
46. Gerrish E, Watkins SL. The relationship between urban forests and income: a meta-analysis. Landscape and Urban Planning. 2018;170:293–308.
47. Hope D, Gries C, Zhu W, Fagan WF, Redman CL, Grimm NB, et al. Socioeconomics drive urban plant diversity. Proceedings of the National Academy of Sciences of the United States of America. 2003;100(15):8788–92.

48. Magle SB, Fidino M, Sander HA, Rohnke AT, Larson KL, Gallo T, et al. Wealth and urbanization shape medium and large terrestrial mammal communities. Global Change Biology. 2021;27(21): 5446–59.
49. Watkins SL, Gerrish E. The relationship between urban forests and race: a meta-analysis. Journal of Environmental Management. 2018;209:152–68.
50. Mars R. Shade [Internet]. 99% Invisible; 2020 [cited 2020 Jan 14]. Available from: https://99percentinvisible.org/episode/shade/
51. Heck S. Greening the color line: historicizing water infrastructure redevelopment and environmental justice in the St. Louis metropolitan region. Journal of Environmental Policy and Planning. 2021;23(5): 565–80.
52. Alvarez Guevara JN, Ball BA. Urbanization alters small rodent community composition but not abundance. PeerJ. 2018;30(6):e4885.
53. Shochat E, Stefanov WL, Whitehouse MEA, Faeth SH. Urbanization and spider diversity: influences of human modification of habitat structure. Ecological Applications. 2004;14(1):268–80.
54. Soanes K, Lentini PE. When cities are the last chance for saving species. Frontiers in Ecology and the Environment. 2019;17(4):225–31.
55. Whyte, K. Indigenous experience, environmental justice and settler colonialism [Internet]. SSRN; 2016. Available from: http://dx.doi.org/10.2139/ssrn.2770058
56. Maathai W. Unbowed: a memoir. New York: Anchor; 2017.
57. Jaffe R. Concrete Jungles: urban pollution and the politics of difference in the Caribbean. New York: Oxford University Press; 2016.
58. Ellis-Soto D, Chapman M, Locke D. Uneven biodiversity sampling across redlined urban areas in the United States [Internet]. EcoEvoRxiv; 2022 [cited 2022 Jul 11]. Available from: https://osf.io/ex6w2
59. Gardiner MM, Burkman CE, Prajzner SP. The value of urban vacant land to support arthropod biodiversity and ecosystem services. Environmental Entomology. 2013;42(6):1123–36.
60. Villaseñor NR, Chiang LA, Hernández HJ, Escobar MAH. Vacant lands as refuges for native birds: an opportunity for biodiversity conservation in cities. Urban Forestry & Urban Greening. 2020;49: 126632.
61. Dunn H. Portland paves way as most eco-friendly city in U.S. [Internet]. KOIN; 2022 Aug 9. Available from: https://www.koin.com/news/portland/report-portland-paves-way-as-most-eco-friendly-city-in-u-s/
62. Dyckhoff T. The five best places to live in the world, and why [Internet]. The Guardian; 2012 Jan 20. Available from: https://www.theguardian.com/money/2012/jan/20/five-best-places-to-live-in-world
63. Grudowski M, Heil N. The new American dream towns [Internet]. Outside; 2005 Aug 1. Available from: https://www.outsideonline.com/adventure-travel/destinations/north-america/new-american-dream-towns/
64. Shandas V, Hellman D. Toward an equitable distribution of urban green spaces for people and landscapes: an opportunity for Portland's green grid. In: Nakamura F (ed.) Green Infrastructure and Climate Change Adaptation. Singapore: Springer; 2022. p. 289–301.
65. Nesbitt L, Meitner MJ. Exploring relationships between socioeconomic background and urban greenery in Portland, OR. Forests. 2016;7(8):162.
66. Gibson KJ. Bleeding Albina: a history of community disinvestment, 1940-2000. Transformative Anthropology. 2007;15(1):3–25.
67. Goodling E, Green J, McClintock N. Uneven development of the sustainable city: shifting capital in Portland, Oregon. Urban Geography. 2015;36(4):504–27.
68. Hare W. Portland's gentrification has its roots in racism. High Country News; 2018 May 28.
69. Lane D. Major lenders aid decline of NE Portland [Internet]. The Oregonian; 1990. Available from: https://www.oregonlive.com/portland/2014/08/major_lenders_discourage_homeo.html
70. Bates LK. Gentrification and Displacement Study: planning implementing an equitable inclusive development strategy in the context of gentrification [Internet]. Urban Studies and Planning Faculty Publications and Presentations; 2013. 95 p. Available from: https://doi.org/10.15760/report-01
71. Chapple K, Thomas T, Zuk M. Urban Displacement Project website [Internet]. 2021 [cited 2022 Oct 16]. Available from: https://www.urbandisplacement.org/
72. Hagerman C. Shaping neighborhoods and nature: urban political ecologies of urban waterfront transformations in Portland, Oregon. Cities. 2007;24(4):285–97.
73. Mahmoudi D, Lubitow A, Christensen MA. Reproducing spatial inequality? The sustainability fix and barriers to urban mobility in Portland, Oregon. Urban Geography. 2020;41(6):801–22.
74. Donovan GH, Prestemon JP, Butry DT, Kaminski AR, Monleon VJ. The politics of urban trees: tree planting

is associated with gentrification in Portland, Oregon. Forest Policy and Economics. 2021;124:102387.

75. Goodling E. Urban political ecology from below: producing a "peoples' history" of the Portland Harbor. Antipode. 2019;53(3):745–69.
76. Triguero-Mas M, Anguelovski I, Connolly JJT, Martin N, Matheney A, Cole HVS, et al. Exploring green gentrification in 28 Global North cities: the role of urban parks and other types of greenspaces. Environmental Research Letters. 2022;17(10):104035.
77. Herron E. Racist housing policies have created dangerous "heat islands" in cities [Internet]. Vice; 2020 Apr 22. Available from: https://www.vice.com/en/article/m7q8v4/racist-housing-policies-have-created-dangerous-heat-islands-in-portland-denver
78. Fahy B, Brenneman E, Chang H, Shandas V. Spatial analysis of urban flooding and extreme heat hazard potential in Portland, OR. International Journal of Disaster and Risk Reduction. 2019;39:101117.
79. Nilon CH. Urban biodiversity and the importance of management and conservation. Landscape Ecology and Engineering. 2011;7(1):45–52.
80. US Census Bureau. U.S. Census Bureau QuickFacts: Maryland [Internet]. Suitland, MD: US Census Bureau; 2022 [cited 2022 Jul 12]. Available from: https://www.census.gov/quickfacts/fact/table/baltimorecountymaryland,baltimorecitymaryland/PST045221
81. Grove M, Ogden L, Pickett S, Boone C, Buckley G, Locke DH, et al. The legacy effect: understanding how segregation and environmental injustice unfold over time in Baltimore. Annals of the American Association of Geographers. 2018;108(2):524–37.
82. DiMento JF, Ellis C. Changing Lanes: visions and histories of urban freeways. Cambridge, MA: MIT Press; 2013. 361 p.
83. Glotzer P. Exclusion in Arcadia: how suburban developers circulated ideas about discrimination, 1890–1950. Journal of Urban History. 2015;41(3):479–94.
84. Glotzer P. How the Suburbs Were Segregated: developers and the business of exclusionary housing, 1890–1960. New York: Columbia University Press; 2020. 320 p.
85. Riley CB, Gardiner MM. Examining the distributional equity of urban tree canopy cover and ecosystem services across United States cities. PLoS ONE. 2020;15(2):e0228499.
86. Schwarz K, Fragkias M, Boone CG, Zhou W, McHale M, Grove JM, et al. Trees grow on money: urban tree canopy cover and environmental justice. PLoS ONE. 2015;10(4):e0122051.
87. Boone CG, Buckley GL, Grove JM, Sister C. Parks and people: an environmental justice inquiry in Baltimore, Maryland. Annals of the American Association of Geographers. 2009;99(4):767–87.
88. Namin S, Xu W, Zhou Y, Beyer K. The legacy of the Home Owners' Loan Corporation and the political ecology of urban trees and air pollution in the United States. Social Science & Medicine. 2020;246:112758.
89. Kinzig A, Warren P, Martin C, Hope D, Katti M. The effects of human socioeconomic status and cultural characteristics on urban patterns of biodiversity. Ecology. 2005;10(1):23.
90. Voelker N, Swan CM. The interaction between spatial variation in habitat heterogeneity and dispersal on biodiversity in a zooplankton metacommunity. Science of the Total Environment. 2021;754:141861.
91. Anderson EC, Avolio ML, Sonti NF, LaDeau SL. More than green: tree structure and biodiversity patterns differ across canopy change regimes in Baltimore's urban forest. Urban Forestry and Urban Greening. 2021;65:127365.
92. Rega-Brodsky C, Nilon C, Warren P. Balancing urban biodiversity needs and resident preferences for vacant lot management. Sustainability. 2018;10(5):1679.
93. Aguilar AG. Peri-urbanization, illegal settlements and environmental impact in Mexico City. Cities. 2008;25(3):133–45.
94. Zérah M–H. Conflict between green space preservation and housing needs: the case of the Sanjay Gandhi National Park in Mumbai. Cities. 2007;24(2):122–32.
95. Anguelovski I, Irazábal-Zurita C, Connolly JJT. Grabbed urban landscapes: socio-spatial tensions in green infrastructure planning in Medellín. International Journal of Urban and Regional Research. 2019;43(1):133–56.
96. Barrows L, Bu L, Calvin E, Krassner A, Quinn N, Richardson J, et al. Growth Management in Medellín, Colombia [Internet]. Medellín, Colombia: Columbia University Graduate School of Architecture, Planning and Preservation; 2016. 70 p. Available from: https://planning-org-uploaded-media.s3.amazonaws.com/document/AICP-Student-Project-Awards-2016-Columbia-Report-rev.pdf
97. Irazábal C. The politics of participation: metropolitan region greenbelt planning [Internet]. International Planning Case Studies Project; 2016. Available from: https://planningcasestudies.org/case-studies-1/the-politics-of-participation-metropolitan-region-greenbelt-planning
98. Johnson S. Green 2.0 2020 NGO and Foundation Transparency Report Card [Internet]. Green 2.0;

2020 [cited 2022 Oct 11]. 101 p. Report No.: 4. Available from: https://diversegreen.org/wp-content/uploads/2021/02/green-2.0-2020-transparency-report-card.pdf
99. Taylor DE. The State of Diversity in Environmental Organizations: mainstream NGOs foundations government agencies [Internet]. Green 2.0; 2014 [cited 2022 Oct 12]. 192 p. Available from: http://orgs.law.harvard.edu/els/files/2014/02/FullReport_Green2.0_FINALReducedSize.pdf
100. Paez T. The fight against privatization of beaches in Puerto Rico [Internet]. Liberation; Newspaper for the Party for Socialism and Liberation; 2022 May 31 [cited 2022 Oct 11]. Available from: https://www.liberationnews.org/the-fight-against-privatization-of-beaches-in-puerto-rico/
101. Narcisse D. Beyond treading water: bringing water justice to America's urban poor. Race, Gender & Class. 2017;24(1–2):27–64.
102. Johnson H. A personal story. St John's Law Review. 1994;9(2):513–18.
103. Bouman MJ. A mirror cracked: ten keys to the landscape of the Calumet region. Journal of Geography. 2001;100(3):104–10.
104. Wali A, Darlow G, Fialkowski C, Tudor M, del Campo H, Stotz D. New methodologies for interdisciplinary research and action in an urban ecosystem in Chicago. Conservation and Ecology. 2003; 7(3):2.
105. Pellow DN. Popular epidemiology and environmental movements: mapping active narratives for empowerment. Humanity & Society. 2003;27(4):596–612.
106. Frazier CM. Repurposing Queens: Excavating a Black Feminist eco-ethic in a time of ecological peril [Internet] [Ph.D. Thesis]. Northwestern University; 2019.
107. ENLACE. Little Village Quality-Of-Life-Plan. Chicago, IL: ENLACE Chicago; 2013. 64 p.
108. LVEJO. About us: history [Internet]. Chicago, IL: LVEJO; 2014 [cited 2022 Jun 14]. Available from: http://www.lvejo.org/about-us/history/
109. LVEJO. 2015–2020 Strategic Plan; industrial legacies, climate adaptions, community resiliency. Chicago, IL: LVEJO; 2014. 4 p.
110. Butler K, Acosta-Córdova J, Muhammad J, Pino J. Little Village Environmental Justice Organization (LVEJO) Guidelines for Future Planning and Development [Internet]. Chicago, IL: LVEJO; 2018 [cited 2022 Jun 14]. 16 p. Available from: https://www.metroplanning.org/uploads/cms/documents/lvejo_guidelines.pdf
111. Maathai W. The Green Belt Movement: sharing the approach and the experience. New York: Lantern Books; 2003.
112. Maathai W, MacDonald M. Hands off our Uhuru Park, green spaces a matter of life and death. Daily Nation; 2019 Oct 30.
113. Green Belt Movement. The Green Belt Movement Annual Report 2011. Green Belt Movement; 2011. 18 p.
114. Fridah. Govt bows to pressure after public outcry to spare Uhuru Park [Internet]. Daily Active; 2019. Available from: https://dailyactive.info/2019/10/24/govt-bows-to-pressure-after-public-outcry-to-spare-uhuru-park/
115. Gathara P. Nairobi should rethink its colonialist approach to urban design [Internet]. Bloomberg CityLab; 2019. Available from: https://www.bloomberg.com/news/articles/2019-11-11/why-a-new-expressway-in-nairobi-is-a-bad-idea
116. Locke DH. Residential housing segregation and urban tree canopy in 37 US cities; data in support of Locke et al. 2021 in npj Urban Sustainability [Internet]. Environmental Data Initiative; 2020 [cited 2022 Sep 25]. Available from: https://portal.edirepository.org/nis/mapbrowse?packageid=knb-lter-bes.5008.2
117. Winling L. Mapping inequality [Internet]. University Libraries, Virginia Tech; 2021 [cited 2022 Oct 18]. Available from: https://data.lib.vt.edu/articles/dataset/Mapping_Inequality/14113463

CHAPTER 3

Human Motivations and Constraints in Urban Conservation

Kelli L. Larson and Jeffrey A. G. Brown[‡]

Introduction

A key challenge to conserving urban biodiversity is garnering public support for managing urban landscapes in ways that maintain and enhance biological diversity, including both plants and animals. Historically, the expansive lawns in neighborhoods, parks, and other urban settings have diminished biodiversity through the cultivation of monocultural turfgrass and associated landscape designs that minimize vegetation diversity and structure (1,2). While these homogeneous landscapes reduce diversity and can have lasting effects on land management (e.g., irrigation and fertilization), modern plant choices and landscape decisions by residents, park managers, local governments, and other entities affect biological diversity in cities. The significant influence of people's decisions on vegetation communities is evident in the introduction of non-native plant species, which has resulted in increased plant species richness in cities compared to relatively natural ecosystems (3,4). Ultimately, the decisions of multiple actors—coupled with social and environmental processes—govern urban landscapes in ways that affect vegetative and animal diversity where most of the human population lives. In addition, people's decisions collectively embody a myriad of options for managing land and other resources in cities, the outcomes of which may lead to extirpation of species and environmental harm, as well as species conservation and the enhancement of biological diversity.

Although urbanization has negatively impacted wildlife in cities, residential yards, parks, and other urban landscapes (e.g., open spaces) can provide beneficial habitat to birds, pollinators, and other wildlife (5). Moreover, human connections to nature in cities—including vegetated landscapes and wildlife—can enhance well-being through improved mental health and other benefits, such as aesthetic appreciation and recreational enjoyment (6). Yet some scholars warn that the extinction of nature experiences in cities may diminish human well-being and appreciation of nature, thereby weakening support for conservation (7). We argue that significant potential exists for urban conservation initiatives that enhance biodiversity while improving the well-being of people and wildlife. The enhancement of biodiversity and the well-being of urban residents should be a central goal in the pursuit of equitable conservation outcomes.

The social sustainability of conservation—defined as the ability of landscapes to provide ecologically beneficial outcomes while maintaining human interest over time (8,9)—is key to success since humans are the dominant species in cities. Public support for, and engagement in, urban conservation are therefore crucial. Achieving social sustainability requires thoughtful consideration and management of perspectives and actions among diverse urban residents. To garner public

[‡] Global Institute of Sustainability and Innovation, Arizona State University, USA

Kelli L. Larson and Jeffrey A. G. Brown, *Human Motivations and Constraints in Urban Conservation*. In: *Urban Biodiversity and Equity*. Edited by: Max R. Lambert & Christopher J. Schell, Oxford University Press. © Oxford University Press (2023). DOI: 10.1093/oso/9780198877271.003.0003 appreciation, conservationists should design efforts around residents' aesthetic and lifestyle preferences since people tend to prioritize landscape beauty and positive sensory experiences alongside recreational and leisure opportunities suited to their lifestyles (e.g., whether they enjoy hiking and jogging or gathering with family and friends outdoors (9,10)). Similarly, public views and behaviors toward urban animals—including residents' and others' values and attitudes that affect the appreciation and tolerance of wildlife in different contexts—must be considered for socially sustainable urban conservation (11,12). Meanwhile, understanding how individuals' actions and societal decisions currently and potentially affect urban landscapes, vegetation, and wildlife can inform and improve the management of human–environment interactions for biological conservation in cities.

The complex and multifaceted factors that influence urban conservation range from how residents and other actors manage individual parcels and local landscapes, whether yards, parks, community gardens, or other open spaces, to broader-scale forces such as municipal-to-state governance and political-economic dynamics (13,14). Accordingly, the toolkit for conservation involves various private and public management strategies that facilitate behavior change, enable policies, and otherwise incentivize or establish the most effective strategies in a given location. Geographically tailored approaches are necessary to address the diverse motivations and constraints of different people and cultures present in urban regions, in addition to the distinct ways in which people interact with specific biomes and environments.

Here, we present insights from transdisciplinary human–environment research to explain how varied, multi-scalar social and ecological forces shape people's perspectives and actions and, in turn, biological diversity in cities. Additionally, we discuss how diverse approaches can enhance urban conservation for the sustained well-being of people and wildlife across the globe. To demonstrate motivations and constraints to conservation at multiple geographic levels, we pay particular attention to residents and households as actors (inclusively defined as homeowners and renters, as well as noncitizens) since their collective actions influence conservation at broader scales. While conservation efforts by municipal, county, and other governments are critically important, we believe focusing on diverse residents is central to socially sustainable conservation that is both effective and equitable.

Factors that motivate and constrain conservation

Four social science concepts are useful for thinking about the range of human influences on conservation in general. First, broadly defined, *attitudinal* factors underscore the value-based interests and ideals that motivate people's actions at individual-to-societal levels, thereby influencing public support and action for conservation (15–17). As explained further below, attitudinal factors encompass personal, social, and environmental values, beliefs, attitudes, intentions, and related constructs in the realm of human judgements and decisions. Second, *institutional* forces are defined as the formal and informal rules that govern human behavior and societal decisions. Whereas formal rules are those codified in law and therefore legally enforceable, informal rules encompass norms and customs that are enforced through social relations (13,18). Third, *structural* forces include aspects of the physical environment and socioeconomic dynamics (including technology) that enable or constrain personal or societal decisions toward conservation outcomes (13,17). In urban systems, structural factors include environmental conditions and infrastructure, as well as market-based and political-economic realities that constrain or control behaviors and associated outcomes. Fourth, *situational* or contextual forces are particular to human–environment interactions in specific places (15,19). Although related to the other forces described herein, situational factors are worth distinguishing since they are crucial to understanding problems and targeting solutions to conservation in particular places (for a specific case example, see the case study).

Attitudinal factors

Commonly defined as positive and negative judgments people hold, attitudes encompass relatively general to specific evaluations about some object or phenomenon (20,21). Herein, we use the phrase attitudinal forces broadly to encompass the entire cognitive hierarchy (Figure 3.1), ranging from general life values and broad-based beliefs, or value orientations (also called worldviews), to more specific attitudes and behavioral intentions (16,22). Values are general principles that influence everyday decisions and transcend situations (23). Related to conservation, important values include altruism and biocentric worldviews that emphasize the rights and protection of nature and wildlife over anthropocentric values that stress utilitarian benefits and impacts on people (24,25). Biocentric values also encompass mutualistic worldviews that stress living in harmony with nature and coexisting with wildlife. Related to values, beliefs about what is right and wrong, or good and bad, further influence people's decisions about biological conservation. Beliefs and attitudes toward behaviors and their outcomes have long been known to affect individuals' actions, including those related to the ease of undertaking actions and perceptions about their consequences (15,26). Overall, a variety of values, beliefs, and attitudes combine to influence human motivations and willingness to undertake conservation and other environmental management activities. However, attitudinal judgments are often steadfast and difficult to change, and due to the other forces influencing behaviors, people's decisions and actions do not always reflect their expressed values and attitudes (21,25).

To facilitate urban conservation, appealing to anthropocentric values related to aesthetics and leisure will likely increase public support and appreciation, particularly in the US and developed contexts (10,27). When promoting naturalistic

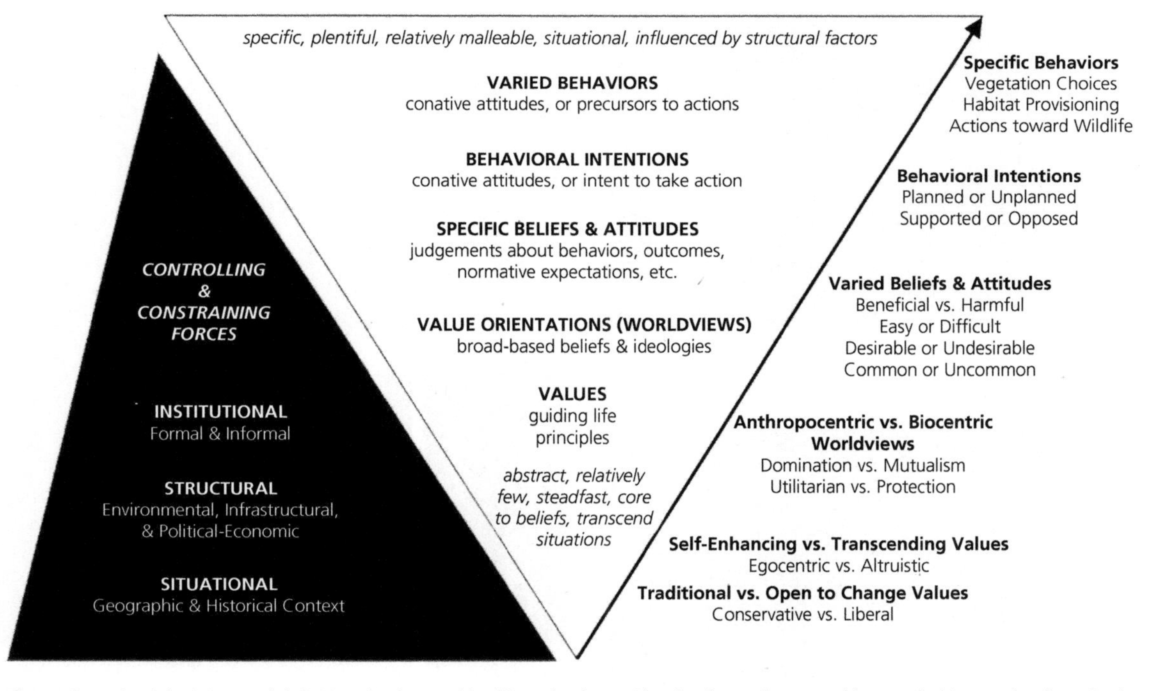

Figure 3.1 Visual depictions and definitions for the cognitive hierarchy along with other forces that control human decisions and actions. On the right, the specific values, beliefs, worldviews, attitudes, and behaviors presented represent core constructs and ideas in the scholarly literature, especially as applied to environmental and biological conservation. On the left, the dark triangle depicts other forces that control and constrain individual and societal behaviors and decisions. The solid base of these controlling factors is a visual reminder of the tenuous relationship between values and actions, and attitudes and behaviors, which, as the cognitive hierarchy triangle suggests, can be easily "tipped" or overcome by other factors.

Source: adapted from Whittaker et al. (22) and Larson et al. (16).

landscapes, their aesthetic appeal is paramount. Although aesthetic tastes vary, research shows that people overwhelmingly appreciate neat, orderly landscapes, or those managed with "cues-to-care" such as linear or curvilinear features, flowering plants, and open vistas (8,28). These and other cues-to-care can garner public appreciation and attention for naturalistic landscapes (for examples, see Figure 3.2 and https://www.allaboutbirds.org/news/tips-to-make-a-wildlife-garden-look-great/). Yet given varied lifestyles and interests, parks and other landscapes should be managed for diverse goals that may require functional landscapes (e.g., lawns and ballfields) and other recreational infrastructure, in addition to naturalistic landscapes inclusive of native plants and other features that enhance biological conservation. Ultimately, providing leisure opportunities associated with urban nature can enhance public support for biological conservation (10). Since urban residents have varied recreational interests, providing diverse nature opportunities in parks and open spaces (e.g., safe lawns and playgrounds for kids to play and trails for hiking and birding) is worthwhile.

For yard management within neighborhoods, many residents prioritize landscapes that support their lifestyle preferences. While some residents manage landscapes and gardens as a leisure pursuit (i.e., hobby), which has been linked to managing landscapes for wildlife, a majority of residents tends to prefer exerting little time and effort in managing their yards (27,29). As a result, conservation efforts in residential neighborhoods must contend with low-maintenance yard priorities, which underscore designs including slow-growing plants that

Figure 3.2 Photos demonstrating various cues-to-care for enhancing public appreciation of nature-based conservation solutions. (a) Signage at a rainwater garden explains how it protects water quality (Maplewood, MN). (b) Bollards delineating an edge and flowering shrubs provide cues of management for a bioretention garden (Detroit, MI). (c) The combination of trees planted in a straight row and incorporation of showy flowers highlight that an urban wetland is actively managed (North Saint Paul, MN). (d) A clearly mown edge along a walkway signals care of a wetland in an urban park (Saint Paul, MN).
Photos courtesy of Joan Nassauer.

minimize pruning and maintenance, as well as low water-use plants that decrease irrigation requirements. Meanwhile, increasing plant variety or abundance and promoting plants that require significant pruning or maintenance are more likely to face resistance among land managers.

Regarding animals, mutualistic values that emphasize human–wildlife coexistence are critical for urban conservation, as is managing tolerable levels of risks when residents have concerns about specific wildlife or aspects of urban nature (11,12). Historically, Americans have tended to emphasize the utilitarian values of wildlife and nature (30). However, value orientations have shifted toward mutualism since the 1970s to emphasize the care and protection of wildlife. With this shift in values, attitudes toward various wildlife have become more positive. For example, one study showed that a higher portion of US residents in 2014 liked bats, vultures, and rats compared to a 1978 study (31). Butterflies, ladybugs, coyotes, and lizards have also garnered more positive attitudes (31). Meanwhile, attitudes toward other species, including raccoons, swans, skunks, and mosquitoes, have become more negative, perhaps due to the perceived nuisances or damage they cause (31). Addressing both real and perceived concerns about safety and health outcomes is paramount for socially sustainable urban conservation,[1] as is overcoming the disgust and other negative sentiments people tend to exhibit toward arthropods, insects, and reptiles (32,33). Addressing public fear and disgust for certain wildlife could include procedural information on how to avoid or deter disliked or potentially harmful wildlife. Meanwhile, highlighting the species that are more valued and appreciated by the public can boost support for conservation.

Institutional forces

Institutions are central to understanding environmental governance, which broadly refers to societal decisions by government and nongovernment actors (34,35). Uncodified, informal institutions are social norms that affect individual behaviors and societal decisions. Significantly influenced by values, norms operate through people's beliefs about socially appropriate or desirable behaviors, which are "enforced" by individuals' propensity to conform and associated social sanctions (e.g., peer pressure; 18). In many urban contexts, societal expectations about maintaining traditional lawns—with their manicured look and hyper-green monocultures of turfgrass—represent one constraint to biodiverse landscapes in neighborhoods and parks (1). More broadly, people may become accustomed to particular plants, including non-native vegetation, and thus want to maintain them, especially if they have historic or cultural significance that reinforces their appeal (18,36). Regardless, norms are very powerful forces that affect human behavior, and when deeply engrained, they can result in habitual actions that are hard to break (15).

Legally enforceable, formal institutions at various levels of governance (see the case study and the "Multi-scalar drivers" section below) also influence vegetation choices and human–wildlife interactions through regulations and restrictions on what people can and cannot do. In many cases, informal rules may be codified through legal channels if acceptance of a social norm is widespread, and they can also help enforce regulations. One example includes US municipal regulations on "weed heights" that commonly require mowing or removal of grass or other urban vegetation when they reach a height of 3–12 inches (9). Historically, weed-height regulations were put into place to control insects, rats, and other "pests," but these regulations also reinforce neat, orderly landscapes including traditional lawns. Similar laws exist for the maintenance of trash in urban areas that are associated with public health and safety, in addition to avoiding nuisance vegetation or wildlife. Ultimately, formal rules are often only effective when they are enforced, whether through legal channels or informal social pressure.

Historic decisions embedded in formal and informal institutions are often barriers to modern conservation goals such as expanding naturalistic landscapes and wildlife habitat, as well as equitably distributing access to nature across metropolitan

[1] Byers et al.'s chapter discusses the strong linkages between wildlife and human health in cities.

Figure 3.3 Example "No Mow May" images circulated by the UK Plantlife organization that promotes individuals and communities circulating their pledges to support bees and other pollinators.
Images courtesy of Plantlife (https://nomowmay.plantlife.org.uk/resources/).

regions (9).[2] In the face of such limits, collective action among government and nongovernment entities may be needed to counter established rules and/or to change social norms. One example that has emerged in recent years is the promotion of "no-mow lawns" or pollinator gardens that do not comply with weed-height ordinances (37). Given the significant potential for residential yards to support bees, butterflies, and other pollinators (5,38), overcoming the institutional obstacle posed by weed-height ordinances is an important urban conservation priority. An example is the rise of "no mow May" campaigns, which originated in the UK through an organization called Plantlife (Figure 3.3) (see https://nomowmay.plantlife.org.uk/) and has spread to the US and elsewhere. This effort relies on social media and messaging to support pollinator lawns by discouraging mowing for the month of May. In some locations, such as the Minneapolis–St. Paul area in Minnesota, conservation advocates have successfully lobbied municipal governments to officially waive weed-height ordinances in the critical month of May to support nectar-hungry pollinators in the spring season (37).

Structural forces

Structural constraints to conservation include broader forces that shape people's decisions, often in ways that limit personal volition (i.e., agency; 16,39). Structural factors are a major reason why environmental values and attitudes often do not lead to conservation actions, as is the case for lawn management (1,16). As Robbins (1) argues, the marketing forces of the lawn industry have led residents to intensively maintain grass—with high inputs of fertilizers, pesticides, and, in some regions, water—in spite of alternative preferences and environmental concerns about lawn management. Such forces are powerful since multibillion-dollar corporations profit from selling turfgrass, chemicals, and machines such as lawnmowers to achieve a golf course-like appearance. In doing so, the industry appeals to normative pressures of maintaining industrial lawns by marketing the status associated with perfect lawns and even the happiness of families who have them. In the face of such powerful forces, counterefforts are crucial including the marketing and promotion of alternative lawns (e.g., clover or pollinator lawns) that require fewer inputs and less intensive management (40).

Beyond lawn-care companies, other industries influence vegetation and landscaping choices. Another major player are plant nurseries, whose business practices constrain the vegetation available to residents, developers, landscape architects, and other consumers. Critical barriers to planting native vegetation include their limited availability in local plant stores, especially prior to the 1990s, as well as higher prices charged for native plants (41,42). The broader horticultural industry further determines which types of vegetation can be cultivated in nurseries and urban environments, while landscape architects influence historic and modern trends that enhance or diminish the appeal of naturalistic landscapes that support wildlife and biodiversity (see the case study on pages 55–56). Furthermore, the real-estate industry—especially developers—have significant control over the

[2] Hoover's and Scarlett's chapter paints in detail the need for equitable urban conservation.

installation of vegetation and urban landscapes, which can affect their management for several years or decades (16,36).

Lastly, landscaping and pest-control services affect plants and animals in urban areas by spraying pesticides to control weeds, insects, and other unwanted biota that can be viewed as messy or otherwise raise concerns among urban residents (43,44). While unwanted vegetation, insects, and other pests are often sprayed with toxic chemicals, unwanted mammals and other animals can be trapped or removed through non/governmental services to either minimize nuisances or reduce the fear associated with certain wildlife (e.g., carnivores and snakes; 45,46). Broad-sweeping actions to control species are also commonly associated with negative health outcomes for people and include both direct and indirect consequences for biodiversity.[3] For example, pesticide campaigns to control mosquitoes, whether to control diseases or nuisances, kill both target and nontarget species (47) and reduce the availability of food for insectivorous species (48). Moreover, the targeting of carnivores, such as gray wolves, coyotes, and foxes, results in increased disease prevalence among wildlife populations and contributes to the rise of destructive species such as feral cats, which are known to kill birds (49,50). Nonlethal pest management strategies, such as trapping and relocating wildlife, can help keep urban wildlife communities intact but often face monetary barriers and may confront opposition if people are unwilling to live with certain species (51).

Overall, structural forces affecting conservation include political-economic dynamics involving markets and the capitalistic drive for profit that enable or constrain decisions, along with marketing and promoting vegetation and wildlife management in ways that impact conservation outcomes. In addition, environmental conditions including historic and existing land use/cover change influence what is possible by determining vegetation composition and habitat structure, which individual landowners may not be able or compelled to change. Environmental, social, and other political factors that represent potential structural drivers or constraints are further discussed below, specifically as situational forces affecting conservation in particular settings.

Situational factors

Situational forces encompass the biophysical, infrastructural, sociocultural, political, economic, and historical circumstances that enable or thwart conservation in particular locations. Although we cannot detail all the ways in which different situational factors affect conservation, we highlight three points to demonstrate how contextual factors influence conservation.

First, legacy effects underscore how past conditions and decisions affect present-day circumstances (14,36,52). Examples we have already discussed include: (1) historic values and traditions (e.g., anthropocentrism and the lawn norm), which are slow to change, that influence public priorities and landscaping practices; (2) previously implemented laws and regulations (e.g., weed-height and pest ordinances) that constrain the modern creation of naturalistic landscapes and wildlife habitat in residential and urban settings; and (3) former landscape installations and vegetation choices that persist well into the future to affect land management and environmental outcomes. Another poignant example of legacy effects is racist redlining policies, which were largely implemented in the early to mid-1900s to segregate residents by race and ethnicity. Ultimately, these housing polices led to investments—including the establishment of parks, trees, and other vegetation—in whiter and wealthier areas (53). The legacies of these policies remain today in widespread urban environmental injustices across American cities. Specific examples include fewer parks and less vegetation in predominately Black and Brown neighborhoods compared to their White counterparts, and, by extension, lower access to recreational opportunities, the shading and cooling effects of trees, and other benefits (54).[4]

[3] Byers et al.'s chapter discusses how management actions for people and/or wildlife can have unintended health consequences.

[4] Hoover's and Scarlett's chapter reviews the extensive inequities in access to urban nature and conservation.

Second, behavior research has underscored the effectiveness of situational approaches in fostering conservation behaviors, particularly through the approach known as community-based social marketing, or CBSM (19). Recognizing that human behaviors are highly dependent on the contexts in which people live, CBSM offers a pragmatic strategy for assessing the motivations (benefits) and constraints (barriers) to desirable and undesirable behaviors with specific conservation outcomes. Depending on the factors that affect behaviors in particular settings, a range of potential strategies can be tailored to increase the positive outcomes and decrease the negative outcomes associated with people's choices and actions. As a whole, CBSM draws from behavioral theory and research while relying on a market segmentation approach that underscores targeting solutions to particular people and contexts.

In targeting strategies to specific contexts, demographic attributes of people and populations are important considerations, as they may affect conservation activities in varied ways. One way to think about demographics, which is consistent with market segmentation approaches, relates to values and priorities that are linked to different life stages and lifestyles; for example, whereas elderly people may prefer low-maintenance, climate-adapted yards, adults with young children may prefer lawns that allow kids to play outdoors (10,36,55). Another conceptual lens through which to examine demographics is personal or social capacity, which may be affected by education and income levels, age or homeownership, race or ethnicity, or other personal and household attributes (15). Regarding personal capacity and other explanations associated with demographic attributes, we caution natural and physical scientists against simplistic thinking about what they represent. For example, in originally describing the "luxury effect," Hope et al. (56) position income as "financial wherewithal," implying that higher biodiversity in wealthier neighborhoods is explained by intentionally increasing vegetation. While economic capacity might be a relevant mechanism to explain biodiversity, such myopic views do not fully recognize the complexities of social-ecological dynamics. Moreover, they often overlook more critical perspectives that underscore the power of status, privilege, and other structural or systematic forces that explain environmental conservation in some areas but not others.

Multi-scalar drivers and interactions in social-ecological systems

The societal forces that impact conservation operate at different geographic levels through individual and collective behaviors and decisions (Figure 3.5). Accordingly, scholars and practitioners must carefully consider how personal and local factors affect regional and broader-scale forces to enable or constrain conservation, in addition to how broader-scale forces affect individual and local-scale conservation decisions and practices. At the individual scale, actors in urban conservation include residents, landowners, activists, business leaders, politicians, government personnel, and others who undertake or influence actions to enhance or support biological diversity. At broader scales, actors include government and nongovernment organizations as well as for-profit companies at local (municipal), regional (e.g., county or metro-wide), state, national, and global levels. To illuminate cross-scale motivations and constraints, we primarily focus on residents and households as actors whose collective actions and voting behaviors can influence conservation, in addition to how institutional, structural, and situational factors affect individual and local-scale decisions.

Although changing individual behaviors presents a unique challenge to conservation, specifically because the ultimate environmental impacts of human behavior largely depend on collective actions undertaken at larger scales, we argue that focusing on individuals' motivations and constraints is crucial for garnering the widespread support and action needed to protect biodiversity and achieve nature conservation in cities and beyond. Yet action and change at the broader scales of local, state, and federal governments are needed, coupled with consideration of the political-economic forces of corporations and nonprofit organizations, to effect change from both the top-down (e.g., through government regulations)

Case study How multi-scalar forces affected a conservation-oriented landscape transition in the desert metropolis of Phoenix, AZ, USA

Figure 3.4 Examples of climate-adapted xeric landscapes in metropolitan Phoenix, AZ, USA.
Photos by author Kelli L. Larson.

Situated in the warm, arid biome of the Sonoran Desert, metropolitan Phoenix, AZ presents an interesting case wherein multiple factors have influenced a shift toward more naturalistic, desert-like landscapes in recent decades. Specifically, residential and other landscapes have shifted away from traditional lawns toward climate-adapted xeric yards with gravel groundcover and low water-use vegetation (Figure 3.4) (17,36). In fact, while approximately 70%–90% of houses built before the 1980s were planted with turfgrass, this number drastically declined by the 2010s, with only 10% of new homes built since then having lawns (57). While government and media sources often suggest that residents' increasing preferences for xeric landscapes have driven this change, empirical research tells a different story in which the decisions of developers and landscape architects are more likely the central drivers of this landscape transition (17,36). As Ferris explains, the shift toward xeric landscapes in greater Phoenix is at least partly due to a "new breed of architects and designers" who emphasized "living with the local environment" and, by extension, designed new landscapes that embraced local vegetation such as palo verde trees (genus *Parkinsonia*), *Agave*, and other desert plants.

The legacies of lawn installation before the 1980s persist today in the existing landscapes of older neighborhoods and developments (36). Multiple factors explain the persistence of lawns, including the cost and effort associated with their conversion. Some residents have also become accustomed to the lush landscapes of yesteryear and still want to maintain lawns (17). In fact, despite the still-common belief—among policy professionals, real-estate personnel, the media, and broader public—that Midwesterners migrating to the Southwest are "bringing" lawns with them, several studies since the early 2000s have shown that long-time Phoenicians are the residents who demand the lush, exotic landscapes to which they have become accustomed (for a review, see 36). In some older neighborhoods, these historic norms are reinforced by historic overlays (designated by the municipal government) that preserve palm trees and their "traditional character," while local residents also sometimes pressure their neighbors to maintain lawns rather than converting their yards to xeric landscapes (18). Broadly, the well-watered landscapes in the "Phoenix oasis"—which have been promoted by booster campaigns declaring, "the desert is a myth" and "do away with the desert"—have left a legacy in which longer-term residents actually prefer and maintain grass more so than newcomers from humid climates (10,36).

Another important finding from Phoenix and elsewhere is that the shift away from high-management lawns has not been driven by residents' individual concerns about drought, water conservation, or environmental impacts (10,16). Instead, biocentric values have been associated with both xeric yards and mesic lawns, wherein some people even view the cultivation of grass in the desert as tending to nature. Moreover, individuals with more biocentric views have been shown to irrigate their yards more frequently in winter months, likely due to their enjoyment of landscaping as a form of outdoor recreation. Given these counterintuitive realities, efforts that assume environmental values, knowledge, or attitudes will lead to conservation action are likely to fail (21,30). As another example, research from Phoenix has shown that neither prioritizing low water-use plants nor increased knowledge about native plants is associated with actual native vegetation diversity in people's yards (58). However, prioritizing native vegetation when making plant choices does significantly influence yard composition, as do beliefs that native (desert) plants belong in the city.

Since aesthetically appealing and low-maintenance yards are the top motivations for many residents overall, conservationists should attend to these priorities (10,29). This approach was recently underscored by a study that found residents who prioritize easy-to-maintain yards are significantly less likely to garden for wildlife. Meanwhile, those who garden for leisure (i.e., as a hobby) are more likely to plant native vegetation and incorporate other

Case study *Continued*

features to support birds, pollinators, and other wildlife (27). Yet wildlife gardeners or people who enjoy landscaping may not always be conservation-minded. Moreover, and as noted above and found elsewhere, residents who enjoy spending time outdoors—including avid gardeners—may actually manage their yards more intensively by irrigating or using more fertilizers and pesticides than residents who desire low-maintenance and passive management of their yards (27,59).

Overall, diverse factors influence the adoption of naturalistic landscapes that support biological conservation. While attitudinal factors may not drive conservation behaviors in many circumstances, personal priorities and social norms do have an impact. Moreover, broader-scale forces, including the decisions of real-estate developers and trends in landscape architecture and the horticultural industry, have been strong drivers of landscape change with conservation benefits in Phoenix, AZ, as well as other regions.

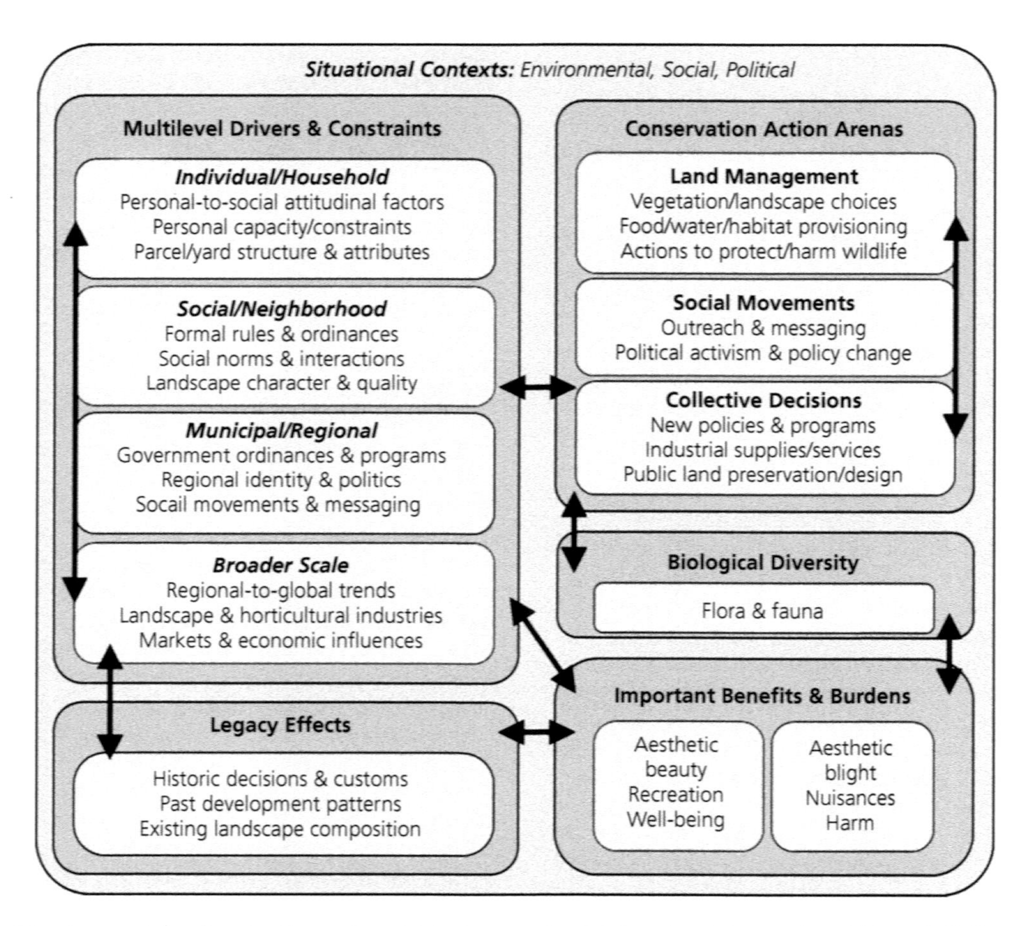

Figure 3.5 Interconnected, multi-scalar drivers and constraints of conservation. This figure depicts the varied societal forces that influence conservation at different scales and in different domains. The intent is to provide an overview, rather than an exhaustive framework. The social benefits and burdens box, in particular, highlights important outcomes for the social sustainability of urban biological conservation.

Source: adapted from Cook et al. (14).

as well as the bottom-up (e.g., through collective action among communities).

Residents and other actors significantly affect conservation through the aggregate effects of individual actions, as well as through collective efforts and organized activism for or against conservation at neighborhood, municipal, and larger scales (13). As detailed above, interpersonal, social dynamics within and beyond households and neighborhoods affect conservation through shared views, social pressures, and common behaviors that influence decisions about urban landscapes and their positive or negative impacts on biological conservation. Social groups that are particularly influential when it comes to personal values, attitudes, and behaviors are families and schools, as well as political and religious groups (60). This is largely because ideals, customs, and habits are often shaped and reinforced through everyday social interactions, especially in childhood or through groups that are inherently ideological (i.e., religious groups and political affiliations). One example is the Judeo-Christian roots of the predominantly anthropocentric worldview in the Western world, which positions people as having dominion and control over nature, which has been indicted in today's ecological crises (61,62). In contrast, scholars have described a more interdependent worldview in which Hispanic and Latinx communities tend to see themselves as more interconnected with nature and, as such, often feel more vulnerable to risks. While some cultural beliefs (i.e., shared views among a group of people) are situated in particular places, socially shared "cultural" views are not necessarily bound to specific locations, especially given the high levels of mobility and telecommunications in modern society. However, many social institutions (i.e., norms and regulations) and structural forces (e.g., political-economic, environmental factors) do indeed operate in specific places and across geographic levels, specifically within neighborhoods, municipalities, and broader governance units that affect conservation in metropolitan regions. Locally, neighborhood dynamics are important to conservation since neighboring residents often influence each other's land management through formal and formal institutions that operate within shared landscapes.

Within urban neighborhoods, residents influence each other through norms and the local activities of neighborhood organizations that affect household-level decisions through stated or implied expectations to maintain the quality and character of neighborhood landscapes. Oftentimes, these local social interactions can lead to homogeneity of landscapes and their management within neighborhoods (63,64). Environmental and social differences across neighborhoods can also affect the wildlife experienced by local residents, along with associated actions taken to either support or thwart their protection. For instance, due to local vegetation structure in yards and nearby parks and preserves, some neighborhoods experience relatively drab generalist birds whereas others experience more diverse birds including colorful and unique specialist species (64). This, in turn, can affect local appreciation of wildlife and continued or further efforts to manage residential landscapes for wildlife (65). Neighborhood organizations such as Homeowners' Associations (HOAs) are also a noteworthy actor in urban conservation since they are increasingly prevalent in urban subdivisions and commonly regulate vegetation and land management practices, often to the detriment of environmental outcomes (66). However, the effects of neighborhood dynamics are highly context-specific; for instance, HOA rules have been linked to traditional lawn management with high chemical inputs (in Baltimore, MD; 64), as well as to more naturalistic landscapes with higher plant and bird diversity (in Phoenix, AZ; 65).

Furthermore, municipal regulations determine what is allowed and prohibited within their jurisdictions, such that residents and neighborhood organizations must comply or face legal repercussions such as fines. The primary goals of municipal ordinances tend to focus on maintaining aesthetics and avoiding nuisances, along with safeguarding public health and safety (9). As discussed above, common regulations in many municipalities limit "weed heights" and require the maintenance of trees to avoid encroachment or harm to nearby residents. Some municipalities also require a minimum amount of vegetation for new developments, which can support biological diversity depending on the specific legal stipulations. In some regions and states, government agencies or other organizations often compile plant lists that guide residents and others about the characteristics of plants, sometimes including details about nativity and whether they attract birds or other wildlife (for details, see 9). While plant recommendations are not compulsory, such guides can be very informative in promoting vegetation with positive conservation outcomes that also address aesthetic and maintenance preferences and priorities based plant attributes (e.g., flowers, growth rates, leaf litter). Overall, municipalities often engage in regulatory and nonregulatory programming that can affect conservation of plants and animals, whether intentionally or not (9). Residents, conservation groups, and other entities can therefore lobby local governments to change regulations and programs that impede conservation, or to implement policies or efforts to support biodiversity conservation (e.g., as seen above in some municipalities waiving weed-height ordinances in the spring to support habitat for pollinating bees).

As also discussed above, situational factors in particular geographic contexts significantly affect conservation, including hydroclimatic conditions, political and economic forces, and historic legacies of past decisions and customs that constrain or enable specific practices or decisions (see the case study above). The implementation of state or national policies and programs may also set priorities and legal expectations for diverse actors in their jurisdictions. The U.S. Endangered Species Act is one national example that mandates protection of plants and animals under threat for extinction. Internationally, the UN Convention on Biological Diversity and the Convention on International Trade in Endangered Species of Wild Fauna and Flora are two examples that may affect urban conservation globally. Generally, urban conservationists can leverage broader-scale policies and programs while also addressing how they might either support or inhibit local efforts.

Social-ecological applications: conservation strategies

As emphasized throughout, conservation efforts are most likely to succeed if they target local areas and otherwise tailor activities to the needs and interests of diverse people. Locations planned for new urban development or redevelopment are a high priority for urban conservation since installed landscapes can set the stage for biodiversity and land management for decades.[5] Neighborhoods or areas near parks and preserves may also be more receptive or relevant as targeted geographic areas, since natural areas and open spaces often serve as wildlife habitat (64,67). Working with conservation-minded politicians, developers, community organizations, and local opinion leaders can further leverage conservation actions in particular places or at broader scales.

Rather than trying to change people's beliefs and opinions, navigating and working with existing values and ideals is more likely to be effective for urban conservation (21,25). For instance, promoting vegetation and landscape designs that are easy to maintain *and* support biological diversity is recommended. This might include marketing or incentivizing native plants or other vegetation that are climate-adapted (to minimize irrigation needs), slow growing (to minimize pruning), or otherwise reduce maintenance efforts. Considering aesthetic ideals, designing naturalistic and wildlife-supporting yards with cues-to-care (Figure 3.2) is also crucial (8,68). Specific tips to foster appreciation of conservation-oriented landscapes include: mowing or maintaining linear or curved edges; arranging plants so they do not get overcrowded and messy; liberally using flowering plants; establishing viewsheds that emphasize color and contrasts; and maintaining features—such as bird boxes and baths, feeders, bat or bee houses (68)—that exhibit the intent to support wildlife. In general, emphasizing the aesthetic and low-maintenance appeal of conservation-oriented landscapes and actions, as well as recreational and other benefits (e.g., colorful birds, pest control, or pollination services), can foster appreciation and support for their adoption.

Similarly, highlighting iconic native wildlife that people appreciate as conservation mascots might enhance support for and actions toward

[5] Pejchar's and Reed's chapter discusses new residential development plans that consider conservation upfront.

conservation (69,70). A good example in America is the bald eagle; the decline of eagle populations in the early 1900s was of particular concern in the US since it is closely associated with residents' sense of patriotism, freedom, and pride. Federal bans in the use of DDT—which significantly led to the decline in eagles—ultimately resulted in the successful conservation of eagles and other species impacted by pesticides (71). This success story underpins the potential to leverage positive and patriotic views of wildlife in urban conservation efforts.

In general, conservation is most likely to be supported when species are charismatic (e.g., large or colorful (72)); iconic in a region (e.g., the giant panda in China); or symbolize a cultural ideal (e.g., edelweiss representing individualism and ruggedness in the European Alps (71,73)). Meanwhile, conservation of disliked or disdained species, which can include carnivores, insects, and herpetofauna, especially snakes, may experience greater opposition or resistance (72). However, the support for particular wildlife may be regionally or culturally specific; as an example, see the support for the conservation of bats in Tanzania (74). When attempting to conserve regionally disliked species, emphasizing the ecosystem services, or benefits, they provide may help foster coexistence (45). However, changing irrational fears or attitudes and behaviors rooted in trauma or deeply engrained ideologies will be difficult if not impossible (12). While creating positive or benign experiences with wildlife (e.g., at zoos or through youth education) may help foster tolerance and coexistence, specific procedural tips (75) on how to deter unwanted wildlife from private property or how to keep people, pets, and property safe will likely be more effective.

Generally, behavioral research has shown that procedural knowledge—that is, on how to undertake conservation behaviors—is far more effective in spurring action than declarative knowledge, which is factual or technical information (75). In fact, many environmental education efforts often fail to change behaviors because they focus more on declarative knowledge than other domains of knowledge that better align with social science theory on what motivates or inhibits human behavior. Two other types of knowledge that are more effective in motivating conservation behaviors include effectiveness and social knowledge (76). Effectiveness knowledge emphasizes the impacts of behaviors and, therefore, underscores behavioral theory's focus on individuals' perceived consequences of behaviors, as well as their sense of self-efficacy (i.e., as their ability to undertake behaviors to effect change; 15,77). Meanwhile, social knowledge relates behavior norms in representing people's understanding or beliefs about which behaviors are common and socially accepted (75,76). Thus, when using educational and outreach approaches, reliance on procedural, effectiveness, and social knowledge regarding conservation behaviors will be more effective than relying on technical knowledge about species or biodiversity.

As indicated earlier, changing or establishing social institutions to support conservation is another important strategy. To change formal institutions, lobbying elective officials, collective political action, or other forms of activism and advocacy that change regulations and policies may be necessary. To change informal institutions, strategic marketing to establish new trends and customs may lead to new norms of diversifying vegetation, planting natives, and supporting mutualism with local wildlife. To develop marketing strategies or otherwise tailor conservation to particular contexts, we strongly recommend following the CBSM approach, which enables evidence-based strategies for targeting messaging to people's existing values and attitudes, in addition to tailoring incentives, infrastructure, and other strategies to specifically overcome the barriers to conservation actions in particular contexts (19). Lastly, partnerships with government agencies at varying levels and community organizations of various types can synergize conservation. Such collective efforts include collaborations among researchers and practitioners to inform the evidence-based solutions needed for urban conservation.[6]

[6] Magle et al.'s chapter details collective efforts among diverse conservation groups to provide management relevant for biodiversity in multiple cities.

Global perspectives

The species in urban regions and how people interact with them shape residents' attitudes and actions toward local conservation of diverse plants and animals (78,79). While some species or taxa are generally viewed positively or negatively across cultures, local cultural values and traditions, coupled with species interactions, influence a community's willingness to protect plants and animals (80–82). To successfully implement conservation initiatives, practitioners must understand and incorporate local perspectives and acknowledge historic and current human–wildlife interactions in distinct regions across the globe. Here, we highlight examples of attitudes toward non-native and invasive plant species, as well as megafauna (i.e., large and charismatic wildlife species), to demonstrate major issues associated with urban conservation outside the US.

In countries including England, Australia, China, and the United Arab Emirates, conservationists are increasingly promoting the use of native plants in residential and public landscapes (83–85). The drive to promote native plants stems from the desire to increase local biodiversity and increase plants that are locally adapted to specific climates and environmental conditions (55,86,88). However, current and historic trade and transportation of species from areas including the South Pacific, South America, and Africa can distort the perception of "nativity" (89,90). For example, in the Thulamela Municipality of South Africa, residents adopted 21 species of non-native plants for a wide variety of uses, with some plants becoming staples as food sources and in traditional medicine (91). In cases where non-native species become staples, whether economically or culturally, efforts to eradicate non-native vegetation and restore native vegetation may receive little support. Thus, it is critical to understand how local communities utilize vegetation so conservation efforts can meet local needs. This may include protecting culturally important non-native plants, especially when they are not invasive or detrimental, along with promoting locally native vegetation.

Globally, many conservation programs focus on protecting megafauna species (e.g., lions and tigers), especially in Sub-Saharan Africa and other regions of the Global South. However, while megafauna often garner international support for conservation, local conflicts with large carnivores or other wildlife can limit the willingness of residents to undergo conservation actions (92–94). Large cats have long been poster species for wildlife conservation (94,95) but can be perceived as risky or dangerous when their distribution overlaps with urban landscapes (92,96). Similarly, conservation target species such as primates may be negatively perceived when they venture into urban landscapes and damage property (97,98). To conserve urban wildlife in the Global South, understanding how species are perceived locally, not just globally, is essential. Moreover, identifying ways to reduce the impacts of wildlife on local livelihoods and property is paramount to just conservation that ensures the well-being of local residents as well as wildlife.

Conclusion

Understanding and overcoming the human motivations and constraints to urban conservation requires robust broad-based knowledge from social scientists including psychologists, sociologists, political scientists, anthropologists, and geographers, among other disciplines and fields. Interdisciplinary research among social and ecological scientists is also necessary to understand how individual and societal decisions affect conservation outcomes, as well as how biodiversity of vegetation and animals in turn impacts the health and well-being of people. To inform urban conservation, we have highlighted the myriad of attitudinal, institutional, structural, and situational forces that enable and impede urban conservation actions and support for biodiversity preservation at various geographic scales and across diverse places. Since human–environmental interactions are complex and often situationally specific, we recommend tailoring urban conservation initiatives to particular people and places. Evidence-based strategies are imperative given that common assumptions about what motivates and constrains choices are often wrong, and the complexity of social-ecological dynamics defies myopic and simplistic solutions. Partnerships among conservationists, researchers, policy-makers, activists, and community groups are most likely to be effective in creating synergies that advance conservation knowledge and practice locally to globally.

References

1. Robbins P. Lawn People: how grasses, weeds, and chemicals make us who we are. Philadelphia, PA: Temple University Press; 2012. 209 p.
2. Blanchette A, Trammell TL, Pataki DE, Endter-Wada J, Avolio ML. Plant biodiversity in residential yards is influenced by people's preferences for variety but limited by their income. Landscape and Urban Planning. 2021;214:104149.
3. Knapp S, Kühn I, Schweiger O, Klotz S. Challenging urban species diversity: contrasting phylogenetic patterns across plant functional groups in Germany. Ecology Letters. 2008;11(10):1054–64.
4. Faeth SH, Saari S, Bang C. Urban biodiversity: patterns, processes and implications for conservation. In: Encyclopedia of Life Sciences [Internet]. Hoboken, NJ: John Wiley & Sons, Ltd.; 2012 [cited 2022 Sep 28]. Available from: https://onlinelibrary.wiley.com/doi/abs/10.1002/9780470015902.a0023572
5. Lerman SB, Narango DL, Avolio ML, Bratt AR, Engebretson JM, Groffman PM, et al. Residential yard management and landscape cover affect urban bird community diversity across the continental USA. Ecological Applications. 2021;31(8):e02455.
6. Soulsbury CD, White PCL. Human–wildlife interactions in urban areas: a review of conflicts, benefits and opportunities. Wildlife Research. 2016;42(7):541–53.
7. Soga M, Gaston KJ. Extinction of experience: the loss of human–nature interactions. Frontiers in Ecology and the Environment. 2016;14(2):94–101.
8. Nassauer JI. Messy ecosystems, orderly frames. Landscape Journal. 1995;14(2):161–70.
9. Larson KL, Andrade R, Nelson KC, Wheeler MM, Engebreston JM, Hall SJ, et al. Municipal regulation of residential landscapes across US cities: patterns and implications for landscape sustainability. Journal of Environmental Management. 2020;275:111132.
10. Larson KL, Casagrande D, Harlan SL, Yabiku ST. Residents' yard choices and rationales in a desert city: social priorities, ecological impacts, and decision tradeoffs. Environmental Management. 2009;44(5):921–37.
11. Frank B. Human–wildlife conflicts and the need to include tolerance and coexistence: an introductory comment. Society and Natural Resources. 2016;29(6):738–43.
12. Pooley S, Bhatia S, Vasava A. Rethinking the study of human–wildlife coexistence. Conservation Biology. 2021;35(3):784–93.
13. Chowdhury RR, Larson K, Grove M, Polsky C, Cook E, Onsted J, et al. A multi-scalar approach to theorizing socio-ecological dynamics of urban residential landscapes. Cities and the Environment. 2011;4(1): 1–19.
14. Cook EM, Hall SJ, Larson KL. Residential landscapes as social-ecological systems: a synthesis of multi-scalar interactions between people and their home environment. Urban Ecosystems. 2012;15(1): 19–52.
15. Stern PC. New environmental theories: toward a coherent theory of environmentally significant behavior. Journal of Societal Issues. 2000;56(3): 407–24.
16. Larson KL, Cook E, Strawhacker C, Hall SJ. The influence of diverse values, ecological structure, and geographic context on residents' multifaceted landscaping decisions. Human Ecology. 2010;38(6):747–61.
17. Wheeler MM, Larson KL, Andrade R. Attitudinal and structural drivers of preferred versus actual residential landscapes in a desert city. Urban Ecosystems. 2020;23(3):659–73.
18. Larson K, Brumand J. Paradoxes in landscape management and water conservation: examining neighborhood norms and institutional forces. Cities and the Environment. 2014;7(1):6.
19. McKenzie-Mohr D. Fostering Sustainable Behavior: an introduction to community-based social marketing. Gabriola Island, BC: New Society Publishers; 2011.
20. Thurstone LL. Attitudes can be measured. American Journal of Sociology. 1928;33(4):529–54.
21. Heberlein TA. Navigating Environmental Attitudes. New York: Oxford University Press; 2012. 228 p.
22. Whittaker D, Vaske JJ, Manfredo MJ. Specificity and the cognitive hierarchy: value orientations and the acceptability of urban wildlife management actions. Society and Natural Resources. 2006;19(6):515–30.
23. Stern PC. Information, incentives, and proenvironmental consumer behavior. Journal of Consumer Policy. 1999;22(4):461–78.
24. Dunlap RE, Van Liere KD, Mertig AG, Jones RE. New trends in measuring environmental attitudes: measuring endorsement of the New Ecological Paradigm: a revised NEP scale. Journal of Social Issues. 2000;56(3):425–42.
25. Manfredo MJ. Who Cares about Wildlife?: social science concepts for exploring human-wildlife relationships and conservation issues. New York: Springer-Verlag; 2008.

26. Fishbein M, Ajzen I. Attitudes towards objects as predictors of single and multiple behavioral criteria. Psychological Review. 1974;81:59–74.
27. Larson KL, Lerman SB, Nelson KC, Narango DL, Wheeler MM, Groffman PM, et al. Examining the potential to expand wildlife-supporting residential yards and gardens. Landscape and Urban Planning. 2022;222:104396.
28. Gobster PH, Nassauer JI, Daniel TC, Fry G. The shared landscape: what does aesthetics have to do with ecology? Landscape Ecology. 2007;22(7):959–72.
29. Larson KL, Nelson KC, Samples SR, Hall SJ, Bettez N, Cavender-Bares J, et al. Ecosystem services in managing residential landscapes: priorities, value dimensions, and cross-regional patterns. Urban Ecosystems. 2016;19(1):95–113.
30. Manfredo MJ, Teel TL, Don Carlos AW, Sullivan L, Bright AD, Dietsch AM, et al. The changing sociocultural context of wildlife conservation. Conservation Biology. 2020;34(6):1549–59.
31. George KA, Slagle KM, Wilson RS, Moeller SJ, Bruskotter JT. Changes in attitudes toward animals in the United States from 1978 to 2014. Biological Conservation. 2016;201:237–42.
32. Landová E, Bakhshaliyeva N, Janovcová M, Peléšková Š, Suleymanova M, Polák J, et al. Association between fear and beauty evaluation of snakes: cross-cultural findings. Frontiers in Psychology. 2018;9:333.
33. Fukano Y, Soga M. Why do so many modern people hate insects? The urbanization–disgust hypothesis. Science of the Total Environment. 2021;777:146229.
34. Adger WN, Brown K, Fairbrass J, Jordan A, Paavola J, Rosendo S, et al. Governance for sustainability: towards a "thick" analysis of environmental decision-making. Environment and Planning: Economy and Space. 2003;35(6):1095–110.
35. Ostrom E. Background on the institutional analysis and development framework. Policy Studies Journal. 2011;39(1):7–27.
36. Larson KL, Hoffman J, Ripplinger J. Legacy effects and landscape choices in a desert city. Landscape and Urban Planning. 2017;165:22–9.
37. Toro ID, Ribbons RR. No Mow May lawns have higher pollinator richness and abundances: an engaged community provides floral resources for pollinators. PeerJ. 2020;8:e10021.
38. Lerman SB, Contosta AR, Milam J, Bang C. To mow or to mow less: lawn mowing frequency affects bee abundance and diversity in suburban yards. Biological Conservation 2018;221:160–74.
39. Roy Chowdhury R, Turner II BL. Reconciling agency and structure in empirical analysis: smallholder land use in the southern Yucatán, Mexico. Annals of the Association of American Geographers. 2006;96(2):302–22.
40. Bormann FH, Balmori D, Geballe GT. Redesigning the American Lawn: a search for environmental harmony. 2nd ed. New Haven, CT: Yale University Press; 2001.
41. Avolio ML, Pataki DE, Trammell TL, Endter-Wada J. Biodiverse cities: the nursery industry, homeowners, and neighborhood differences drive urban tree composition. Ecological Monographs. 2018;88(2): 259–76.
42. Hooper VH, Endter-Wada J, Johnson CW. Theory and practice related to native plants: a case study of Utah landscape professionals. Landscape Journal. 2008;27(1):127–41.
43. Marion WR. Urban wildlife: can we live with them? In: Crabb AC, Marsh RE (eds.) Proceedings of the Vertebrate Pest Conference 13. Davis, CA: University of California, Davis; 1988. p. 34–8. Available from: https://escholarship.org/uc/item/9156k01t
44. Baker SE, Maw SA, Johnson PJ, Macdonald DW. Not in my backyard: public perceptions of wildlife and "pest control" in and around UK homes, and local authority "pest control". Animals. 2020;10(2):222.
45. Soulsbury CD, White PCL. Human–wildlife interactions in urban areas: a review of conflicts, benefits and opportunities. Wildlife Research. 2015;42(7):541–53.
46. Bateman HL, Brown JA, Larson KL, Andrade R, Hughes B. Unwanted residential wildlife: evaluating social-ecological patterns for snake removals. Global Ecology and Conservation. 2021;27:e01601.
47. Hoang TC, Pryor RL, Rand GM, Frakes RA. Use of butterflies as nontarget insect test species and the acute toxicity and hazard of mosquito control insecticides. Environmental Toxicology and Chemistry. 2011;30(4):997–1005.
48. Mitra A, Saha P, Chaoulideer ME, Bhadra A, Gadagkar R. Chemical communication in *Ropalidia marginata*: Dufour's gland contains queen signal that is perceived across colonies and does not contain colony signal. Journal of Insect Physiology. 2011;57(2):280–4.
49. Crooks KR, Soulé ME. Mesopredator release and avifaunal extinctions in a fragmented system. Nature. 1999;400(6744):563–6.
50. Gehrt SD, Wilson EC, Brown JL, Anchor C. Population ecology of free-roaming cats and interference competition by coyotes in urban parks. PLoS ONE. 2013;8(9):e75718.
51. McManus JS, Dickman AJ, Gaynor D, Smuts BH, Macdonald DW. Dead or alive? Comparing costs and benefits of lethal and non-lethal human–wildlife conflict mitigation on livestock farms. Oryx. 2015;49(4): 687–95.

52. Grimm NB, Faeth SH, Golubiewski NE, Redman CL, Wu J, Bai X, et al. Global change and the ecology of cities. Science. 2008;319(5864):756–60.
53. Schell CJ, Dyson K, Fuentes TL, Des Roches S, Harris NC, Miller DS, et al. The ecological and evolutionary consequences of systemic racism in urban environments. Science. 2020;369(6510):eaay4497.
54. Larsen L, Harlan SL. Desert dreamscapes: residential landscape preference and behavior. Landscape and Urban Planning. 2006;78(1–2):85–100.
55. Yabiku ST, Casagrande DG, Farley-Metzger E. Preferences of landscape choice in a southwestern desert city. Environmental Behavior. 2008;40(3):382–400.
56. Hope D, Gries C, Zhu W, Fagan WF, Redman CL, Grimm NB, et al. Socioeconomics drive urban plant diversity. Proceedings of the National Academy of Sciences of the United States of America. 2003;100(15):8788–92.
57. Ferris K. Changing landscape: Phoenix grows into distinctively desert city [Internet]. Phoenix, AZ: Arizona Municipal Water Users Association; 2015. Available from: https://www.amwua.org/blog/changing-landscape-phoenix-grows-into-distinctively-desert-city
58. Wheeler MM, Larson KL, Bergman D, Hall SJ. Environmental attitudes predict native plant abundance in residential yards. Landscape and Urban Planning. 2022;224:104443.
59. Templeton SR, Yoo SJ, Zilberman D. An economic analysis of yard care and synthetic chemical use: the case of San Francisco. Environmental Resource Economics. 1999;14(3):385–97.
60. Putnam RD. Bowling Alone: the collapse and revival of American community. New York: Simon and Schuster; 2000. 550 p.
61. White L. The historical roots of our ecologic crisis. Science. 1967;155(3767):1203–7.
62. Manfredo MJ, Teel TL, Dietsch AM. Implications of human value shift and persistence for biodiversity conservation. Conservation Biology. 2016;30(2): 287–96.
63. Polsky C, Grove JM, Knudson C, Groffman PM, Bettez N, Cavender-Bares J, et al. Assessing the homogenization of urban land management with an application to US residential lawn care. Proceedings of the National Academy of Sciences of the United States of America. 2014;111(12):4432–7.
64. Andrade R, Franklin J, Larson KL, Swan CM, Lerman SB, Bateman HL, et al. Predicting the assembly of novel communities in urban ecosystems. Landscape Ecology. 2021;36(1):1–15.
65. Warren PS, Lerman SB, Andrade R, Larson KL, Bateman HL. The more things change: species losses detected in Phoenix despite stability in bird–socioeconomic relationships. Ecosphere. 2019;10(3):e02624.
66. Carr MF, Boyd Kramer D. Homeowners' associations: barriers or bridges to more sustainable residential development? Landscape and Urban Planning. 2022;224:104419.
67. Andrade R, Larson KL, Hondula DM, Franklin J. Social–spatial analyses of attitudes toward the desert in a southwestern U.S. city. Annals of the American Association of Geographers. 2019;109(6):1845–64.
68. Nassauer JI. Care and stewardship: from home to planet. Landscape and Urban Planning. 2011;100(4):321–3.
69. Hayden D, Dills B. Smokey the bear should come to the beach: using mascot to promote marine conservation. Social Marketing Quarterly. 2015;21(1): 3–13.
70. Butler P, Green K, Galvin D. The principles of pride: the science behind the mascots [Internet]. Arlington, VA: Rare; 2013. Available from: https://rare.org/wp-content/uploads/2019/02/Rare-Principles-of-Pride.pdf
71. Seasholes B. The bald eagle, DDT, and the Endangered Species Act: examining the bald eagle's recovery in the contiguous 48 states [Internet]. 2007 Jul 1 [cited 2022 Sep 28]; Available from: https://policycommons.net/artifacts/1175255/the-bald-eagle-ddt-and-the-endangered-species-act/1728384/
72. Ducarme F, Luque GM, Courchamp F. What are "charismatic species" for conservation biologists? BioSciences Masters Reviews. 2013. 8 p.
73. Schirpke U, Meisch C, Tappeiner U. Symbolic species as a cultural ecosystem service in the European Alps: insights and open issues. Landscape Ecology. 2018;33(5):711–30.
74. Bowen-Jones E, Entwistle A. Identifying appropriate flagship species: the importance of culture and local contexts. Oryx 2002;36(2):189–95.
75. Frisk E, Larson KL. Educating for sustainability: competencies & practices for transformative action. Journal of Sustainability Education. 2011;2.
76. Kaiser FG, Fuhrer U. Ecological behavior's dependency on different forms of knowledge. Applied Psychology. 2003;52(4):598–613.
77. Ajzen I. Perceived behavioral control, self-efficacy, locus of control, and the theory of planned behavior. Journal of Applied Social Psychology. 2002;32(4): 665–83.
78. Aronson MFJ, Nilon CH, Lepczyk CA, Parker TS, Warren PS, Cilliers SS, et al. Hierarchical filters determine community assembly of urban species pools. Ecology. 2016;97(11):2952–63.

79. Dunn RR, Gavin MC, Sanchez MC, Solomon JN. The pigeon paradox: dependence of global conservation on urban nature. Conservation Biology. 2006;20(6):1814–16.
80. Doherty JF, Ruehle B. An integrated landscape of fear and disgust: the evolution of avoidance behaviors amidst a myriad of natural enemies. Frontiers in Ecology and the Environment. 2020;8:564343.
81. Weinstein SB, Buck JC, Young HS. A landscape of disgust. Science. 2018;359(6381):1213–14.
82. Frynta D, Marešová J, Řeháková-Petrů M, Šklíba J, Šumbera R, Krása A. Cross-cultural agreement in perception of animal beauty: boid snakes viewed by people from five continents. Human Ecology. 2011;39(6):829–34.
83. Özgüner H, Kendle AD, Bisgrove RJ. Attitudes of landscape professionals towards naturalistic versus formal urban landscapes in the UK. Landscape and Urban Planning. 2007;81(1):34–45.
84. Peterson MN, Thurmond B, Mchale M, Rodriguez S, Bondell HD, Cook M. Predicting native plant landscaping preferences in urban areas. Sustainable Cities and Society. 2012;5:70–6.
85. du Toit MJ, Kotze DJ, Cilliers SS. Landscape history, time lags and drivers of change: urban natural grassland remnants in Potchefstroom, South Africa. Landscape Ecology. 2016;31(9):2133–50.
86. Alam H, Khattak JZK, Ppoyil SBT, Kurup SS, Ksiksi TS. Landscaping with native plants in the UAE: a review. Emirates Journal of Food and Agriculture. 2017;29(10):729–41.
87. Shaw A, Miller KK, Wescott G. Australian native gardens: is there scope for a community shift? Landscape and Urban Planning. 2017;157:322–30.
88. Narango DL, Tallamy DW, Marra PP. Native plants improve breeding and foraging habitat for an insectivorous bird. Biological Conservation. 2017;213:42–50.
89. Shackleton RT, Le Maitre DC, van Wilgen BW, Richardson DM. Use of non-timber forest products from invasive alien *Prosopis* species (mesquite) and native trees in South Africa: implications for management. Forested Ecosystems. 2015;2(1):16.
90. Sinclair JS, Lockwood JL, Hasnain S, Cassey P, Arnott SE. A framework for predicting which non-native individuals and species will enter, survive, and exit human-mediated transport. Biological Invasions. 2020;22(2):217–31.
91. Semenya SS, Tshisikhawe MP, Potgieter MT. Invasive alien plant species: a case study of their use in the Thulamela Local Municipality, Limpopo Province, South Africa. University of Venda; 2012. Available from: http://univendspace.univen.ac.za/handle/11602/1380
92. Bredin YK, Lindhjem H, van Dijk J, Linnell JDC. Mapping value plurality towards ecosystem services in the case of Norwegian wildlife management: a Q analysis. Ecological Economics. 2015;118: 198–206.
93. Ripple WJ, Abernethy K, Betts MG, Chapron G, Dirzo R, Galetti M, et al. Bushmeat hunting and extinction risk to the world's mammals. Royal Society Open Science. 2016;3(10):160498.
94. Entwistle A, Dunstone N. Priorities for the Conservation of Mammalian Diversity: has the panda had its day? Cambridge: Cambridge University Press; 2000. 478 p.
95. Macdonald C, Gallagher AJ, Barnett A, Brunnschweiler J, Shiffman DS, Hammerschlag N. Conservation potential of apex predator tourism. Biological Conservation. 2017;215:132–41.
96. Marchini S, Macdonald DW. Mind over matter: perceptions behind the impact of jaguars on human livelihoods. Biological Conservation. 2018;224: 230–7.
97. Hoffman TS, O'Riain MJ. Monkey management: using spatial ecology to understand the extent and severity of human–baboon conflict in the Cape Peninsula, South Africa. Ecology and Society. 2012; 17(3):13.
98. Brotcorne F, Huynen MC, Deleuze S, Antoine-Moussiaux N, Wandia N, Poncin P. The necessity of a One Health perspective for managing urban primates. University of Liège; 2019. Available from: https://orbi.uliege.be/handle/2268/267986

CHAPTER 4

Conservation on the Urban Fringe

Sustaining Biodiversity and Advancing Equity in Suburban Ecosystems

Liba Pejchar[‡] and Sarah E. Reed[§]

Introduction

Suburban development is among the greatest threats to biodiversity, particularly in high-income nations, yet communities on the urban fringe are also home to diverse human (1) and wildlife communities (2). After population growth in cities slowed in the early 21st century, former agricultural and natural lands were transformed into sprawling housing developments in the US and beyond (3). Suburbs (periurban and exurban areas—defined, respectively, as moderately dense residential and commercial development at the edge of urban centers, and as low-density rural development more distant from cities (4))—are understudied yet increasingly important components of the greater urban ecosystem (5,6). Although suburbs take many different forms across the globe (3), these areas often have the capacity to support a large fraction of native biodiversity (2), including human-sensitive species, and they are also home to increasingly diverse human populations (1).

Similarly to denser cities, suburban systems generate altered plant and animal communities and inequitable access to nature;[1] however, housing density is generally lower in suburbs, population growth remains rapid, and ecological structure, function, and species assemblages differ in suburbs (7). For example, species richness often peaks midway along the urban–rural development gradient, because suburbs can be home to both human-sensitive and human-adapted species (8). Because of these characteristics (higher species richness, lower development density, ongoing rapid growth), there are more opportunities for innovative land-use planning to retain and restore habitat for biodiversity and provide environmental benefits to a diverse human population in suburbs relative to cities (9).

Conserving biodiversity and providing equitable and affordable housing are often portrayed as conflicting goals in suburban communities (10), perhaps because of a history of racist and classist land-use policy, income inequality, and the premium associated with living adjacent to environmental amenities (11). Moving forward, understanding and advancing innovative policies, tools, and strategies for conserving and managing biodiversity for nature and people in suburban ecosystems will be essential for meeting society's linked conservation and equity goals. Our objectives here are to (1) use a social equity lens to evaluate the ways that existing tools and strategies for suburban biodiversity conservation help resolve or exacerbate historic injustices in housing and environmental quality, and (2) inform practice (e.g., land-use planning and land management) and guide applied research to fill knowledge gaps at the intersection of conservation and equity. In the following sections, we describe historic and current trends and drivers of suburban population and housing growth

[‡] Colorado State University, USA

[§] Robert and Patricia Switzer Foundation, USA

[1] Larson's and Brown's chapter outlines how individuals' actions and societal decisions affect urban landscapes, vegetation, and wildlife.

Liba Pejchar and Sarah E. Reed, *Conservation on the Urban Fringe*. In: *Urban Biodiversity and Equity*. Edited by: Max R. Lambert & Christopher J. Schell, Oxford University Press. © Oxford University Press (2023). DOI: 10.1093/oso/9780198877271.003.0004 and changing demographics. Next, we summarize the current state of knowledge on the relationship between suburban development patterns, biodiversity, and human access to environmental benefits. These sections are followed by a social-ecological synthesis of the major policy tools and strategies for achieving biodiversity and equity goals in the urban fringe. Finally, we explore how these largely US-based tools and strategies may or may not apply in a global context, and we recommend a path forward for reconciling the dual goals of nature conservation and equitable access to housing and nature, particularly for those groups historically marginalized and excluded from suburban communities.

Drivers of suburban and exurban population growth

Suburbs have expanded around cities over the past half-century to become a dominant land-use type in the US (12) and globally (3). Conservation and equity along the urban–rural development gradient cannot be considered in isolation because cities and suburbs are linked by multiple interactions and flows, including human migration, economies, and ecosystem processes (13). In contrast to urban growth patterns elsewhere in the world, in the US population growth in large cities has slowed, stalled, or even reversed in recent decades and shifted to the suburbs (14,15) (Figure 4.1), where land consumption per capita is much greater than in urban areas (12–16). Exurban development in particular is the fastest-growing type of land use in the US and now encompasses at least 25% of the private land area (17).

Land-use policies and lending practices have shaped patterns of human settlement and open space, leading to ongoing legacies of environmental injustice in suburbs. Neighborhood classifications for Federal Housing Administration (FHA) mortgage loan guarantees, known as "redlining," denied government-insured mortgages to residents of predominantly Black and low-income neighborhoods. After World War II, these discriminatory lending practices combined with Department of Veterans Affairs mortgages and FHA construction loans that explicitly permitted racial discrimination to create layers of economic disadvantage and segregation in suburban communities and reinforced historic neighborhood covenants that restricted the purchase of suburban homes by specific racial, ethnic, religious, or national groups.

The Fair Housing Act was passed in 1968 to prohibit discrimination in the rental, sale, or financing of housing on the basis of race, religion, or national origin. Yet, racist and classist real-estate and lending practices continued. These included block busting (i.e., persuading owners to sell at a low price out of fear of racial or class turnover), steering (i.e., influencing a buyer's choice of community based on race,

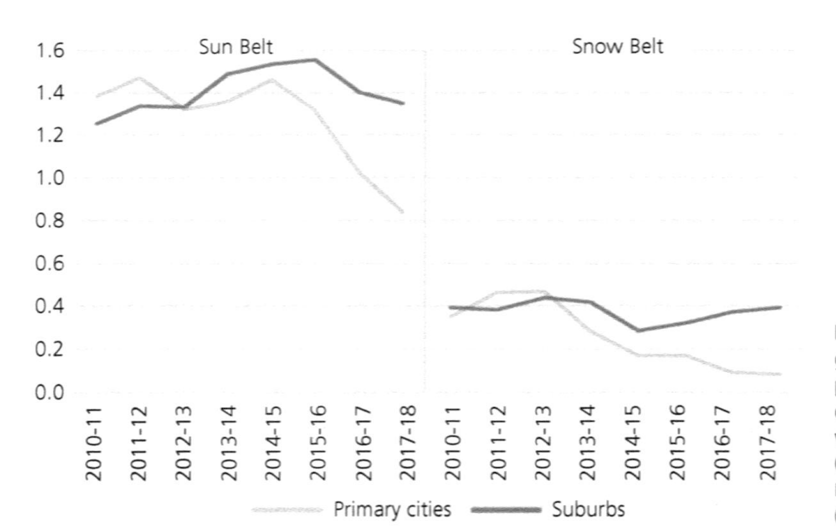

Figure 4.1 Average proportional annual growth for cities and suburbs with populations of over 1 million in US census districts in the Sun Belt (southern and western states) and the Snow Belt (midwestern and northeastern states). Reproduced with permissions from Brookings (16).

religion, or other characteristics), inequitable home valuation, and subprime lending. Suburban governments effectively locked in the segregated settlement patterns established by restrictive covenants and discriminatory lending practices by adopting exclusionary zoning laws (18). As a result of these policies and practices, today three-quarters (75%) of developable land area is zoned for single family detached homes, and investments in parks and green space remain focused in White and wealthier neighborhoods. This leaves communities of color and poor communities with less access to environmental amenities (19).

Despite their segregated history, today's suburbs are no longer neighborhoods dominated by upper-income White families. Instead, they are increasingly racially, ethnically, and economically diverse. Over the decade from 2010 to 2019, there was a 15% increase in people of color (relative to a 5% increase in White residents) in suburban communities (1). People of color now make up nearly 30% of the suburban population, and more students of color attend suburban schools than schools in central cities (20). The suburbs are also home to the fastest-growing areas of concentrated poverty in the US (21). Whereas prior waves of migration to the suburbs were drawn by amenities such as open space, parks, and schools, some more recent migrants may be displaced by gentrification and increasingly prohibitive housing costs in major cities. In addition, remote work has become more mainstream, making commute time less important, and increasing opportunities to prioritize remote locations and outdoor amenities (22). Rapid social and economic changes associated with the Covid-19 pandemic may further contribute to population growth and land-use change in the suburbs.

As suburbs become more dense (1), and exurban development in particular becomes more prevalent and widespread, planned growth to sustain vibrant natural and human communities is essential. Yet, people don't have full control over where to live. Such decisions are necessarily filtered through the actions of developers, policymakers, planners, real-estate agents, and market forces (23). Understanding how development patterns (e.g., housing density, configuration) affect biodiversity (e.g., species richness, assemblages, relative abundance) and environmental (in)justices will be critical to evaluating the strengths and shortcomings of existing policies and developing new strategies to achieve conservation and equity goals.

Development patterns, biodiversity, and human access to environmental benefits

Most research on the relationships between urbanization, biodiversity, and environmental justice focuses on large cities. Yet, a large proportion of the population, and most of the land undergoing residential and commercial development, are in peri-urban and exurban areas, beyond the urban core (6). Development in these areas may have a disproportionate impact on biodiversity and access to environmental benefits because exurban development covers an order of magnitude greater land area and tends to be in closer proximity to protected lands (6,24). Although species diversity is just one of many metrics of interest, the high levels of landscape and habitat heterogeneity (e.g., gardens, public green spaces) in neighborhoods between the countryside and the urban core are often associated with peak levels of plant and animal species richness (25). Nonetheless, despite supporting more native species relative to urban areas, the evidence that suburban development alters species assemblages is overwhelming (6,26).

The effects of suburban development patterns on plant and animal communities

Direct habitat loss is the most obvious impact of development (e.g., conversion of agricultural and natural systems into buildings, associated infrastructure, lawns, and gardens), replacing native species or crops with those more likely to be suited to built environments. Although habitat loss is the greatest threat to biodiversity globally (27), habitat loss is less extensive in suburban and especially exurban areas relative to cities because housing densities are lower (4). Thus, other indirect effects of residential and commercial development are likely to be equally or more important to biodiversity. These include fragmentation and edge effects (e.g., habitat perforation in exurban landscapes), the addition of resource subsidies, altered or disrupted species interactions, and wildlife–human conflict (17).

Even the addition of single homes in a forest or grassland can change the composition of the wildlife community, creating an ecological effect zone of approximately 200 m around each house, a distance that appears to be relatively consistent among ecosystem types (28,29). Within this housing effect zone, habitat generalists are favored over specialists (17,24,30,31), exotic species over native species, and early successional over longer-lived, later-maturing, and less fecund species (6). As such, suburban landscapes are less likely to support native human-sensitive species even in comparison to other human-dominated lands such as ranches (32).

Both the density and configuration of housing development affect native plant and animal communities (26). For example, periurban and exurban development vary in regard to development density and configuration, and thus biodiversity responses across this gradient (e.g., woodland bird richness and abundance) can also vary (33). Cluster or "conservation development" may have fewer impacts than houses spread evenly, even if overall housing density is the same (34). The relationship between housing density and biodiversity can also be nonlinear, resulting in sudden declines at particular thresholds of development beyond which much of the original biota may be lost (35).

Land stewardship and anthropogenic subsidies in suburban neighborhoods

In addition to development design and configuration, stewardship of suburban lands can also have direct and indirect impacts on plant and animal communities. For example, suburban landscapes with more trees and fewer mowed lawns support higher bird diversity in the northeastern US (36). More diverse plant communities provide habitat for a wider array of species and greater resilience to pests and disease (37). Thus, when suburban developments are structurally diverse and include native plant communities, wildlife may thrive in unexpected ways. For instance, some native pollinators may be resilient to the built environment in suburbs and exurbs, accessing resources from both native and non-native plants (38). Similarly, others have found no evidence for negative effects of suburban development on mammalian carnivore and herbivore diversity, richness, and occupancy relative to undeveloped natural areas (39). These patterns could be explained by heterogeneous habitat types and structural diversity, as well as the widespread addition of resource subsidies in suburban systems.

Resource subsidies can alter plant and animal communities and increase gamma species richness (the total number of species in a landscape, not just a patch), particularly in otherwise resource-poor environments (40). Such subsidies include water features (e.g., birdbaths), nesting sites (e.g., artificial cavities), supplemental food (e.g., bird feeders, ornamental plants), and accidental feeding (e.g., garbage). These resource subsidies tend to benefit human commensalists (e.g., grey squirrels, raccoons). Together with the increased prevalence of household pets (e.g., cats, dogs) and exotic vegetation from landscaped gardens, subsidies can negatively affect human-sensitive species through interspecific interactions such as higher predation pressure, competitive exclusion, and nest parasitism (2,4). Suburban and exurban development can also affect trophic interactions in nearby undeveloped ecosystems. For example, more introduced plants and irrigation in neighborhoods in the arid and resource-limited Chihuahuan desert led to spatial spillover and bottom-up effects on lagomorphs, leading to more herbivory, and more mesopredators in nearby natural areas (41).

Suburban human–wildlife interactions and equitable access to nature

Suburban landscapes arguably experience more human–wildlife interactions than either urban or rural areas because these areas are prime human habitat and also provide habitat for many animals. Wildlife managers are thus challenged with both sustaining native biodiversity and minimizing the threat of "overabundant" wildlife in these landscapes (7). For example, moose regularly wander into suburban Boston, and mountain lions move through neighborhoods at the wildland–urban interface in Colorado and California. Human presence and wildlife policies may provide some prey species with a haven from predation or hunting (39). Wildlife in suburban areas can harm property, create road hazards, and spread disease, and yet they can also be sources of delight and inspiration for residents (42).[2]

Altered species assemblages and ecological interactions have feedback effects on human communities, impacting health and well-being (2,43). For example, bird diversity and residents' happiness with their neighborhoods were spatially correlated, suggesting that habitat where birds and people thrive could be compatible (44). Historic land uses and land-use policies can also have legacy effects for both nature and people (45). For example, the current abundance of pollinators in an exurbanizing landscape was affected by past land use, leading to the slow loss of species through extinction debt (46). Similarly, racist housing policies have led to long-lasting and ongoing inequities in access to environmental benefits and exposures to environmental disamenities in suburban and exurban ecosystems. Residents of lower socioeconomic status neighborhoods are less likely to be able to enjoy diverse plant and bird communities (47).

Environmental amenities and disamenities in suburbia

Environmental justice has largely focused on the distribution of disamenities (e.g., toxic waste sites), but more recently has expanded to include environmental amenities, such as parks and other green spaces (19). Such open spaces provide demonstrated physical and mental health benefits, enhanced social interaction and security, and ecosystem services such as reduced urban heat island effects, reduction in air and water pollutants, flood control, often reduced light and noise pollution, provision of plant and wildlife habitat, and sometimes local food production. Suburbanization has only been linked to questions of environmental justice within the last several decades (48). Some have proposed that the ability of White people to enjoy the privilege of clean, safe, and relatively inexpensive living environments in the suburbs is a prime example of environmental racism (49).[3]

Although suburbia is becoming more racially and ethnically diverse, these areas are far from environmentally equitable. Urban exclusion and housing crises are driving people into the suburbs where they can experience greater exposure to natural disasters and climate risks (e.g., wildfires and infectious diseases (13,50)). For example, "starter home" neighborhoods on the fringe of cities meant to benefit lower-income families are disproportionately located near preexisting unwanted land uses such as environmental hazards and heavy manufacturing, transmission lines, and busy roadways, thus putting already vulnerable communities at greater risk (23). Racial and ethnic identity and socioeconomic status are also predictors of greater exposure to light pollution across the urban–rural gradient (51). Achieving just access to environmental benefits and protection from environmental disamenities must include both distributional and procedural equity. It is critical that historically marginalized communities have a voice at the table if policy tools and strategies are to meet both conservation and equity goals in suburbs.

Social-ecological applications: evaluating policy tools and strategies for meeting conservation and equity goals

Given the limited funding available for land protection and the long legacy of housing discrimination and environmental injustice, there is a critical need to examine innovative policies and incentives for integrating conservation into land-use planning, development, and land stewardship of new and existing developments. Suburbs and exurbs, like cities, present opportunities to design human-dominated habitats that are suitable for

[2] Byers et al.'s chapter discusses the interplay between human and wildlife disease in urban areas. Larson's and Brown's chapter outlines human motivations for urban conservation, especially in residential areas.

[3] Hoover's and Scarlett's chapter reviews the deep history of environmental racism—especially around biodiversity—in urban areas.

nature and people. Yet, many land-use mechanisms (e.g., zoning ordinances, deed restrictions, neighborhood covenants) have afforded Whiter and wealthier neighborhoods with disproportionate environmental benefits while protecting them from environmental injustices at the expense of people of color and low-income communities (23). Here, we describe various land-use policies and planning tools that were designed to achieve either conservation or equity goals and consider the ways in which they could be modified to both sustain biodiversity and advance environmental justice (see Case study 1).

Overlay districts

An overlay district, or overlay zone, is a zoning district that is layered over existing zoning districts and implements additional regulations to protect or promote special features that cross multiple zones or jurisdictions for new developments or redevelopments. For example, a conservation overlay district identifies conservation targets and supplements the underlying zoning standards with additional requirements that are designed to protect those targets (53,54). Developers of properties within a conservation overlay district may be required to preserve certain natural features or conduct environmental assessments to avoid or mitigate potential impacts. Overlay districts may also be used to advance housing equity goals. In areas at risk of gentrification and displacement of low-income or longstanding communities, an affordable housing overlay zone awards density bonuses and expedited review to developers in exchange for constructing a high proportion of units at below-market rates (55). To achieve both biodiversity and equity goals, conservation and affordable housing overlay districts could be designed and implemented in coordination with one another to ensure that affordable units are located in close proximity to environmental benefits.

Transfer of development rights

Transfer of development rights (TDR) programs allow landowners to buy, sell, or transfer development rights from one property to another (56,57). The goal of TDR is to reduce or eliminate development potential in areas that are high priority for conservation (e.g., where high levels of biodiversity or rare species occur), while directing new development to areas where infrastructure and services already exist. Less commonly, TDR programs can also be used to preserve affordable housing. For example, an existing affordable housing site

Case study 1 Habitat certification programs are a promising tool for engaging diverse communities in restoring and sustaining native plant and animal communities in suburban systems

Overview: Habitat certification (or community stewardship) programs are a means of engaging the local community in the conservation of habitat for native plants and animals (Figure 4.2). Agencies or organizations that lead these programs typically provide a combination of monetary incentives, professional guidance, and/or public recognition in exchange for planting and stewarding native plant communities. These programs have the potential to meet conservation challenges associated with suburban and exurban development by enhancing habitat for native plants and animals, and they give residents opportunities to connect with local biodiversity.

Critical analysis: Although community engagement in such programs is not often examined through a diversity and equity lens, there is evidence that place-based conservation programs that recruit residents in urbanizing areas can have lasting benefits for local conservation efforts (52). To meet equity goals, programs should employ budget equity analysis or similar tools to ensure equitable investments among neighborhoods by racial or economic composition (65). To work toward environmental justice, programs could disproportionately focus on communities that may not have preexisting experience with or positive attitudes toward conservation, but have been historically underrepresented or marginalized in conservation and planning decision-making.

Figure 4.2 Audubon's Habitat Heroes program provides guidance on improving at-home bird habitat which can enhance habitat for human-sensitive species. (a) Programs such as Audubon's Habitat Heroes provide guidance on how to steward suburban and exurban spaces to enhance habitat for wildlife, (b) More human-sensitive bird species use private certified natural areas relative to other neighborhood open space Data from Jimenez et al. (74).

may sell its unused development capacity to generate revenue to preserve the site and continue operations (58); although this application of TDR was innovated in urban areas, the concept could be extended to expand housing equity in suburban areas. Further, these TDR approaches can be merged to simultaneously achieve conservation in high-biodiversity areas and generate housing units and operating revenue in areas in need of affordable housing. In all cases, TDR programs are most likely to be successful in communities where support for conservation is high, demand for housing is high, and alternatives for development are low.[4]

Conservation or cluster development

Clustering is a common design technique used to conserve land within new residential subdivisions and minimize the negative ecological impacts of housing and associated infrastructure (Figure 4.3). In a conservation or cluster development (CD), homes are built on smaller lots and clustered together, allowing for the remainder of the property to be protected as open space. Clustering should reduce the ecological impacts of residential development because the ecological effect zones, or the areas around individual homes where the wildlife community is affected, would overlap (59,60), reducing fragmentation and increasing the amount and connectivity of habitat for many species.

Despite the promise of this approach for meeting conservation objectives, we are not aware of any examples of conservation or cluster developments that were designed to advance equity goals. On the contrary, they have been criticized as private conservation strategies accessible only to wealthy residents, and our own research demonstrated significant increases in the value of homes located in conservation developments relative to homes in other, comparable rural developments (61). Yet, there is no reason why a conservation development could not incorporate affordable units or be designed with a specific intention of making environmental benefits accessible to historically marginalized communities. Likewise, although currently conservation developments are more common in rural settings (62), the design technique could be applied to generate similar environmental and amenity benefits in higher-density suburban communities.

[4] Larsen's and Brown's chapter outlines how conservation efforts are most likely to succeed if they target local areas and tailor activities to the needs and interests of diverse people.

Inclusionary zoning

Inclusionary zoning is a planning tool that requires or incentivizes the inclusion of affordable housing units in new or renovated development. Intended as an antidote to the segregation caused by exclusionary zoning and its associated policies, inclusionary zoning aims to create opportunities for low-income households to avoid displacement by gentrification or provide them with access to high-amenity neighborhoods (63). On its own, inclusionary zoning may make only passive contributions toward advancing environmental justice, because it depends on people to move to enjoy environmental benefits. However, when applied in combination with efforts to acquire open space, improve air and water quality, and restore soil health in "eco districts" or "green zones," inclusionary zoning offers potential to meet conservation and equity goals in redeveloped neighborhoods (64).

Budget equity analysis

Several large cities, including Baltimore, Minneapolis, and Seattle, have begun to implement spatially explicit analyses of their cities' capital investments in amenities such as public spaces and recreation by neighborhoods' racial and economic composition. Scholar Sheryll Cashin argues that such practices will help to reverse "opportunity hoarding" and ensure consistent and equitable distribution of public funding and opportunities among communities (65). Budget equity analyses have been piloted in urban areas, but the principles are easily applicable to address environmental injustices in nearby suburban and exurban areas. Particularly if capital budgets include resources for investment in conservation via land acquisition, habitat restoration, and land stewardship, then neighborhood analyses could help to ensure that the greatest investments to sustain biodiversity are made in communities with proportionately fewer environmental amenities. In contrast to many of the other tools listed in this section, budget equity analysis is especially well-suited to advance conservation and environmental justice within existing suburban neighborhoods.

Figure 4.3 (a) Conservation or cluster development (CD) is an alternative to exurban sprawl that concentrates homes on a small portion of a parcel and preserves the rest of the land as open space. (b) Although CD has potential to provide housing and access to environmental amenities to diverse populations, CDs are typically only affordable to high-income families, a lost opportunity to meet joint conservation and equity goals. Images courtesy of C. Farr (a) and S. Bombaci (b).

Global perspectives

Although many of the strategies and tools outlined in the previous section exist in a US context, the challenges faced by the rapid growth of suburban development are global in relevance (3). Just as in the US, in addition to driving environmental degradation through conversion of natural lands, how these developments are distributed and constructed can lead to inequities in regard to both exposure to environmental hazards and lack of access to environmental amenities. Approximately two-thirds of all Canadians live in suburban areas which are growing far more rapidly compared to city centers (37). Suburbs are also expanding in the developing world where the poor are being displaced from city centers due to rising prices, and suburban sprawl is increasingly widespread to meet the needs of the emerging middle and upper classes (48).

Some US-based tools have been implemented in other countries, and the US could learn from other countries' policy innovations for achieving conservation and equity goals. For example, three Italian cities have adopted a TDR approach to control urban expansion (66), and inclusionary zoning has begun to contribute affordable units to the suburban housing supply in South Australia (67). Yet, just as in North America, the world continues to struggle with designing and evaluating planning tools that meet these multiple objectives. For example, the UK requires that all new housing developments result in no net loss of biodiversity, and they are now considering adopting more stringent measures that could require net gains in habitat for biodiversity (68). These policies are progressive from a conservation perspective, but equitable outcomes for people are not always included in social and environmental impact assessments (69). Similarly, Sweden has pioneered the creation of high-density and climate-friendly neighborhoods, but although successful in preserving open space and achieving clean energy goals, they have fallen somewhat short in supporting native biodiversity, and housing prices remain inaccessible for lower-income families (70). Although suburbia may take different forms globally, from sprawling single family homes to higher-density peri-urban neighborhoods, rapid and often unplanned growth remains a common and nearly universal challenge to protecting species and rectifying past and ongoing injustices.

Conclusion

With the rapid expansion of suburban development, society faces a formidable challenge. We must sustain habitat for biodiversity in human-dominated environments, and we must provide safe and affordable housing with equitable access to nature for all people. Although biodiversity conservation and social justice are often framed as mutually exclusive or conflicting goals, it is past time to move beyond either/or thinking to develop a more expansive and equity-centered land ethic for human-dominated ecosystems (see Case study 2). We must harness innovative tools such as upzoning, inclusionary zoning, and budget equity analysis to reallocate resources and create networks of housing and open space that both sustain nature and engage people, to the benefit of biodiversity and human communities. For example, a recently developed screening tool identifies communities with high social marginalization and low access to protected open space (71). It shows that areas prioritized for conservation according to environmental justice criteria are markedly different than areas prioritized based on conventional conservation criteria. Incorporating environmental justice criteria into land-use planning processes could thus fundamentally shift patterns of future land conservation and development. We must also build transdisciplinary teams that include diverse disciplinary experts, boundary organizations and experts, and community leaders, to avoid making decisions based on siloed thinking (72). For example, conservation scientists often recommend concentrating development close to cities to minimize the development footprint (73). Yet, such developments can exacerbate environmental injustices when affordable housing is placed in close proximity to industry, and far from green space (23). When "infill"-oriented development is proposed, it must be paired with sufficient investments in local built (e.g., schools, community centers) and natural (e.g., parks, greenways) amenities to avoid replicating historic injustices and exacerbating environmental vulnerability.

Key gaps in knowledge for achieving these goals will require transdisciplinary research to overcome. For example, we should use carefully designed experiments to evaluate the mechanisms driving changes in species abundance and relative composition as a function of development density and distribution, and across a gradient of socioeconomic conditions. We must also consider the role of connectivity in suburbs (e.g., greenbelts) for people and for biodiversity (30). How can we design such features to maximize use and movement, and minimize harm (e.g., human–wildlife conflict, safety concerns)? Backyard and neighborhood habitat enhancement programs can provide benefits for nature and people (74) (Case study 1) but these programs are typically designed for homeowners. How can we develop impactful incentive programs for suburban renters? Finally, in a future defined by climate change, more study is also needed to link research on temporal and spatial patterns of natural disasters to environmental and human health outcomes along the urban–rural gradient (50). Similarly, recent alliances among climate change, housing, and racial justice activists are calling for upzoning (aka YIMBY) to provide more abundant housing, with a smaller environmental footprint (75). Integrating such policy and planning efforts more explicitly with biodiversity conservation goals could expand support for these initiatives from communities concerned about the loss of green space, and help achieve equity and conservation along the urban fringe.

Case study 2 Three-way land deal to conserve land and build affordable housing: an approach for achieving equity and conservation goals?

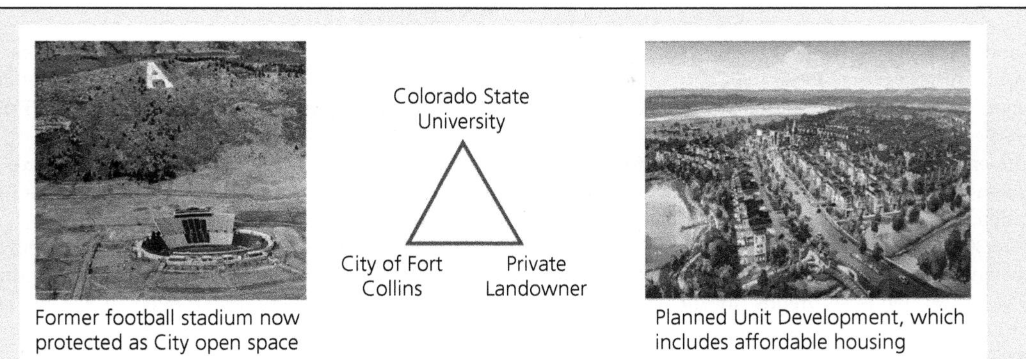

Figure 4.4 A three-way land deal that conserves biodiversity and expands affordable housing, often viewed as conflicting goals in urbanizing landscapes.
Images courtesy of Colorado State University, the TB Group.

Overview: Colorado State University (CSU) planned to build affordable housing on the site of its former football stadium. However, residents organized a ballot initiative, compelling the City to instead purchase and protect this land as open space, arguing its location adjacent to the Rocky Mountain foothills offers scenic beauty, recreational opportunities, and habitat for plants and animals. The City agreed, with the provision that CSU use the funds from the sale to build affordable housing elsewhere in the City (Figure 4.4).

Critical analysis: Although framed as a "win-win" for conservation and equity, the motivations for this project are more complex, and the gains somewhat unclear. Many of the groups that organized to lobby the City to retain the property in open space could be perceived as motivated primarily by NIMBY concerns about increased traffic and lost outdoor recreation amenities for current residents in adjacent neighborhoods. Further, the parcel does not have high biodiversity value and was not an acquisition priority for the City. However, the land does have potential to contribute to restorative justice if the City responds to calls from Indigenous groups to return this unceded treaty land back to native stewardship.

References

1. Anderson D. The suburbs have 15% more people of color than a decade ago versus 5% more white people, likely contributing to Democrats' 6-point suburban gain in 2020. Redfin News; 2020 [cited 2022 May 18]. Available from: https://www.redfin.com/news/suburbs-demographics-more-diverse-liberal/
2. Marzluff JM, Clucas B, Oleyar MD, DeLap J. The causal response of avian communities to suburban development: a quasi-experimental, longitudinal study. Urban Ecosystems. 2016;19(4):1597–621.
3. Hamel P, Keil R (eds.). Suburban Governance: a global view. Toronto: University of Toronto Press; 2015.
4. Bock CE, Bock JH. Biodiversity and residential development beyond the urban fringe. In: Esparza AX, McPherson G (eds.) The Planner's Guide to Natural Resource Conservation: the science of land development beyond the metropolitan fringe. New York: Springer; 2009. p. 59–84.
5. Theobald DM. Land-use dynamics beyond the American urban fringe. Geographical Review. 2001;91(3):544–64.
6. Hansen AJ, Knight RL, Marzluff JM, Powell S, Brown K, Gude PH, et al. Effects of exurban development on biodiversity: patterns, mechanisms, and research needs. Ecological Applications. 2005;15(6):1893–905.
7. DeStefano S, DeGraaf RM. Exploring the ecology of suburban wildlife. Frontiers in Ecology and the Environment. 2003;1(2):95–101.
8. Marzluff JM. Welcome to Subirdia: sharing our neighborhoods with wrens, robins, woodpeckers, and other wildlife. New Haven, CT: Yale University Press; 2014.
9. Clergeau P, Savard JPL, Mennechez G, Falardeau G. Bird abundance and diversity along an urban–rural

gradient: a comparative study between two cities on different continents. Condor. 1998;100(3):413–25.
10. Briechle KJ. Conservation-Based Affordable Housing: improving the nature of affordable housing to protect place and people. Arlington, VA: The Conservation Fund; 2006 [cited 2022 Oct 25]. 103 p. Available from: https://www.conservationfund.org/images/resources/Conservation-Based-Affordable-Housing-Study-all-9-06-lo-res.pdf
11. Immergluck D, Balan T. Sustainable for whom? Green urban development, environmental gentrification, and the Atlanta Beltline. Urban Geography. 2018;39(4):546–62.
12. Brown DG, Johnson KM, Loveland TR, Theobald DM. Rural land-use trends in the conterminous United States, 1950–2000. Ecological Applications. 2005;15(6):1851–63.
13. Jokisch BD, Radel C, Carte L, Schmook B. Migration matters: how migration is critical to contemporary human–environment geography. Geography Compass. 2019;13(8):e12460.
14. Jackson KT. Crabgrass Frontier: the suburbanization of the United States. New York: Oxford University Press; 1987.
15. Frey WH. Big city growth stalls further, as the suburbs make a comeback. Brookings Report; 2019 [cited 2022 May 18]. Available from: https://www.brookings.edu/blog/the-avenue/2019/05/24/big-city-growth-stalls-further-as-the-suburbs-make-a-comeback/
16. Theobald DM. Landscape patterns of exurban growth in the USA from 1980 to 2020. Ecology and Society. 2005;10(1):32.
17. Gilbert NA, Ferguson PF. Disturbance, but not the right kind: exurban development does not create habitat for shrubland birds. The Wilson Journal of Ornithology. 2019;131(2):243–59.
18. Rugh JS, Massey DS. Stalled integration or end of the segregated century? Du Bois Review. 2014;11(2): 205–32.
19. Boone CG, Buckley GL, Grove JM, Sister C. Parks and people: an environmental justice inquiry in Baltimore, Maryland. Annals of the Association of American Geographers. 2009;99(4):767–87.
20. Lewis-McCoy RL. Inequality in the Promised Land: race, resources, and suburban schooling. Stanford, CA: Stanford University Press; 2014.
21. Murphy AK, Allard SW. The changing geography of poverty. Focus. 2015;32(1):19–23.
22. Glennon MJ, Kretser HE. Exurbia east and west: responses of bird communities to low density residential development in two North American regions. Diversity. 2021;13(2):42.
23. Currie MA, Sorensen J. Repackaged "urban renewal": issues of spatial equity and environmental justice in new construction, suburban neighborhoods, and urban islands of infill. Journal of Urban Affairs. 2019;41(4):464–85.
24. Fraterrigo JM, Wiens JA. Bird communities of the Colorado Rocky Mountains along a gradient of exurban development. Landscape and Urban Planning. 2005;71(2–4):263–75.
25. Nitoslawski SA, Duinker PN. Managing tree diversity: a comparison of suburban development in two Canadian cities. Forests. 2016;7(6):119.
26. Pejchar L, Reed SE, Bixler P, Ex L, Mockrin MH. Consequences of residential development for biodiversity and human well-being. Frontiers in Ecology and the Environment. 2015;13(3):146–53.
27. Foley JA, DeFries R, Asner GP, Barford C, Bonan G, Carpenter SR, et al. Global consequences of land use. Science. 2005;309(5734):570–4.
28. Glennon MJ, Kretser HE. Size of the ecological effect zone associated with exurban development in the Adirondack Park, NY. Landscape and Urban Planning. 2013;112:10–17.
29. Odell EA, Knight RL. Songbird and medium-sized mammal communities associated with exurban development in Pitkin County, Colorado. Conservation Biology. 2001;15(4):1143–50.
30. Goad EH, Pejchar L, Reed SE, Knight RL. Habitat use by mammals varies along an exurban development gradient in northern Colorado. Biological Conservation. 2014;176:172–82.
31. Suarez-Rubio M, Leimgruber P, Renner SC. Influence of exurban development on bird species richness and diversity. Journal of Ornithology. 2011;152(2):461–71.
32. Maestas JD, Knight RL, Gilgert WC. Biodiversity across a rural land-use gradient. Conservation Biology. 2003;17(5):1425–34.
33. Merenlender AM, Reed SE, Heise KL. Exurban development influences woodland bird composition. Landscape and Urban Planning. 2009;92(3–4):255–63.
34. Pejchar L, Morgan PM, Caldwell MR, Palmer C, Daily GC. Evaluating the potential for conservation development: biophysical, economic, and institutional perspectives. Conservation Biology. 2007;21(1):69–78.
35. Farr CM, Pejchar L, Reed SE. Subdivision design and stewardship affect bird and mammal use of conservation developments. Ecological Applications. 2017;27(4):1236–52.
36. Belinsky K, Ellick T, LaDeau S. Using a birdfeeder network to explore the effects of suburban design on invasive and native birds. Avian Conservation and Ecology. 2019;14(2):2.

37. Nitoslawski SA, Duinker PN, Bush PG. A review of drivers of tree diversity in suburban areas: research needs for North American cities. Environmental Reviews. 2016;24(4):471–83.
38. Koyama A, Egawa C, Taki H, Yasuda M, Kanzaki N, Ide T, et al. Non-native plants are a seasonal pollen source for native honeybees in suburban ecosystems. Urban Ecosystems. 2018;21(6):1113–22.
39. Parsons AW, Forrester T, Baker-Whatton MC, McShea WJ, Rota CT, Schuttler SG, et al. Mammal communities are larger and more diverse in moderately developed areas. ELife. 2018;7:e38012.
40. Bock CE, Jones ZF, Bock JH. The oasis effect: response of birds to exurban development in a southwestern savanna. Ecological Applications. 2008;18(5):1093–106.
41. DaVanon KA, Howard LK, Mabry KE, Schooley RL, Bestelmeyer BT. Effects of exurban development on trophic interactions in a desert landscape. Landscape Ecology. 2016;31(10):2343–54.
42. Clergeau P, Mennechez G, Sauvage A, Lemoine A. Human perception and appreciation of birds: a motivation for wildlife conservation in urban environments of France. In: Marzluff JM, Bowman R, Donnelly R (eds.) Avian Ecology and Conservation in an Urbanizing World. Boston, MA: Springer; 2001. p. 69–88.
43. Liu J, Dietz T, Carpenter SR, Alberti M, Folke C, Moran E, et al. Complexity of coupled human and natural systems. Science. 2007;317(5844):1513–16.
44. Oleyar MD, Greve AI, Withey JC, Bjorn AM. An integrated approach to evaluating urban forest functionality. Urban Ecosystems. 2008;11(3):289–308.
45. Schell CJ, Dyson K, Fuentes TL, Des Roches S, Harris NC, Miller DS, et al. The ecological and evolutionary consequences of systemic racism in urban environments. Science. 2020;369(6510):eaay4497.
46. Cusser S, Neff JL, Jha S. Land use change and pollinator extinction debt in exurban landscapes. Insect Conservation and Diversity. 2015;8(6):562–72.
47. Kinzig AP, Warren P, Martin C, Hope D, Katti M. The effects of human socioeconomic status and cultural characteristics on urban patterns of biodiversity. Ecology and Society. 2005;10(1):23.
48. Leichenko RM, Solecki WD. Consumption, inequity, and environmental justice: the making of new metropolitan landscapes in developing countries. Society and Natural Resources. 2008;21(7):611–24.
49. Pulido L. Rethinking environmental racism: White privilege and urban development in Southern California. Annals of the Association of American Geographers. 2000;90(1):12–40.
50. Greenberg M. Seeking shelter: how housing and urban exclusion shape exurban disaster. Sociologica. 2021;15(1):67–89.
51. Nadybal SM, Collins TW, Grineski SE. Light pollution inequities in the continental United States: a distributive environmental justice analysis. Environmental Research. 2020;189:109959.
52. Jimenez MF, Pejchar L, Reed SE. Tradeoffs of using place-based community science for urban biodiversity monitoring. Conservation Science and Practice. 2021;3(2):e338.
53. McElfish JM Jr. Nature-Friendly Ordinances: local measures to conserve biodiversity. Washington, DC: Environmental Law Institute; 2004.
54. Duerksen C, Snyder C. Nature-Friendly Communities: habitat protection and land use planning. Washington, DC: Island Press; 2013.
55. Tziganuk A, Irvine B, Cook-Davis A, Kurtz LC. Exclusionary zoning: a legal barrier to affordable housing. Phoenix, AZ: ASU Morrison Insitute for Public Policy; 2022. 25 p. Available from: https://morrisoninstitute.asu.edu/sites/default/files/exclusionary_zoning_legal_barrier_to_affordable_housing.pdf
56. Kaplowitz MD, Machemer P, Pruetz R. Planners' experiences in managing growth using transferable development rights (TDR) in the United States. Land Use Policy. 2008;25(3):378–87.
57. Pruetz R, Standridge N. What makes transfer of development rights work?: success factors from research and practice. Journal of the American Planning Association. 2008;75(1):78–87.
58. Nelson AC, Pruetz R, Woodruff D. The TDR Handbook: designing and implementing successful Transfer of Development Rights programs. Washington, DC: Island Press; 2012.
59. Theobald DM, Miller JR, Hobbs NT. Estimating the cumulative effects of development on wildlife habitat. Landscape and Urban Planning. 1997;39(1): 25–36.
60. Odell EA, Theobald DM, Knight RL. Incorporating ecology into land use planning: the songbirds' case for clustered development. Journal of the American Planning Association. 2003;69(1):72–82.
61. Hannum C, Laposa S, Reed SE, Pejchar L, Ex L. Comparative analysis of housing in conservation developments: Colorado case studies. Journal of Sustainable Real Estate. 2012;4(1):149–76.
62. Mockrin MH, Reed SE, Pejchar L, Salo J. Balancing housing growth and land conservation: conservation development preserves private lands near protected areas. Landscape and Urban Planning. 2017;157: 598–607.

63. Schuetz J, Meltzer R, Been V. 31 flavors of inclusionary zoning: comparing policies from San Francisco, Washington DC, and suburban Boston. Journal of the American Planning Association. 2009;75(4):441–56.
64. Fitzgerald J. From eco-districts to green justice zones. Planetizen; 2020 [cited 2022 Jun 23]. Available from: https://www.planetizen.com/blogs/111005-eco-districts-green-justice-zones
65. Cashin S. White Space, Black Hood: opportunity hoarding and segregation in the age of inequality. Boston, MA: Beacon Press; 2021.
66. Colavitti AM, Serra S. The transfer of development rights as a tool for the urban growth containment: a comparison between the United States and Italy. Papers in Regional Science. 2018;97(4):1247–65.
67. Gurran N, Gilbert C, Gibb K, van den Nouwelant R, James A, Phibbs P. Supporting Affordable Housing Supply: inclusionary planning in new and renewing communities. Melbourne, Australia: Australian Housing and Urban Research Institute Limited; 2018. Available from: http://www.ahuri.edu.au/research/final-reports/297
68. zu Ermgassen SO, Baker J, Griffiths RA, Strange N, Struebig MJ, Bull JW. The ecological outcomes of biodiversity offsets under "no net loss" policies: a global review. Conservation Letters. 2019;12(6):e12664.
69. Bull JW, Baker J, Griffiths VF, Jones JPG, Milner-Gulland EJ. 2019. Ensuring no net loss for people and biodiversity: good practice principles. SocArXiv. Available from: https://doi.org/10.31235/osf.io/4ygh7
70. Austin G. Case study and sustainability assessment of Bo01, Malmö, Sweden. Journal of Green Building. 2013;8(3):34–50.
71. Sims K, Lee LG, Estrella-Luna N, Lurie M, Thompson JR. Environmental justice criteria for new land protection can inform efforts to address disparities in access to nearby open space. Environmental Research Letters. 2022;17(6):064014.
72. Deziel NC, Shamasunder B, Pejchar L. Synergies and trade-offs in reducing impacts of unconventional oil and gas development on wildlife and human health. BioScience. 2022;72(5):472–80.
73. Gude PH, Hansen AJ, Jones DA. Biodiversity consequences of alternative future land use scenarios in Greater Yellowstone. Ecological Applications. 2007;17(4):1004–18.
74. Jimenez MF, Pejchar L, Reed SE, McHale MR. The efficacy of urban habitat enhancement programs for conserving native plants and human-sensitive animals. Landscape and Urban Planning. 2022;220:104356.
75. Sommer L. Why sprawl could be the next big climate change battle. National Public Radio; 2020 [cited 2022 Jun 20]. Available from: https://www.keranews.org/2020-08-06/why-sprawl-could-be-the-next-big-climate-change-battle

CHAPTER 5

Portland's Conservation Organizations

Acknowledging Racial Inequity and Responding with Community-Informed Solutions

Laura Guderyahn[‡] and Mary Logalbo[§]

What you do to the land, you do to the people and what you do to the people, you do to the land. This is why we cannot separate social equity from ecological restoration.

Movement Generation (1)

Introduction

Environmental health and access to natural areas affect us all, but who benefits and who is harmed across the landscape are not equally felt across all communities. Portland, OR is an important place to explore this issue given its recognition as a city leading in its sustainability efforts, yet plagued with a racist history that has contributed to stark inequitable outcomes today (2,3). Communities of color in Portland have disproportionately experienced poor health outcomes and local environmental risks including elevated air pollution levels and urban heat island impacts (4–6). Relatedly, one's race has been a strong determinant of if one will own land or have a say in how land is managed in Portland (7). To undo these unjust outcomes, the policies, practices, and systems that got us here must be examined, deconstructed, and redesigned with voices of those most marginalized centered to craft better solutions for our future. Proactively dismantling barriers and welcoming all in caring for our land is imperative to ensuring long-term environmental health for all.

While conservation organizations work to center equity, inclusion, and justice in their work, they must ask how they can respond to the legacy of the past and ensure the story is different moving forward. New approaches are required as organizations realize the need to prioritize equity and justice alongside their preexisting conservation goals. Beyond the justice imperative of centering equity, conservation organizations recognize that conservation efforts are most successful through an inclusive and just approach. Here, we first center our work in the broader national and international context of how racist and classist systems in other areas of the world have similarly determined how land, homes, natural resources, and power have been made available to dominant cultural communities and withheld from marginalized communities. Following this overview, we use an example to spotlight how our two Portland conservation organizations are working to embed equity in our city's conservation movement. We present a true account of successes, failures, challenges, and hopes. Finally, we conclude with the business case for equity and a call to action from wherever the reader might sit.

History of racist human–land relationships in Oregon

The historical relationships between land and marginalized people are still impacting conservation organizations' work and the communities

[‡] City of Portland Parks & Recreation, USA

[§] West Multnomah Soil & Water Conservation District, USA

Laura Guderyahn and Mary Logalbo, *Portland's Conservation Organizations*. In: *Urban Biodiversity and Equity*. Edited by: Max R. Lambert & Christopher J. Schell, Oxford University Press. © Oxford University Press (2023). DOI: 10.1093/oso/9780198877271.003.0005 served in Portland today. Legal, social, and institutional factors have determined how land, homes, and natural resources have been made available to White communities and withheld from communities of color over the course of the past 200 years (7). Indigenous communities, the original caretakers and inhabitants of the land we are discussing, have survived centuries of actions focused on land grab-induced displacement. Despite Indigenous peoples having lived and cared for the land for thousands of years, the US claimed sovereignty and right of ownership and control over Oregon using the Christian framework of the Doctrine of Discovery and its declaration that "the principle of discovery gave European nations an absolute right to New World lands" (8). Before Oregon was even declared a state, the Donation Land Claim Act (DLCA) of 1850 was passed allowing White settlers to claim large acreage parcels in the Oregon Territory, rapidly displacing Indigenous communities who had been living on this land for generations (9). Prior to enacting this legislation, lawmakers authorized commissioners' treaty negotiation powers to extinguish Indian title, remove tribes, "and leave the whole of the most desirable portion open to white settlers" (10). The DLCA was the only federal land-distribution act in US history that specifically limited land grants by race, creating a long-standing legacy of social inequality in Oregon (9). Additional legislation followed, further eroding the ability of Native Americans to own and benefit from the contiguous acres of land these communities had lived on and cared for over millennia (7). As the majority of the fertile land in Oregon transitioned into the hands of homesteaders, the fertile Willamette Valley transitioned to farmed land. Additional legislation such as the Dawes Act (or General Allotment Act of 1887) abolished the idea of communal land which was integral to Native American culture, and promoted "the idea that the true route to civilization involved an incentive to work that was only possible through the individual ownership of property" (11). In 1857, Oregonians voted to ban slavery, but also to enact a clause that prohibited Black people from being in the state, owning property, or entering into contracts (12). Although Asian Americans were able to establish themselves as landowning farmers and ranchers out west, an Alien Land Law of 1923 was signed by Oregon's 17th governor (a card-carrying member of the Ku Klux Klan), which "prohibited immigrants from owning land in their own name" (11). This was mainly directed at Japanese ranchers and farmers (13). Although the Japanese American Citizens League was eventually able to repeal this land law through a Supreme Court ruling in their favor, they were largely unable to buy back their farms after returning from Japanese Internment Camps in the wake of World War II (14).

As cities such as Portland grew, New Deal agencies and policies helped create patterns of spatial segregation that persist today (3). One example is federally backed housing programs that created suburban housing along with "redlining" maps to indicate where loans were "safe to insure," resulting in the mass production and sale of homes to White individuals with the requirement that none of the homes were sold to African-Americans (15). Historical racist policies like redlining have had pronounced, lasting impact on communities of color in various neighborhoods in Portland.[1] The histories noted are only a small sample of examples showcasing how opportunity to build land-based wealth, power, and access to services were handed to White people and systematically denied to Black, Indigenous, and other People of Color.

Global perspectives—conservation's deep colonial history and continued role in disparities

Oregon is not alone in having a history of racist policies, practices, and systems that created significant disparities in who owns, manages, and benefits from land today (16). Rather, it is situated in a broader history of colonialism. In most cases, the act of colonialism explicitly included the belief that the colonizing culture has a religious and political supremacy over the native population, justifying one country taking full or partial political control of another country and occupying it with settlers for purposes of profiting from its resources and economy (17). This "Age of Exploration" began in the

[1] Hoover's and Scarlett's chapter dissects the ongoing legacy of historical racist policies that impact communities and urban environments today, including those in Portland.

1400s and was marked by the specific intention of European countries to explore the world to find more efficient trade routes, spread Christianity, and control priority resources like spices and precious metals. The ultimate goal was to increase the colonizing country's wealth and power on the world stage.

In 1419, Portugal colonized the North African territory of Ceuta, prompting Spain to begin colonizing as well (18). In 1492, Christopher Columbus landed in the Bahamas, marking the beginning of Spanish colonialism. Battling each other for new territories to exploit, Spain and Portugal went on to colonize and control Indigenous lands in the Americas, India, Africa, and Asia. This pattern of land grabs through colonialism flourished during the 17th century with the establishment of the French and Dutch overseas empires, along with the English overseas possessions, which would later become the sprawling British Empire, spanning the globe to cover nearly 25% of the Earth's surface at the peak of its power in the early 1900s (19). When Spain and Portugal were permanently weakened after the beginning of decolonization in the late 1700s, Great Britain, France, the Netherlands, and Germany made South Africa, India, and Southeast Asia the targets of their colonial efforts.

Despite the power of colonizers who claimed lands that were already home to Indigenous peoples, resistance is an integral part of the story of colonialism. Even before decolonization, Indigenous people on all continents staged violent and/or nonviolent resistance to those seeking takeovers of their ancestorial homelands (20).

After World War II, nationalist movements for self-determination won victories worldwide, and the world's political map rapidly transformed as former colonies and protectorates became independent states. Despite the overthrow of colonizing countries, the resulting disparities in land access, representation, and services persist to this day around the world. Further, modern-day versions of global land grab policies persist. Corporations and states including the US, European nations, and the rising "BRICS" economies of Brazil, Russia, India, China, and South Africa are all involved in the purchasing of significant tracts of overseas land (21). As a result, millions of hectares of farmland, as well as forests and peatlands, are being rapidly converted into large mono-cropped plantations of cash crops, to the detriment of local landowners (22). The disproportionate ownership of land by dominant-culture communities, and the resulting accumulation of intergenerational wealth and use of intensive agriculture drive the disparities between industrialized and developing economies today (23).

The land wealth disparity within and between societies has become increasingly inequitable, with communities with land access able to take advantage of and benefit from various conservation programs, such as land trusts, conservation easements, private reserves, and other incentives (24). This, along with inequities driving who gets to work, lead, and make decisions in the conservation and environmental sectors, perpetuates service disparities between communities when it comes to accessing and benefiting from environmental services such as park spaces for recreation, safe drinking water, and clean air to breathe (25). It is instrumental that we recognize the consequences of global colonial practices on the genesis of conservation and the inherent inequities embedded in conservation thought and action. The environmental justice movement seeks to address these disparities and will be achieved when "everyone enjoys the same degree of protection from environmental health hazards, and equal access to the decision-making process to have a healthy environment in which to live, learn, and work" (26).

Portland's conservation agencies

Many organizations advance conservation in the Portland Metro Area. Here we detail the mission of our two organizations, the urban conservation work we do, and our growth into equity and justice-centered conservation. The equity work done by our organizations has not been linear and is ever evolving. The authors' views, thoughts, and perspectives of this journey do not necessarily encompass all that has been or is being done by these organizations, and do not necessarily represent the views or positions of the organizations they represent.

Portland Parks & Recreation (PP&R)

The mission of Portland Parks & Recreation (PP&R) is to help Portlanders play—providing the safe places, facilities, and programs which promote physical, mental, and social activity. We get people, especially kids, outside, active, and connected to the community. The vision that guides our actions is:

> Portland's parks, public places, natural areas, and recreational opportunities give life and beauty to our city. These essential assets connect people to place, self, and others. Portland's residents treasure and care for this legacy, building on the past to provide for future generations. (27)

The PP&R 2017–2020 Strategic Plan (28) acknowledges the growth and change that have occurred in Portland City since 2000 and sets forth specific initiatives that would allow PP&R to meet the rising need for a consistent and culturally responsive park and natural area system through:

- Establishing, safeguarding, and restoring the parks, natural areas, public places, and urban forest of the city, ensuring that these are accessible to all;
- Developing and maintaining excellent facilities and places for public recreation and community building;
- Providing dynamic recreation programs and services that promote health and well-being for all;
- Partnering with the community we serve.

Biodiversity and conservation goals were developed through the creation of a Natural Areas Restoration Plan in 2010 (updated in 2015) (29). As a first step to creating these goals, PP&R inventoried and surveyed the vegetation of its natural area parkland from 2003 to 2008. This monumental effort identified vegetation community characteristics such as dominant and invasive plant species, management concerns, and overall ecological health to inform park management and citywide natural resource planning. A total of 1072 surveys were conducted covering 8213 acres. From these surveys, the ecological health of each park unit was rated, and a cumulative score was developed for each park unit that combined the vegetation assessment with a series of weighted factors, such as:

- Ecological health of the riparian area from surveys conducted in 2007 by PP&R;
- Wildlife Assessment completed between 2003 and 2008 by PP&R;
- Special habitat types identified by the City of Portland Bureau of Environmental Services Terrestrial Ecological Enhancement Strategy (TEES);
- Presence of active volunteer/stewardship groups.

The resulting management priority matrix focuses land management activities and resources where they can provide the biggest ecological "lift" in ecosystem health and function. Accordingly, as part of its work, the Land Stewardship Division of PP&R actively restores and manages habitat and biodiversity throughout the city.

West Multnomah Soil & Water Conservation District (the District)

The District's mission is to provide resources, information, and expertise to inspire people to actively improve air and water quality, fish and wildlife habitat, and soil health. We are a publicly funded special district conservation organization directed by an elected board. To accomplish our work, we collaborate with many partners, including private landowners, tenants, schools, culturally specific organizations, nonprofits, and other government organizations including PP&R. Our service area includes Multnomah County west of the Willamette River (including part of the City of Portland), all of Sauvie Island including the Columbia County portion of the island, and a portion of the Bonny Slope region of the Tualatin Mountains in Washington County. We offer assistance with invasive plant management, establishing native plants, livestock management, project funding, wildlife habitat restoration, soil health enhancement, workshops, and conservation planning for healthy forests, farms, streams, and gardens.

The following vision guides what we aim to achieve:

> All people in our district are informed and confidently engaged in the long-term caring for and giving back to the land. Everyone has the opportunity to connect or reconnect with the land, especially those who have been displaced from or deprived of land. People's engagement

and connection to the land ensures clean water, clean air, healthy soil, and diverse habitats, for thriving communities, fish and wildlife. (30)

Working in partnership

In 2003, PP&R became the nation's first park system to be Salmon-Safe Certified, an accreditation that links development and land management practices with protection of agricultural and urban watersheds. The priorities of this certification (irrigation efficiency, stormwater management, pesticide reduction, conservation of native biodiversity, and streamside and wetlands area management) have guided PP&R in reaching the desired outcome of protecting and enhancing the biodiversity and ecological health of our natural areas, provided direction for near- and long-term actions, and established management priorities for the more than 8000 acres of natural areas and 4500 acres of developed park lands that PP&R oversees. Just a few years later, residents of Portland in the District service area voted to establish a permanent property tax base to meet the demand for natural resource education and assistance to protect clean water and wildlife habitat. The work of the District complements PP&R's priorities by implementing conservation work with private and public landowners across PP&R's properties and beyond, spanning into more rural areas of our county. Soil and water conservation districts across the US—born out of the need to foster local solutions to national conservation issues such as the Dust Bowl of the 1930s—have a tradition of serving rural communities and private owners of farms and ranches (31). The origins of the District are no different; however, as our service area urbanized and our understanding of environmental justice grew, the need to embed new strategies focused on equitable outcomes and deconstructing systems that have led us to what we face today has come into focus.

Equity in Portland's conservation work

While achieving conservation and biodiversity goals remain primary land management objectives of the District and of the Land Stewardship Division within PP&R, acknowledging and addressing the racial disparities that shape who participates in and reaps the benefits from parks, natural areas, and private lands has been determined as the lens through which our work must be prioritized and decisions made to ensure better outcomes for all. A significant amount of capacity building and education was needed for the District and PP&R to operationalize our equity work. To start this work our staff and leadership needed to deepen their understanding of equity along with the past and present issues at play. This capacity building necessitated staff and leadership trainings, organizational assessments, and a solid internal and external commitment to equity that included publicly adopted statements declaring this commitment. Both the District and PP&R have decided it is important to center racial equity in our project prioritization and planning with the recognition that doing so helps us both focus on addressing racial disparities and increasing our organizations' overall strength and capacity to better serve other marginalized communities. After grounding our organizations in foundational understandings of and commitment to racial equity, we then implemented newly established equity-centered goals and plans including forming teams or committees to move this work forward, more robust community outreach and engagement, partnership development, and tracking and accountability measures. The more our organizations dove into this work, the more we realized that, to make the changes that were needed, we needed a long-term approach that continuously strengthened community relationships.

In 2011, Portland's Office of Equity and Human Rights (OEHR) was created by City Ordinance to reduce disparities amongst Portland's communities and to provide guidance and assistance to City staff as they develop methods to achieve equitable outcomes and services. OEHR works closely with PP&R's Land Stewardship Division staff (responsible for implementing biological and conservation goals within Portland's park portfolio) to engage communities of color in decision-making, understand the root causes of existing disparities, and identify how our work can reduce these disparities. The District's equity journey started with several impactful staff and board trainings in 2014, followed by the formation of a Diversity, Equity, and Inclusion (DEI) Committee, and the later adoption of a

Racial Equity Statement and DEI principles. In 2021, the District adopted a Long Range Business Plan that centered on DEI with new and diverse perspectives incorporated into the plan (31). Key to this step was reaching out to individuals and communities that the District has not historically worked with. This outreach initiated important new relationships with representatives of underserved and other marginalized communities—relationships the District will strive to strengthen through implementation of the plan. We recognize that while we have made progress toward our goals, we still have a long way to go.

The story of each organization's journey is detailed below.

PP&R

In practice, the biodiversity and conservation goals set by PP&R are combined with the City's Racial Equity Toolkit (RET), which ensures that staff integrate explicit consideration of racial equity when prioritizing recreation, conservation, and biodiversity projects. Using a data-driven approach and an equity lens, the RET ensures that our actions and decisions are designed to achieve equitable outcomes; and helps us engage communities of color in decision-making, understand the root causes of existing disparities, and identify how our work can reduce these disparities.

Some ways the RET has been realized on the ground in land conservation projects include the creation of the Community Benefits Agreement (CBA) and Community Equity and Inclusion Plan (CEIP) to increase minority and female participation in construction trades. These documents apply to restoration and conservation projects that require any level of construction. Additionally, the Community Opportunities and Enhancement Program (COEP) provides workforce development and technical help to businesses through grants. The program's two goals are (1) to increase the number of people of color and women in the trades and (2) to remove barriers for construction firms owned by people of color and women. Both goals seek to increase the number of people of color and women in public contracting. The City of Portland also encourages qualified businesses to become certified by Business Oregon. Business Oregon's COBID program certifies businesses owned by minorities, women, or service-disabled veterans. They also certify emerging small businesses. The City may buy goods and services with COBID-certified businesses without competitive solicitation. It applies to contracts with an estimated cost of $150,000 or less. Direct contracting takes less time to put contracts in place and supports COBID firms. Several culturally and economically specific restoration contracting firms have taken advantage of this program allowing staff to more easily conduct business with them directly (i.e., Verde NW, R Franco Restoration, and Wisdom of the Elders).

In addition, in recognition of a shared interest, and the essential value of working with the Native community, a unique council of interested Native community members advise PP&R about Native cultures, lifeways, and needs. The Native American Community Advisory Council's (NACAC's) role is to advise and collaborate about park development, land management, bureau policy, and projects. The NACAC provides a forum to discuss values such as sacredness of land, the importance of spiritual and cultural connection to heritage, and the recognition that Indigenous peoples have been here since time immemorial. The NACAC is comprised of Native American community members, local federally recognized tribal members, and members of Native American community-based organizations throughout the region. The work of this partnership is to foster collaboration between and among Native communities and the bureau by developing authentic relationships and strengthening community ties. For example, in partnership with NACAC and other Native community groups, the PP&R's Land Stewardship Division is working to reestablish First Food biological communities and cultural resources throughout Portland's parks and natural areas.

The District

The mission and vision shared above were adopted in 2021 and born of an intensive year-and-a-half-long planning process that sought to center DEI. This planning process was informed by hundreds of survey responses, focus groups, and interviews with a wide range of stakeholders. We took a novel approach for this plan's development by deliberately reaching out to underserved and marginalized communities to gather input on how to plan for our future and taking intentional equity lens pauses. A team of eight Community Engagement Liaisons were hired to learn from community members, providing translation services as needed, within the following communities determined to be the most prevalent racial and ethnic communities, other than White communities, residing in our service area according to available school and census data (31): Arabic, African American, Chinese, Latinx, Native American, Slavic, and Vietnamese. In addition, a diverse advisory committee with representation from community members, partners, and culturally specific organizations was convened to make recommendations to our elected Board of Directors as we developed this plan. This newly adopted mission, vision, and plan emphasizes an important growth in what we see as our organizational focus and aim. The intentionality of including those displaced from or deprived of land in our vision is an important step towards answering specific community requests for increased access to land and its care. This has not been a role the District has historically proactively filled as our efforts have largely been focused on privately owned larger properties and the people who own them.

To get to this point, we first needed to lay substantial groundwork in DEI with our staff and board. As was noted in the introduction, this work was initially sparked and shaped by impactful trainings delivered by equity authorities including OEHR, Center for Diversity & the Environment, Coalition of Communities of Color, and Capacity Building Partnerships. These trainings led us through important learning as individuals, as an organization, and as part of various systemic and cultural frameworks. Joining organizational equity-focused cohorts, hosting an equity-centered internship, and developing an organizational equity lens were also important steps in building our capacity for this work. Critical points in publicly committing to this work include the board resolution to form a staff and board DEI Committee and adoption of the following racial equity statement:

Vision—The West Multnomah Soil & Water Conservation District is a culturally inclusive organization that welcomes and engages people of color in all facets of our organization, activities and programs. The Board and staff of the District, the customers we serve, the contractors we hire and the people who benefit from our work resemble the racial diversity found within our service territory. The District reaches out to communities of color to determine their conservation priorities for the purpose of enhancing livability through healthy soil, clean water and diverse habitats. We willingly share with others our experiences in pursuing racial equity.

Need—Our District has taken the initiative to review the history of racial disparity in Oregon, and how this history persists in the form of unconscious biases and cultural barriers that contribute to disparities in how we work, whom we work with and whom we serve. We recognize that gaining the perspectives of, and working with, communities of color will increase our organization's overall strength. By working proactively and deliberately to be equitable and inclusive, we will be more successful in our work.

Accountability—We will hold ourselves accountable to racial equity by addressing disparities when found, and by developing, implementing, tracking and reporting on Specific, Measurable, Achievable, Realistic, and Time-Bound racial equity goals. Priorities include fully understanding the demographics of the communities we serve, developing new and lasting partnerships with communities of color and organizations that represent them, and recruitment and retention of persons of color on our Board, staff and supporting committees.

An example: Connect SW PDX Listening and Action Pledge Project

In 2016, the District was setting out to implement its equity goals locally on-the-ground and conducted a geospatial analysis of the urban conservation programming offered by the District. In doing so we were disappointed to find that one of our service area's most racially diverse neighborhoods in West Portland had never received any direct assistance from the District. Another geospatial data analysis, solely looking at District programming activities offered in 2018, found only one project that was active in areas that were historically redlined (Figure 5.1). The majority of our 2018 projects were found in the "best" (green) neighborhoods with others found in "still desirable" (blue) neighborhoods.

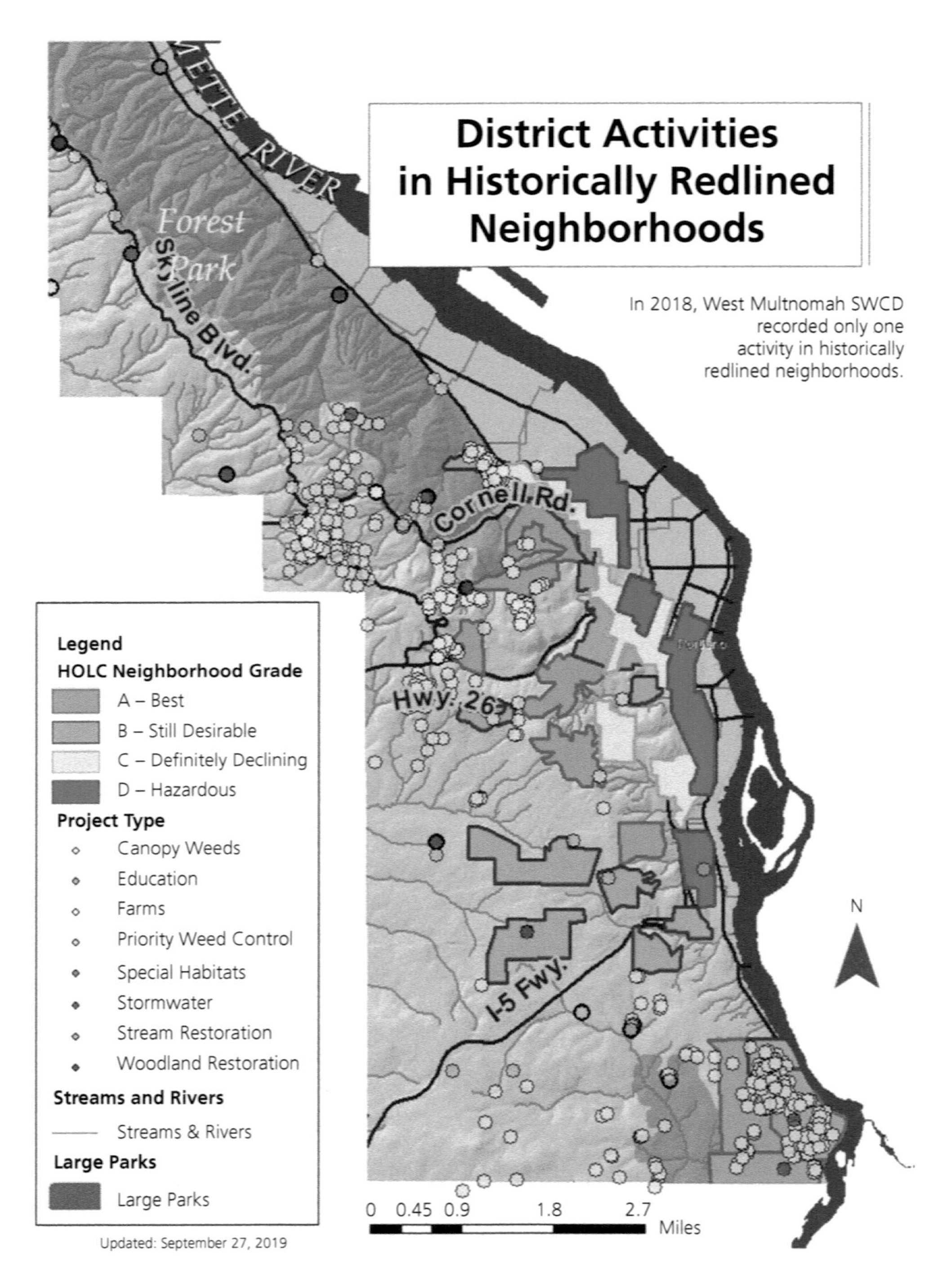

Figure 5.1 Map of conservation activities performed by the West Multnomah Soil & Water Conservation District overlaid against historical Home Owners Loan Corporation (HOLC) redlining neighborhood boundaries. Of the urban communities served, almost all were in formerly green- or blue-lined neighborhoods which were and still are predominantly White and wealthier.

These analyses were an important first step in identifying and understanding that our organization has perpetuated racial inequality and that we have much work to do to better ensure equitable and inclusive outcomes for all communities in our District.

To start dismantling the systems that led to the District's inequitable distribution of services, we sought out new and existing partners, including PP&R, to cocreate a process that helped us understand how to better serve people in this neighborhood. At that time, the District and PP&R staff co-chaired the West Willamette Restoration Partnership (WWRP), a coalition of engaged volunteers, community groups, and organizations working to conserve and enhance forests and natural areas of southwest Portland. The WWRP provided a strong network of conservation partners and hired a skilled consultant coordinator from the Samara Group to deliver a community-focused outreach and engagement project. This project was initiated to expand the reach of the WWRP partners with a neighborhood campaign called "Connect SW PDX" (Figure 5.2). The West Portland Park neighborhood was selected as the focal area with an aim toward engaging and partnering with members of historically underserved communities (with an emphasis on reaching diverse communities of color in alignment with partner organizations' racial equity goals) and identifying project/program opportunities in an area of ecological importance (this neighborhood encompasses the majority of the Falling Creek Watershed, a headwaters tributary to the Tryon Creek Watershed) while identifying opportunities to amend partner offerings to be more culturally responsive. At the outset, the Connect SW PDX Listening Project, funded by the Creating Community Natural Area Stewardship Nature in Neighborhoods (NIN) Metro Grant, sought to identify key environmental health issues of concern in a defined region of southwest Portland and connect residents to resources to address those concerns. The planned framework for the project was based on the Northwest Institute for Community Enrichment Listening Project model for outreach and engagement, which includes: collaborative development of a specialized engagement and education plan, a coordinated group of volunteers to engage with residents ("stakeholders"), asking interested residents to answer a set of listening questions, and to pledge to a personal action that connects them to a local organization or resource (32).

When starting this process, the WWRP quickly realized we were not sure how to best reach the community members we were attempting to better serve, nor did we have the necessary mechanisms in place to reimburse community members whose time and input we sought. Our WWRP team initially planned to implement the project through a volunteer door-to-door effort. However, we quickly realized that the two-year timeframe we had to deliver the project and the lack of an established presence from our WWRP network in this neighborhood presented substantial barriers for us to fully engage residents. In addition, we had not addressed how to manage language barriers, which is particularly complex given over 120 languages are spoken in Oregon (33) and Portland's role as a Sanctuary City makes it home to diverse refugee and immigrant communities. We realized we were ill-equipped to serve our communities and needed help. We reached out to our partners for ideas. These early challenges emphasized our initial mistake of not inviting community members and leaders to the table at the onset of project planning. Doing so would have helped us craft a more effective plan to start with.

We were fortunate to discover a program called Community Engagement Liaison services (CELs) founded by Ping Khaw. Ping was instrumental in helping us redesign our project and worked with the CELs and participating organizations to ensure everything went smoothly throughout the entirety of the project. "CELs are expert bridge-builders . . . Liaisons provide effective avenues for authentic public participation" (34). CELs not only connected us with foreign-language translators from within the community, they also helped us redesign our outreach plan and provided critical insight. Ping helped demonstrate to us how current distrust in the government and our lack of perceived legitimacy among Portland's underserved communities could sink the project. In addition to language barrier challenges, CELs confirmed that we were

Figure 5.2 Connect SW PDX wrap-up celebration. Photo courtesy of Ping Khaw, Community Engagement Liaison Services (CELs).

unwise to try door-to-door canvassing given the racially charged political environment of 2016 that could make individuals reluctant to participate. Instead of a knock on the door from a government stranger, trusted local community leaders attending community events in and around their neighborhoods were able to deliver the project's goals and provide translation services—undoubtedly getting more robust and authentic information from those we most hoped to hear from. Community liaisons included staff from culturally specific organizations like the Immigrant & Refuge Community Organization and Wisdom of the Elders. The CELs program further provided the WWRP network with a missing financial mechanism for reimbursing community leaders when participating agencies had no protocols set up for reimbursing community members for their time. The involvement of Ping and the CELs was pivotal because they helped us meaningfully engage underserved communities in our service area to identify how we could best support their needs.

Our cocreated process landed on a "listening project and action pledge" that provided opportunity for community members to offer us input and provided conservation actions that community members could take with partnering conservation organizations. The Listening Project & Action Pledge was provided in Vietnamese, Arabic, and Somali as well as English as these were determined to be some of the most prevalent limited English proficiency languages in the focal area according to 2010–2014 American Community Survey and 2012 Oregon Department of Education Language Use Survey data. The survey and pledge were designed through a collaborative meeting with project partners on March 15, 2017 and a meeting with Community Engagement Liaisons on April 6, 2017. Partners and liaisons were invited to participate to provide input on questions that conservation organization project partners would like to ask with an aim at better serving and understanding the West Portland Park community. Partners met to compile their knowledge unique to the neighborhood as well as their organizational goals, which were then used to guide the questions in the survey and pledge. This first proactive step into an underserved neighborhood far exceeded all our goals.

Aside from this project's direct outcomes (see the Case study), we experienced a realization that the intentionality put into this project and much more need to be done for both PP&R and the

District to fully engage underserved community members. From analyzing our community listening sessions, surveys, and organizational assessments it was clear that our conservation organizations needed additional mechanisms to both hear from and elevate voices of diverse community members and leaders within our organization to better embed equity in our organizations. As WWRP organizations move forward in these efforts, we have taken with us learning about how we must slow down and intentionally plan with marginalized communities from the onset to craft effective community engagement strategies. Most recently in 2021, the District adopted a 5-Year Long Business Plan informed by community input gathered from a diverse advisory committee comprised of partners, community members, landowners, and community leaders that helped shape our newly declared mission, vision, strategic directions, and goals (31). Some of the CELs that helped with the Connect SW PDX project were rehired to conduct additional community survey work and provide direct input on the plan and marketing materials. This process and its resultant plan aim to "embed equity and inclusion in all that we are and all that we do" (31).

As both of our organizations move forward in implementing our equity-centered plans, we have outlined goals, tactics, and measures of success guiding our steps toward more equitable hiring and procurement practices, fostering a welcoming work environment, diversifying partnerships, continued community engagement, convening advisory committees to better inform our work, community-informed communications, and programming responsive to marginalized community member requests. Programming responsive to these requests includes new initiatives to provide access to and connection with the land including school and community garden resources, promotion of resilient environments and frontline communities in the face of climate change such as those most vulnerable to urban heat island impacts, and support of opportunities that recognize and apply Traditional Ecological and Cultural Knowledge to better steward the land. Partnership development with and support of community-based and culturally specific organizations like HAKI, a local community organization created to be a resource of education, advocacy, and empowerment for East African immigrants (35), and Wisdom of the Elders, an organization that records, preserves, and shares oral history, cultural arts, language concepts, and traditional ecological knowledge of exemplary Native American elders, storytellers, and scientists in collaboration with diverse institutions, agencies, and organizations, have been found to be critical in extending our reach to communities we aim to better serve (36).

Case study Outcomes of Connect SW PDX Listening and Action Pledge Project

More than 100 people participated in the listening project, providing key information on how to better serve their communities, and 49 participants pledged on-the-ground steps to preserve and restore ecological habitat. The demographic data collected showed we reached 26 residents that speak a language other than English at home and that 28% reported a racial or ethnic identity other than White. Many partners used translation services, some for the first time ever, to respond to community members that pledged conservation actions. A neighborhood that was once without any District conservation projects has now seen implementation of multiple projects and the establishment of important relationships with community members and leaders. Projects in this neighborhood included youth environmental education programming, school garden support, riparian habitat restoration planning and implementation, and a stormwater management workshop. The learning from this project has helped PP&R and the District develop additional plans and strategies informed by the communities most impacted as we've moved deeper into our equity work. Strategies include developing community-informed communications plans, collaborating on discussion panels to discuss Indigenous Traditional Ecological Knowledge and "invasive" species management (Figure 5.3), and prioritizing additional support and funding for culturally specific educational programming and partners.

Case study *Continued*

Figure 5.3 Discussion panel on Invasive Species & Traditional Ecological Knowledge (TEK). TEK provides another in-depth and intimate way of knowing a place, its environmental needs, and how people may interact, value, and see the environment. This locally rooted and historically passed-down knowledge presents challenges that confront the western science way of managing, seeing, and talking about the environment and immense opportunities to work with different ways of knowing to achieve better outcomes. Left to right: Wisdom of the Elders staff Alvey Seeyouma, West Multnomah Soil & Water Conservation District staff Mary Logalbo, Portland Parks & Recreation staff Janelle St. Pierre, and host representative Emily Bosanquet from the Pacific Northwest College of the Arts.

Conclusion

These stories highlight important first steps that our organizations have taken on the long road of embedding equity in a way that focuses on better outcomes for those that have been historically marginalized. As we remain dedicated to conservation and environmental stewardship, we have learned the importance of shifting our approaches to be successful at addressing the equity issues we face. The following are lessons learned from Portland partners' conservation equity and inclusion efforts that have proven critical to moving forward on our equity journeys—we are still learning, and this list continues to grow:

(1) Community co-creation. Successful community engagement plans involve community cocreation from the start. Doing so will result in better outcomes and may save time in the long run. Provide community member stipends or incentives to value all the work you are asking them to do and to assist with financial burdens and/or barriers to participate.

> In its simplest terms, community engagement seeks to better engage the community to achieve long-term and sustainable outcomes, processes, relationships, discourse, decision-making, or implementation. To be successful, it must encompass strategies and processes that are sensitive to the community-context in which it occurs.
>
> Penn State University Center for Community Development (see (37))

(2) Building capacity for change. A firm commitment coupled with understanding is required to step into equity work; a need to understand the past and present must be realized before moving forward. Systems and culture change can only come about when we understand what they are and how they impact us at various levels. Individuals and organizations should work together to create trainings and build momentum with the help of professionals in order to "harness the power of racial & ethnic diversity to transform the U.S. environmental movement by developing leaders, catalyzing change within institutions, and building alliances" (Center for Diversity & the Environment; (38)).

(3) Apply an equity lens. Examine decisions by asking about assumptions, those who are most impacted, desired and anticipated outcomes, and diverse perspectives.

> An Equity and Inclusion Lens is like a pair of glasses. It helps you see things from a new perspective ... When we consider the range of equity and inclusion issues, we take action to eliminate barriers so that everyone can benefit.
>
> Nonprofit Association of Oregon (39)

(4) Relationship development. Lasting relationship development requires time, vulnerability, respect, humility, authenticity, and intentionality.

> Dominator culture has tried to keep us all afraid, to make us choose safety instead of risk, sameness instead of diversity. Moving through that fear, finding out what connects us, reveling in our differences; this is the process that brings us closer, that gives us a world of shared values, of meaningful community.
>
> bell hooks (40)

(5) Slow down and remain adaptable. Flexibility in planning and timelines that allow for full incorporation of perspectives and provide space to incorporate the same will result in better outcomes—this work is urgent, but it also requires time to carry it out well.

(6) Accountability and following up are critical to gaining and keeping trust/engagement. Having a plan but not showing results can further damage relationships/trust. Provide transparent and accessible ways those interested can track your progress on equity and inclusion goals. If you haven't made progress, openly explain your challenges.

(7) Allyship, power sharing, and stepping down. Questions from dominant culture that we are asking and still working on, include (1) how to ensure power is shared/shifted, (2) what makes a good ally, and (3) when to step down?

> Global biodiversity declines are increasingly recognized as profound ecological and social crises. In areas subject to colonialization, these declines have advanced in lockstep with settler colonialism and imposition of centralized resource management by settler states. Many have suggested that resurgent Indigenous led governance systems could help arrest these trends while advancing effective

and socially just approaches to environmental interactions that benefit people and places alike.

Artelle et al. (41)

The continued need to make progress on equity and inclusion goals to diversify staff and leadership, ensure equitable access to conservation service benefits, and realize systemic change continues. We understand this work will need to continue over the long term, always evolving, with deliberate thought and evaluation to realize the outcomes we wish to see. We also recognize we do not exist in a vacuum and to be successful equity must be addressed at local, regional, national, and international levels. Not until major strides are made on all these fronts will accountability and best conservation outcomes for ALL be realized. Koffi Dessou, Equity and Operations Manager for the City of Portland, taught our organizations the business case for equity and we hope that we can leave you with the same. Valuing diversity and embedding equity and inclusion in one's organization lead to a stronger organization with better outcomes. Research has illustrated how the most diverse companies are likely to outperform less diverse counterparts (42). Despite this evidence, conservation organizations across the US are consistently represented by low levels of racially diverse staff and board members, with racial diversity lowest on boards and highest among relatively new employees, seasonal workers, and interns (43). The Portland-based Center for Diversity & the Environment "grew out of the belief that everyone has a place in the environmental movement" and is rooted in the belief that "diversity strengthens and enriches our work and makes the environmental movement more relevant than ever" (38). The District and PP&R have found our conservation work strengthened, enriched, and made more relevant as we have deepened our equity and inclusion work. We hope all readers can engage in this valuable work from wherever they sit given the local, national, and international challenges the environmental community must now face and the understanding that we are all needed to realize success.

If we truly want to win on climate and the environment, that means all voices must help drive the process.

Mustafa Ali, Hip Hop Caucus (44)

References

1. Movement Generation Justice & Ecology Project. From banks and tanks to cooperation and caring: a strategic framework for a just transition. Movement Generation; 2016. Available from: https://movementgeneration.org/wp-content/uploads/2016/11/JT_booklet_Eng_printspreads.pdf
2. Goodling E, Green J, McClintock N. Uneven development of the sustainable city: shifting capital in Portland, Oregon [Internet]. Portland State University; 2015. Urban Studies and Planning Faculty Publications and Presentations 107. Available from: https://pdxscholar.library.pdx.edu/usp_fac/107
3. Hughes J. Historical context of racist planning: a history of how planning segregated Portland [Internet]. Portland, OR: City of Portland Bureau of Planning and Sustainability; 2019. Available from: https://www.portland.gov/bps/planning/history-racist-planning-portland
4. Curry-Stevens A, Cross-Hemmer A, Coalition of Communities of Color. Communities of Color in Multnomah County: an unsettling profile. Portland, OR: Portland State University; 2010. Available from: https://www.coalitioncommunitiescolor.org/research-and-publications/cccunsettling profile
5. Profita, C. Study: more people of color live near Portland's biggest air polluters. Oregon Public Broadcasting; 2020 Apr 29. Available from: https://www.opb.org/news/article/oregon-portland-study-people-of-color-polluted-neighborhoods-redlining-covid-19/
6. Hoffman JS, Shandas V, Pendleton N. The effects of historical housing policies on resident exposure to intra-urban heat: a study of 108 US urban areas. Climate. 2020;8(1):12. Available from: https://doi.org/10.3390/cli8010012
7. Keith I. Whose land is our land? Spatial exclusion, racial segregation, and the history of the lands of western Multnomah County [Internet]. West Multnomah Soil & Water Conservation District; 2019. Available from: https://wmswcd.org/wp-content/uploads/2020/10/Whose-Land-is-Our-Land_ReviewDraft_9-24-19.pdf
8. Miller R. Oregon, Indigenous Nations, Manifest Destiny, and the Doctrine of Discovery [Internet]. Confluence; 2019. Available from: https://www.confluenceproject.org/library-post/robert-miller-oregon-indigenous-nations-manifest-destiny-and-the-doctrine-of-discovery/
9. Coleman CR. "We'll all start even": White egalitarianism and the Oregon Donation Land Claim Act.

Oregon Historical Society. 2019;120(4):414–37. Available from: https://www.ohs.org/oregon-historical-quarterly/back-issues/upload/05_Coleman_We-ll-all-start-at-even_OHQ-Winter-2019_120_4_web.pdf
10. Robbins WG. Oregon Donation Land Law. In: Oregon Encyclopedia [Internet]. Portland State University and the Oregon Historical Society; 2022. Available from: https://www.oregonencyclopedia.org/articles/oregon_donation_land_act/
11. Robbins WG. New assaults on Indian land. The Oregon History Project. The Oregon Historical Society; 2014. Available from: https://www.oregonhistoryproject.org/narratives/this-land-oregon/political-and-economic-culture-1870-1920/new-assaults-on-indian-land/
12. Streckert J. Echoes of the Klan. Portland Mercury; 2017 Nov 15. Available from: https://www.portlandmercury.com/feature/2017/11/15/19472650/echoes-of-the-klan
13. Buck AK. Alien Land Laws: the curtailing of Japanese agricultural pursuits in Oregon [Master's Thesis]. [Portland, OR]: Portland State University; 1999. 109 p. Available from: https://doi.org/10.15760/etd.5872
14. Nokes G. Black exclusion laws in Oregon. In: Oregon Encyclopedia [Internet]. Portland State University and the Oregon Historical Society; 2022. Available from: https://www.oregonencyclopedia.org/articles/exclusion_laws/
15. Rothstein A. The forgotten history of how our government segregated the United States [Internet]. The Zinn Education Project, Rethinking Schools and Teaching for Change, Huffington Post; 2017. Available from: https://www.zinnedproject.org/if-we-knew-our-history/forgotten-history-government-segregated-united-states/
16. Ray R, Perry A, Harshbarger D, Elizondo S, Gibbons A. Homeownership, racial segregation, and policy solutions to racial wealth equity [Internet]. Washington, DC: The Brookings Institution; 2021. Available from: https://www.brookings.edu/essay/homeownership-racial-segregation-and-policies-for-racial-wealth-equity/
17. Malpas J. Colonialism. In: Zalta EN (ed.) The Stanford Encyclopedia of Philosophy [Internet]. Winter 2012 ed. Stanford, CA: Stanford University; 2012. Available from: https://plato.stanford.edu/archives/win2012/entries/colonialism/
18. Correia R. Three graphics that explain Portuguese colonialism [Internet]. The Interpreter; 2021 May 28. Available from: https://www.interruptor.pt/artigos/graficos-para-compreender-o-colonialismo-portugues
19. Khilnani S. Cruel Britannica. The New Yorker; 2022 Apr 4.
20. Tussing N. African resistance to European colonial aggression: an assessment. Africana Studies Student Research Conference; 2017 Feb 12. Available from: https://scholarworks.bgsu.edu/cgi/viewcontent.cgi?article=1056&context=africana_studies_conf
21. 5th BRICS Summit. General background [Internet]. Cape Town: Government of South Africa; 2022. Available from: https://www.gov.za/events/fifth-brics-summit-general-background
22. Zoomers A, Kaag M. The global land grab as modern day corporate colonialism. The Conversation; 2014 April 25.
23. Shenk MK, Borgerhoff Mulder M, Beise J, Clark G, Irons W, Leonetti D, et al. Intergenerational wealth transmission among agriculturalists: foundations of agrarian inequality. Current Anthropology. 2010;51(1):65–83. Available from: https://doi.org/10.1086/648658
24. Parker D. Distributional effects of conservation easements. In: Costello C (ed.) Distributional Effects of Environmental Markets: insights and solutions for economics. Bozeman, MT: Property and Environment Research Center; 2019. p. 33–40.
25. Woodford J. The power of storytelling to create representation: an interview with Dawood Qureshi. Conservation Careers; 2022 May 9.
26. EPA. Environmental justice [Internet]. Environmental Protection Agency; 2022. Available from: https://www.epa.gov/environmentaljustice
27. PP&R. The vision. Portland, OR: Portland Parks & Recreation; 2020. Available from: https://www.portland.gov/sites/default/files/2020/ppr-2020-vision.pdf
28. PP&R. Portland Parks & Recreation 2017–2020 strategic plan. Portland, OR: Portland Parks & Recreation; 2017. Available from: https://www.portlandoregon.gov/parks/article/706102
29. PP&R. Portland Parks & Recreation natural areas restoration plan. Portland, OR: Portland Parks & Recreation; 2010. Available from: https://www.portlandoregon.gov/parks/article/323540
30. City of Portland Bureau of Planning and Sustainability. Salmon-safe certification [Internet]. Portland, OR: Bureau of Planning and Sustainability; 2022. Available from: https://www.portland.gov/bps/scg/scg-dashboard/salmon-safe-certification
31. WMSWCD. Long Range Business Plan 2021-2025. West Multnomah Soil & Water Conservation District; 2021. Available from: https://wmswcd.org/wp-content/uploads/2021/07/WMSWCD-LRBP2021-25-final-7.2.21.pdf

32. The Northwest Institute for Community Enrichment. Effecting change [Internet]. Portland, OR: Northwest Institute for Community Enrichment; 2022. Available from: https://ecapdx.weebly.com/northwest-institute-for-community-enrichment.html
33. Zarkhin F. Five languages you probably didn't know Oregonians spoke at home (searchable database). The Oregonian/OregonLive; 2015 November 28. Available from: https://www.oregonlive.com/pacific-northwest-news/2015/11/five_languages_you_probably_di.html
34. Khaw P. Why we need CELs [Internet]. Portland, OR: Community Engagement Liaison services (CELs); 2021. Available from: https://celsservices.com/why-we-need-cels/
35. HAKI. Haki Community Organization [Internet]. Portland, OR: HAKI; 2022. Available from: https://hakicommunity.org/
36. Wisdom of the Elders. About Wisdom of the Elders/mission and vision [Internet]. Portland, OR: Wisdom of the Elders; 2022. Available from: https://wisdomoftheelders.org/mission-and-vision/
37. Wilson C. Grassroots community outreach and engagement [Internet]. Michigan State University, MSU Extension LeadNet; 2018 October 10. Available from: https://www.canr.msu.edu/news/grassroots-community-outreach-engagement
38. CDE. Center for Diversity & the Environment [Internet]. CDE; 2022. Available from: https://www.cdeinspires.org/
39. Nonprofit Association of Oregon. 2019. Equity & inclusion lens guide. Nonprofit Association of Oregon; 2019. Available from: https://nonprofitoregon.org/sites/default/files/NAO-Equity-Lens-Guide-2019.pdf
40. hooks b. Teaching Community: a pedagogy of hope. New York: Routledge; 2003.
41. Artelle KA, Adams MS, Bryan HM, Darimont CT, Housty J, Housty WG, et al. Decolonial model of environmental management and conservation: insights from Indigenous-led grizzly bear stewardship in the Great Bear Rainforest. Ethics, Policy & Environment. 2021;24(3):283–323. Available from: https://doi.org/10.1080/21550085.2021.2002624
42. Dixon-Fyle S, Dolan K, Hunt V, Prince S. Diversity wins: how inclusion matters [Internet]. New York: McKinsey & Company; 2020. Available from: https://www.mckinsey.com/featured-insights/diversity-and-inclusion/diversity-wins-how-inclusion-matters
43. Taylor DE. The State of Diversity in Environmental Organizations: mainstream NGOs foundations government agencies [Internet]. Green 2.0; 2014. 192 p. Available from: https://orgs.law.harvard.edu/els/files/2014/02/FullReport_Green2.0_FINALReducedSize.pdf
44. Gewin V. Why diversity in sustainability matters, and what you can do [Internet]. GreenBiz; 2018. Available from: https://www.greenbiz.com/article/why-diversity-sustainability-matters-and-what-you-can-do

SECTION 2

Innovative Approaches for Understanding and Prioritizing Equitable Urban Conservation

In some cases, conservation science and practice in cities can build on existing approaches from nonurban conservation. However, the unique social-ecological situation of cities relies on applying these approaches in new ways or innovating new methods. This section underscores how centering equity is an essential lens for conserving urban biodiversity.

CHAPTER 6

The Role of Urban Tree Canopies in Environmental Justice and Conserving Biodiversity

Dexter H. Locke‡, J. Morgan Grove‡, and Steward T. A. Pickett§

Introduction

Urban trees are an important part of biodiversity in urban areas, from cities to periurban suburbs and exurbs. Trees are often the base or underlying ecological support for other biota since they provide habitat, food, and influence other environmental conditions like temperature and moisture. Urban trees provide diverse ecosystem services for people, too, such as providing shade and cooler temperatures, aesthetic value, increases in sense-of-place and social cohesion, reductions in crime, and improvements to physiological and psychological health and well-being (1–3). In forested biomes, urbanization usually means replacing trees with roads, buildings, and other "gray" infrastructure. In biomes that are not naturally forested, urbanization can mean the addition of trees such as in Salt Lake City, UT, parts of Los Angeles County, or Sacramento and Davis, CA. Managing urban forests for societal and species' needs is therefore highly regionally and city-specific.

Urban trees are foundational for urban biodiversity conservation. They provide habitat for a diverse range of species, from birds and insects to mammals, reptiles, and amphibians. By providing food, shelter, and nesting sites, urban trees can increase the abundance and diversity of other species in urban areas, and thus enhance alpha and beta biodiversity. In fact, in the early 1990s the former director of the Iowa Department of Natural Resources commented on how cities and suburbs provided some of the only habitat for wildlife in midwestern states like Iowa and Illinois because expansive agriculture was inhospitable (4). The connectivity of urban tree canopy (of any type), and urban forest patches in particular, may create ecological variation that contributes to diverse biological community composition (5,6). Overall, urban trees are essential for conserving biodiversity by providing valuable habitat and resources for wildlife (7–9).

The spatial distribution of urban trees is neither uniform nor random. This is because individual people, organizations, and societal rules and norms shape where new trees are planted, where trees are removed, and which trees are protected (10). The fields of urban and community forestry have the goal of discerning the spatial distribution of urban trees, the complexity of managing urban trees, and which communities do or do not steward urban trees and forests.

A guiding theme from this body of work is documenting how urban trees are unevenly distributed, and at different scales. For instance, at the parcel scale, most urban land is residential. At the neighborhood scale, higher-income areas have more tree canopy cover than lower-income areas (11,12), and predominantly White neighborhoods have more canopy cover than predominantly Black

‡ Northern Research Station, USDA Forest Service, USA
§ Cary Institute of Ecosystem Studies, USA

Dexter H. Locke, J. Morgan Grove, and Steward T. A. Pickett, *The Role of Urban Tree Canopies in Environmental Justice and Conserving Biodiversity.*
In: *Urban Biodiversity and Equity.* Edited by: Max R. Lambert & Christopher J. Schell, Oxford University Press.
DOI: 10.1093/oso/9780198877271.003.0006

neighborhoods (13). A long-term view reveals the legacies of institutionalized racism (particularly the policy of "redlining") (14) on present-day distribution of urban vegetation generally (15) and urban tree canopy cover in particular (16–19). There are, of course, nuances in how these relationships among income, wealth, and race at the household, neighborhood, and municipal scales unfold. Regardless, these patterns demonstrate important interactions among income, wealth, race, and canopy cover which typically lead to higher-income areas having more trees and communities of color having fewer. How urban trees are managed has important implications not only for conservation but also for distributional and procedural environmental injustice.[1] Indeed, urban trees were the first and are probably the best understood dimensions of inequities in urban conservation.

We have four goals here: (1) introduce and describe urban forestry, (2) explain why other disciplines might be interested in urban forestry, (3) illustrate how wildlife biologists can use urban forestry in their studies, and (4) encourage conservationists to think mechanistically about how urban vegetation, environmental justice, and urban wildlife may interact. Here we describe urban forestry, which requires us to differentiate among different parts of an urban forest: street trees, forested patches, and public vs private ownership. This differentiation is important because these different components of the urban forest are managed by different individuals and organizations, with diverse and sometimes competing interests and authorities. We also explain why other urban biodiversity researchers and practitioners may be interested in urban forestry, measuring urban forests, and investigating how urban vegetation relates to species distributions, composition, and biology. Finally, we illustrate how history matters and how urban forests and environmental justice might look in the future.

[1] Hoover's and Scarlett's chapter discusses the differences between distributional and procedural (in)justice in urban conservation.

Urban forests are spatially and socially heterogeneous: ownerships and management vary over small geographic areas

Urban forestry is the practice of planting, managing, and conserving trees and other vegetation in urban and suburban settings. It ranges from a disciplinary to a transdisciplinary field that can draw on botany, horticulture, urban planning, geographic information science, environmental sociology, and diverse other sets of expertise. Urban forestry aims to increase environmental quality by providing shade and mitigating heat, controlling stormwater runoff, beautifying the landscape, reducing crime, and providing opportunities for community stewardship. In doing so, this practice creates a healthier and more sustainable urban environment for people and, often, other species.

Urban forestry is a critical component of urban ecology, which is the study of how organisms interact with the built environment, including people.[2] Urban forestry is gaining popularity among elected officials via the rise of "million trees" campaigns, citywide tree canopy goals (often in response to increasing weather extremes), and other ambitious tree planting programs (20–25). For example, Beijing added 50 million trees in just four years (26). The entire urban forest consists of individual trees and tree clusters managed by different entities, for different purposes, and with different outcomes (27). We emphasize that an urban "forest patch" is an aggregation of trees, also called a forested natural area or urban woodland. But a forest patch may comprise a mixture of trees in a park or several trees in abutting residential properties, yet the various trees in this latter kind of patch occur on different parcels and thus have different managers. In urban areas, most of the land is privately owned and residential. Thus, conservation goals must be aligned with the relevant authority and desires of those who can manage particular urban trees.

Residential lands: As are the cities and suburbs that harbor them, urban forests are heterogeneous in many ways (Figure 6.1). In forested biomes where

[2] Lambert's and Schell's chapter briefly details the history of urban ecology and conservation.

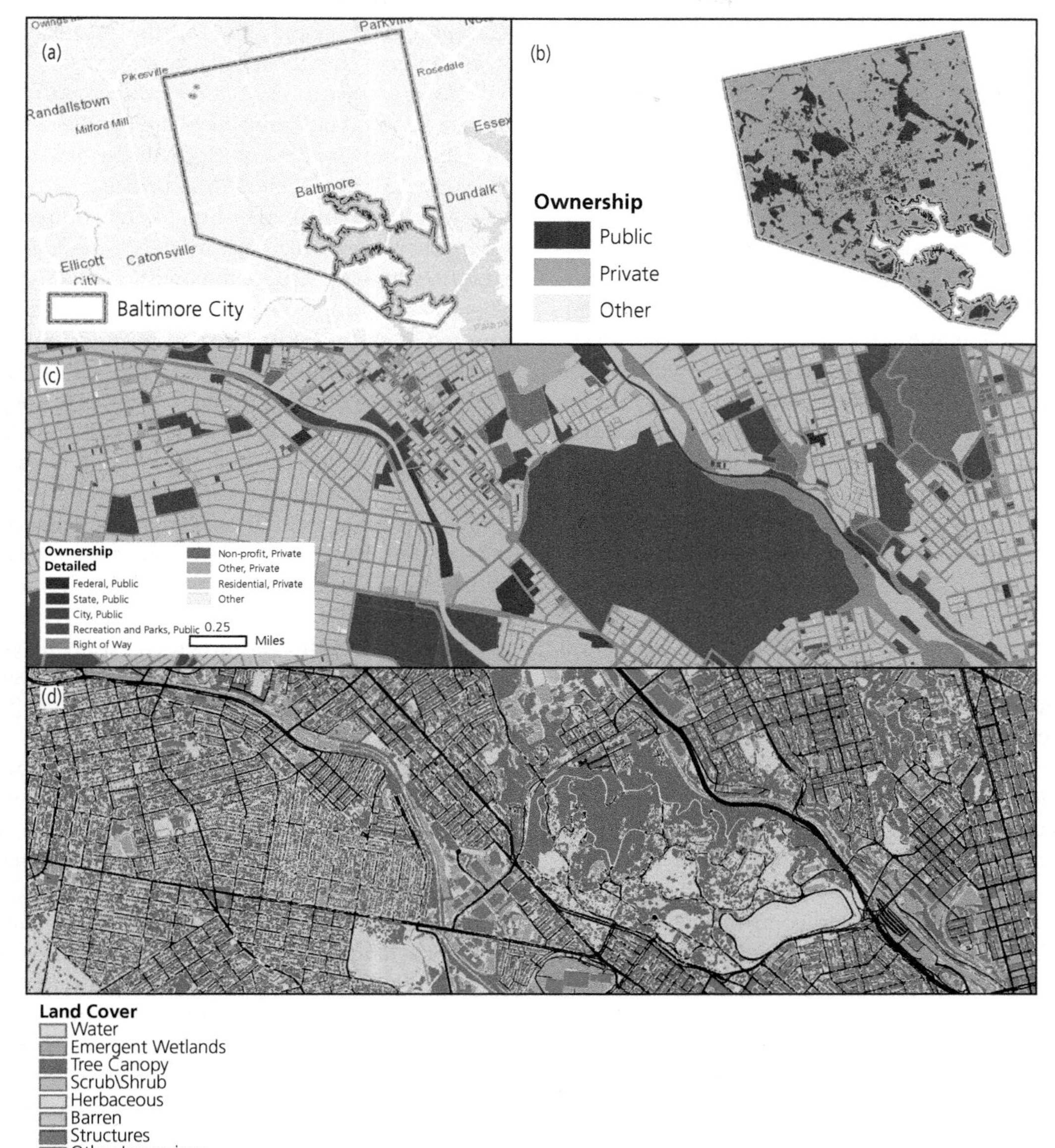

Figure 6.1 Within the City of Baltimore (a), each parcel of land has a distinct owner. In Baltimore, like most urban areas, most of the land is privately owned (b). There are diverse types of public and private land (c), and different owners have different motivations, capacities, and interests. Land ownership says nothing about land cover. High-resolution, high accuracy, 12-class land cover from the Chesapeake Conservancy and partners shows the spatial heterogeneity of land cover (d). The mixes of ownership and cover define constraints and opportunities for urban forestry.

urbanization has occurred, trees live on all different types of land with varied types of ownerships. One of the most clear and consistent findings from dozens of tree canopy assessments in North America using high-resolution (≤ 1 m^2) high-accuracy ($\geq 95\%$) tree canopy maps, is that (1) most land is private residential, (2) most of the tree canopy is private residential, and (3) most of the opportunities for additional tree canopy are also on private residential land. In New York City, for example, there are approximately 750,000 private residential landowners or "forest owners" (28). Each owner may have different motivations, capacities, and interests for tree management, which raises exciting questions about human motivations for urban nature[3] and also complicates managing urban ecosystem services and biodiversity.

Street trees: Street trees have been heavily studied by urban foresters and arboriculturalists (29). These are also the components of urban forests that people probably most commonly experience. Street trees are species diverse because tree populations often consist of locally native species and imported horticultural varieties (30,31). In contrast to residential land, which is managed by numerous landowners, there tends to be one or few organizations charged with managing street trees in a given municipality. The agency responsible for street trees is usually a city's Parks Department, but it is not uncommon to be located within Departments of Transportation or Public Works. Each agency has its own mandate, history, and competing internal priorities and vies for funding with other departments. Understanding how a particular city's government and governance operate is essential to managing street trees for conservation goals. Our experience shows that convincing a mayor or parks department commissioner of the value of street trees is an entirely different task than trying to understand tens of thousands—or hundreds of thousands—of individual residential landowners and their management objectives for their front and back yards. Without people, there would be no street trees. The entire lifecycle of street trees from seed or clone, to a nursery, to a sidewalk, and ultimately until it is removed via chainsaw, chipped, and hauled away is done by and for people. These attributes make street trees a unique component of urban forests.

Parks, open spaces, and forested natural areas: Trees in parks and other open spaces are another important and often-studied part of the urban forest. Many parks have isolated trees or clusters of trees in grassy open areas with little or no competition from other trees which can influence the management needs with respect to conditions like water. There is growing interest in individual trees' and forested areas' performance from a stormwater management perspective (32,33). Forested natural areas, sometimes called urban forest patches, natural area forests (34), and/or urban woodlands (35), bear many similarities to forests in nonurban areas with respect to species composition and structure (35,36). Forested natural areas have spontaneous regeneration, dead material is left to decompose, and, although located within the built environment, they are largely left undisturbed (34,35). Forested natural areas face threats particularly from invasive understory plants and vines (but also pest insects and pathogens) and pressure from herbivory (e.g., from deer and other ruminants).

Patches can and do cross multiple property boundaries, representing additional complexity for management. In the greater Boston area, for example, there are more backyard forested patches and a greater forest patch area than in all public parks combined, in part because there is so much residential land (6,37). From an organismal perspective, a backyard forest patch is a single functional unit. But from a management perspective, in the case of a patch crossing multiple parcel boundaries (like several abutting vacant lots) the threat to development is not the same. A contiguous patch of forest in an Olmsted-designed park like Druid Hill Park in Baltimore City does not face the same development pressures or risk as do two dozen smaller parcels owned by 24 different people—even though as forested patches they may both have similar size, species, structure, and age. Further, multiple levels of government agencies own and manage open spaces and/or forest patches within a city. Visitors to patches on New York City parkland find a sense of refuge, place attachment,

[3] Larson's and Brown's chapter reviews how different human motivations shape how society manages urban habitats and biodiversity.

and experiences with nature (38). Whereas, those living adjacent to some forest patches in Baltimore City are largely ambivalent about patches. The use and experience of urban forest patches warrant further attention among researchers and practitioners. Forested natural area governance is complex and varies geographically (39), especially with multi-parcel forested natural areas. A growing trend in urban conservation are urban land trusts which can buy land, sell the development rights, and protect forested areas from being developed.

Trees on residential properties, street trees, and trees in parks (both in natural areas and on lawns) with different owners and associated motivations reflect just a few different examples for how trees' site-level contexts, ownership, and management interact to provide a different biological template for other organisms. The intersections of ownership, management, and biological context explain, in part, the social and spatial heterogeneity found in US cities' urban forests.

Environmental injustice

The spatial distribution of the urban forest not only relates to insects, birds, mammals, and other wildlife, but also to people. As briefly mentioned earlier, one of the most clear and consistent findings from urban forestry is that (1) higher-income areas have a higher tree canopy cover than nearby lower-income areas (11,12) and (2) communities of color tend to have less tree canopy cover than White areas of cities and suburbs (13). Although such patterns are increasingly being demonstrated in other aspects of urban biodiversity (40), urban forestry has paved the way in understanding the linkages between class and racial inequities and urban nature. Because of the diverse benefits and ecosystem services provided by trees to people in urban areas, the uneven distribution of canopy coinciding with segregating communities along multiple indicators of social status, race, and ethnicity in urban areas produces a distributional environmental injustice.[4]

[4] Hoover's and Scarlett's chapter discusses different forms (e.g., distributional vs procedural) of justice and its relationship to urban environments, including trees.

Understanding patterns of environmental injustice requires understanding processes of environmental injustice over time. Trees do not mature instantly (10), and substantial research shows that historic demographic data predict the present distribution of tree canopy cover (41,42). Bluntly, we do not get to recreate the urban forest every few years; addressing environmental injustice today will matter today and tomorrow. The uneven distribution of trees and urban vegetation in general (14), street trees (30), and tree canopy cover (16–19) today may be an example or result of structural racism (14), as the distribution relates strongly to the history of the racist government practice of redlining and disinvestment in certain neighborhoods. This disparity demonstrates a lack of access to the benefits of urban trees, such as improved air quality, thermal comfort, and stormwater retention, for those living in lower-income neighborhoods. Likewise, household income and wealth do not occur instantly, and the long-term interactions among race, income, and wealth and their consequences for ecological systems need to be understood to address contemporary inequities in urban trees. To address these issues, some state and city governments have begun to invest in programs to promote equitable tree planting and species diversification in order to ensure environmental justice for all their citizens. Yet a recent analysis of more than 100 urban forest management plans in the US shows most plans only briefly mention environmental justice (43).

Along with tree canopy in general and street trees, the distributions of parks and open spaces—and their urban forest patches—also vary by the socioeconomic status of residents. For instance, studies across 10 US cities found that income and education showed strong positive relationships with urban woody vegetation like trees, whereas tree cover was lower in neighborhoods with more Latino, Black, and Indigenous households, and areas with lower percentages of individuals without high school degrees (44). A national synthesis of all publicly accessible open space shows that (1) areas with higher home value have more park access, except in the Northeast; (2) areas with lower income inequality have greater park access; and (3) higher unemployment is associated with greater park access (45).

Given the inequities in the street tree distribution, tree canopy cover, and park access, and the recognition of multiple benefits of urban trees, the 3–30–300 rule has been proposed. Briefly, the 3–30–300 rule proposes that nature exposure from seeing nature, exposure from living among vegetation, and accessing and recreating in green space each provide benefits for human well-being. Therefore everyone should be able to see, "3 trees from every home, school, and place of work," live in a neighborhood with 30% tree canopy cover, and be within 300 meters of the nearest park (46). There is a need to develop the methods for measuring these heuristics and implementing the guidelines, aside from measuring tree canopy at the neighborhood level, which is now well-worked out.

Global perspectives—climate change and the world's urban forests

Human-caused climate change is an ongoing threat to Earth's life support system. Urban forests are poised to help buffer society and biodiversity from some impacts of climate change such as extreme heat and flooding. Yet those urban forests themselves are also threatened by climate change. Research has found that 70% or more of urban tree species found in cities on every continent except Antarctica are at risk from altered temperature and precipitation patterns (47). The situation is worse for equatorial or low-latitude cities where potentially all urban tree species are at risk. The climate threat is active and present given that this research found that over half of urban tree species are currently already exceeding their natural temperature and precipitation regimes and 20%–40% of urban forest tree species around the world are expected to be at enhanced risk from climate change in the places where they are planted by 2050. This threat underscores an urgent need within cities and across cities and regions globally to invest in climate-resilient urban forests. Although urban forests are touted for their ability to buffer impacts of climate change like extreme heat, those same stressors may themselves threaten the world's urban tree canopies. Yet, there are also new frameworks for urban forestry that use climate analog species to help planners plant trees that are more resilient to changing climate conditions (48). These climate analogs can begin accounting for the future temperature, precipitation, and pest/pathogen stressors brought with climate change while also considering aesthetic and biodiversity needs. Such climate analogs may not necessarily be tree species that are native to a region and so planting climate analogs in an urban forest may conflict with some conservation goals that focus on native biodiversity. Conservation practitioners working with urban planners and community groups will have to grapple with a focus on native tree species that may succumb to climate changes and the reality that a non-native climate analog tree may better persist. Even so, depending on the geographic region, there may be opportunities for using assisted migration to plant "native" species from more climate-resilient portions of their range.[5] Ultimately, in many cases around the globe, a tree is likely better than no tree in supporting urban biodiversity and society. Given the need for cities to be hospitable to most of humanity and to play an active role in regional and global conservation efforts, a climate-resilient urban forest is more imperative than ever.

Putting trees on the map

Managing the urban forest and all the biodiversity and communities reliant on those trees necessitates understanding where those trees are, and where they are not. The clear and consistent findings with respect to tree canopy at the neighborhood scale and income and race would not be possible without tremendous innovations in mapping tools. The advent of high-resolution (≤ 1 m^2), high-accuracy ($\geq 95\%$) land cover mapping (Figures 6.1 and 6.2) has allowed urban forestry to produce incredible maps of the urban tree canopy and prioritize equitable conservation measures (49–54). Until these recent innovations, urban forestry—and indeed urban ecologists more broadly—relied on coarse, "mid-resolution" mapping data. Prior mid-resolution mapping data were often at resolutions of 30 m^2. In practice, this coarse resolution made

[5] Spotswood et al.'s chapter discusses translocations and assisted migration as a powerful tool for urban conservation.

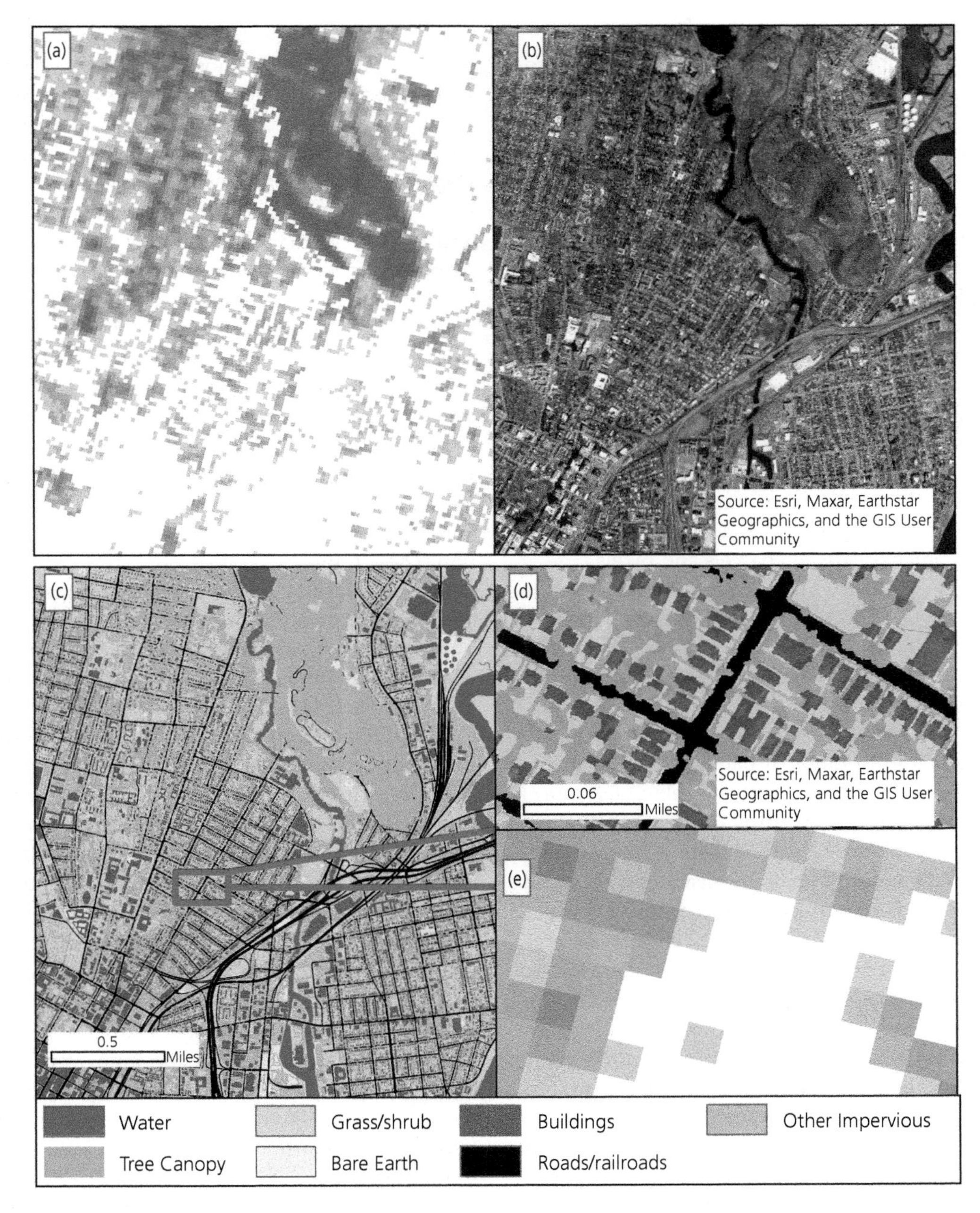

Figure 6.2 New Haven, CT: mid-resolution, Landsat-derived (30 m^2) estimates of tree canopy from the Multi-Resolution Land Characteristics Consortium (https://www.mrlc.gov/data/type/land-cover) miss key patches of urban tree canopy (a) when compared to imagery (b). High-resolution (1 m^2), high-accuracy ($\geq$95%) land cover created by the University of Vermont for the South Central Regional Council of Governments (https://scrcog.org/) has higher categorical resolution as well (c). Zooming in (d) shows small and large street trees planted by the Urban Resources Initiative (https://uri.yale.edu/) and maintained by the City of New Haven, and trees in abutting residential backyards. The same area (e) when mapped with 30-meter data simply fails to capture human- and wildlife-relevant tree canopy.

it impossible to meaningfully map vegetation at a practical scale for management.

Take for example trees along a backyard fence, then a raised bed garden, followed by grass, then the building footprint, a grassy front yard, a sidewalk with street trees, and then the road. All of that spatial heterogeneity—ecologically and socially—would be represented by one single number in prior datasets, despite being several distinct management zones. Worse, often a 30-m^2 pixel covers multiple owners' properties. Whereas these prior, coarse data could not map where vegetation was and where it was needed, newer mapping tools provide trees, other vegetation, and various built features like houses or roads at the parcel scale. As described earlier, most trees and most opportunities for tree management are on residential land—which are relatively small urban parcels—and so prior data could not match the management units due to size mismatches between parcels and pixels.

The now-industry standard land cover data are created by a handful of universities and private companies, and their partnerships. Examples include Virginia Tech (55), UC Davis (37,56,57), the University of South Florida (58–62), and American University (63–66). Private firms routinely creating high-resolution urban tree canopy maps include PlanIt Geo (https://planitgeo.com/) and Earth Define (https://www.earthdefine.com/). The Spatial Analysis Lab at the University of Vermont (UVM) is the leader in this area (https://www.gisforscience.com/chapter11/v2/) and has completed 87 land cover and/or tree canopy change maps, covering 53 million acres, with 1.4 trillion classified pixels (https://uvm.maps.arcgis.com/apps/webappviewer/index.html?id=fe15424caceb4dd294be816581ec2d77).

For example, the standard data from the UVM have seven classes: tree canopy, grass/shrub, bare soil, water, buildings, transportation (roads/railroads), and other impervious surfaces (parking lots, sidewalks, etc.). The Chesapeake Bay Program's freely available watershed-wide maps span roughly 200 jurisdictions at a 1-meter resolution (https://www.chesapeakeconservancy.org/conservation-innovation-center/high-resolution-data/lulc-data-project-2022/). UVM is a core partner on that project. Those data contain trees over roads, trees over buildings, and trees over other impervious surfaces so that the user can flexibly combine tree canopy subclasses or impervious subclasses to match the analytical needs of the questions asked. Government agencies also create similar data. For example, the Washington Department of Fish and Wildlife has produced 1-m resolution data for most of western Washington and some of eastern Washington that allow for mapping contemporary and changes in individual trees and tree canopies in cities, suburbs, and rural areas (https://hrcd-wdfw.hub.arcgis.com/pages/canopy).

Further, many cities now have street tree censuses, often for the first time (i.e., New York City, New Haven, Baltimore City, Washington D.C.). Unlike high-resolution tree canopy data, these census data contain lat/long points of individual trees and contain data about tree species, size, and condition. These digital inventories are tied to municipal databases and may have volunteer engagement in collecting the data. For instance, in the 2015–16 New York City street tree census (called "TreesCount!"), over 2000 volunteers walked over 11,000 miles to map over 660,000 trees across over 130,000 blocks; these volunteers completed one-third of the city's census (https://www.nycgovparks.org/trees/treescount). Such repositories are helping produce platforms like OpenTrees which are invaluable public engagement tools and which capitalize on data from over 200 sources and 20 countries currently, although these data are heavily biased to North America, Europe, and Australia (https://opentrees.org) (Figure 6.3).

The push for open data enables anyone to access entire street tree inventories with a few clicks of a mouse, which allows for innovative social and conservation insights. For example, research using Baltimore's freely accessible tree census data found that street trees include 59 unique species that otherwise would not be there (31). However, these data also showed that formerly redlined neighborhoods comprise street trees that have lower species diversity and have fewer old trees (30). Although these data in Baltimore show local organizations are dedicated to ameliorating the distributional injustice, there is still a need for continued effort that diversifies the trees planted,

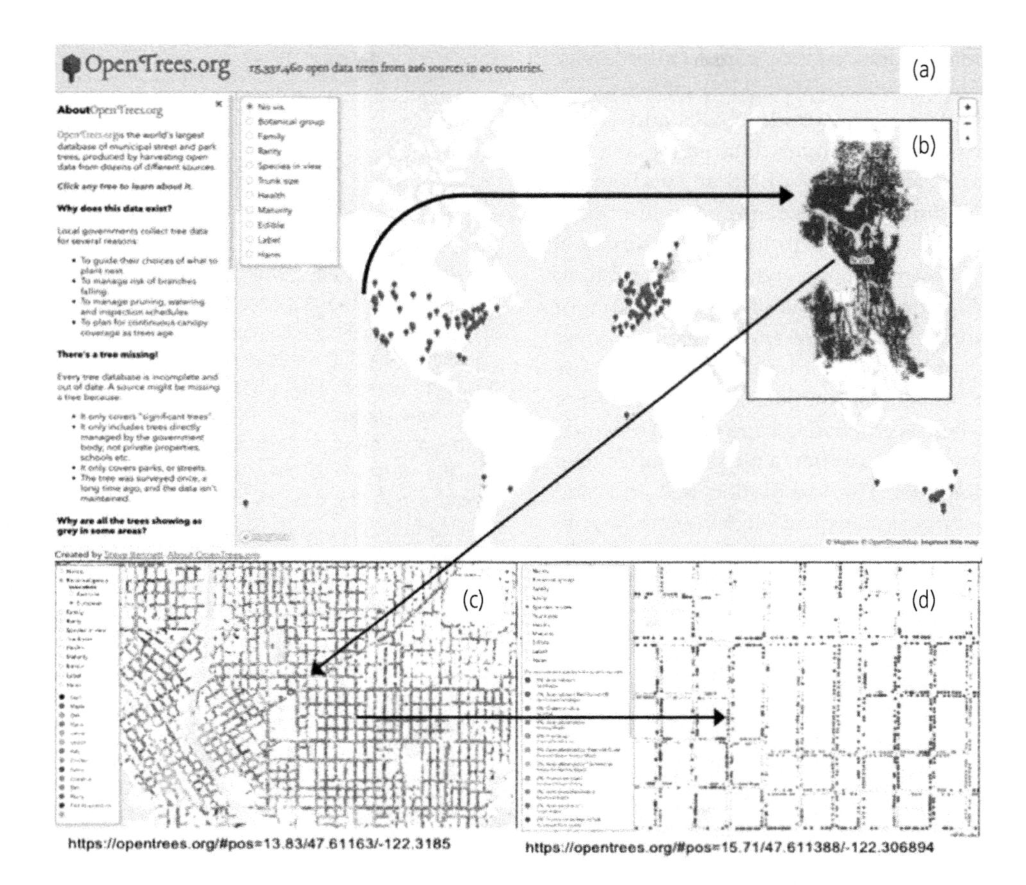

Figure 6.3 Opentrees.org aggregates and serves standardized municipal street tree inventories (a). The map of Seattle, WA (b) can be symbolized by genus (c) and species (d) among other attributes.

both for the biodiversity these urban forests can support but also to enhance the resiliency of the urban forest in marginalized communities against pests and extreme weather, and in other cities, too.

These new tree census and land cover datasets have been used by municipalities for goal setting, monitoring, planning, prioritizing planting, and other purposes (21,67,68). Beyond addressing interesting questions in urban ecology, these mapping innovations have been instrumental in developing our foundational understanding of cities as social-ecological systems, particularly by demonstrating inequities in the distribution and quality of urban tree canopies across wealth and racial boundaries (11–13). Further, if not for urban tree and other vegetation data, studying and managing urban biodiversity more broadly would be substantially more difficult given that trees provide essential habitat conditions for much urban biodiversity.

Urban tree maps facilitate broader urban conservation planning

Urban wildlife biologists can use high-resolution, urban landscape tree and vegetation censuses in at least two helpful ways when designing and analyzing projects. At the design stage, a project area can be divided into different levels of tree canopy and/or impervious surface cover. It is simple and straightforward to obtain the area and percentage of any land cover class (e.g., tree, other vegetation, roads) or combinations of land cover classes within various zones such as parcels, districts, plots, watersheds, or conservation areas. Geographic information system (GIS) software such as ESRI's or free software like QGIS and R can summarize these land cover data easily and are valuable for a multiplicity of urban biodiversity projects including citizen science,[6] for informing sampling plans, or for prioritizing conservation actions. For example, researchers may wish to deploy camera traps in areas of high, medium, or low tree canopy cover. Tree and other land cover summaries can be added in after spatial data are collected or conservation actions are underway. For example, if there is an existing camera trap network, land cover summaries within a range of buffer sizes can be added later. High-resolution tree and land cover data support statistical blocking (e.g., by literal neighborhood blocks with different tree levels) and adding covariates that contextualize habitats.

Case study

In 2006, the Maryland Department of Natural Resources developed the Urban Tree Canopy (UTC) program to encourage cities to conserve and plant new trees to improve the State's waterways and the health of Chesapeake Bay. Baltimore City wanted to participate in the UTC program. Before Baltimore City could establish a goal, however, the City needed to quantify the amount of existing and possible canopy cover for the entire City. For this analysis, the City defined possible canopy cover as areas that were not existing tree canopy cover and were not roads, buildings, or water. The City also needed to know how much of the existing tree canopy was in public ownership, particularly parks and public rights of way in the case of street trees. These questions required detailed, parcel-level analysis. Because of the fine-grain spatial heterogeneity of cities, mid-resolution, remotely sensed data (30 m) were insufficient for this analysis.

Through the Baltimore Ecosystem Study (BES, https://baltimoreecosystemstudy.org/), new methods for combining high-resolution, remotely sensed imagery (>1 m) and LiDAR data were developed to produce highly accurate and visually representative data at the parcel level (69). Visually representative data and parcel-level analysis were two important requirements that decision makers had for the development of these methods. The data product needed to be visually representative at a parcel level in order to convince decision makers and the general public that the data were realistic and therefore legitimate because the data corresponded to how they thought of and perceived their environment. A tree should look like a tree (round, but not too round) and certainly not a pixel (blocky). Summarizing the land cover data within parcels was important because a parcel is the basic unit of decision making and management: parcels are legally binding. The City needed to know how much of the existing and possible urban tree canopy was under its management and how much was associated with other types of ownership or management.

The results from the new land cover map showed canopy cover for the City was 20%, but that the cover varied from 1% to 85% across different neighborhoods. Given the combination of existing and possible canopy cover by neighborhood, some neighborhoods could ultimately have a canopy cover as high as 97% and as low as 29%. In terms of ownership patches, public ownership—street trees and parks—contained only 20% of the existing canopy patches and 14% percent of the possible canopy; while private ownership contained 80% of the existing and 85% of the possible canopy cover. Thus, the City's urban tree canopy goal would be severely limited if it depended only upon conserving existing trees and planting new trees on public lands.

With this information, the City knew the extent of existing and possible canopy cover and its distribution by ownership type and by neighborhood. Yet, the City did not know what the canopy goal should be. Recent watershed research in the region had indicated watershed health declined significantly when existing canopy cover fell below 43% (70). Given that the City's existing canopy cover was 20%, and in consideration of how to "sell" its canopy goal, the City has set a goal of nearly doubling its canopy cover from 20% to 40% by 2036.

Social-ecological applications: who wants urban trees and how to give them away

As described earlier, high-resolution land cover analyses revealed that most land is private residential, most tree canopy cover is private and residential, and most of the opportunities for additional tree canopy are also on private residential land. As well, at the neighborhood level, higher-income and predominantly White areas have more

[6] Perkins et al.'s chapter details how to implement equitable participatory science for urban biodiversity, which often uses GIS tools.

tree canopy cover than their lower-income, communities of color counterparts (11–13). In recognition of these two facts, many municipalities (71) have begun giving free trees away. The rationale is to target residential lands, and by being free, the intention is to reduce the barriers to access in lower-income areas.

Research from Baltimore (72), Washington, D.C. (60), and New York City (73) shows that these well-intentioned programs ultimately end up delivering free or reduced-cost trees to high-income areas with abundant tree canopy on residential land. Put another way, the outreach strategies were extremely effective in reaching one social group, but not others. It is clear that in some areas people do not want trees and/or distrust the government (74,75). In New Haven, CT, the number one reason for requesting a street tree was to replace an old one recently removed (76). In Philadelphia, however, distribution was more even (77), possibly because the City's tree canopy goal was 30% *per neighborhood*. The reasons for participating—or not—vary (71), and this line of research has helped develop best practices for tree distribution programs (78). In some cases, a simple phone call reminder can greatly boost follow-through (79). Each of these programs was young at the time of analysis and has been extremely responsive to these analyses, and the administering groups have changed accordingly. Reanalyses with updated data are needed to evaluate the effectiveness of programmatic changes inspired by these summaries coproduced with urban natural resources managers.

Conclusions and looking forward

In ecology, focus on trees is associated with biomes having sufficient rainfall to support these large, usually water-demanding plants. Urban areas are different from the ecological "addresses" of concern with, study of, and management of trees and forest, however. This is because people in cities and urban places count trees among their important structuring and functional components. Even in typically arid climates and droughty soils, people have often supplied extra moisture or found ways to modify the habitat to support trees. This urban "love affair" with trees means that people have successfully reaped the benefits of arboreal plants in an extraordinary number and variety of cities, suburbs, towns, and villages worldwide, ranging from seacoasts to mountain ranges and tropics to taiga.

The literature and experiences summarized herein yield several insights about trees, some of which point to their relevance to other disciplines within urban sciences, such as wildlife ecology. Other insights show the significant role that trees and urban tree canopy have in setting policy and satisfying goals of sustainability and resilience. We cite eight insights about urban trees and forest.

(1) Urban trees, singly or in stands of several to many trees, are an important contribution to the biodiversity of cities, suburbs, towns, and exurbs. Because people import species not originally present in the native biomes in which cities are located, the tree species richness of the urban forest may be greater than that of surrounding forests and woodlands. Such imported trees may have been chosen for cultural familiarity or significance, or for their ability to thrive in the highly modified environments of cities, including soil compaction and low aeration, soil drought, and atmospheric conditions of pollution, extreme heat, or intense vapor pressure deficits.

(2) All trees in an urban place, regardless of whether planted or tended or not, are part of an urban forest. Of course, an urban forest is usually highly fragmented, with individual specimen trees isolated in yards, parks, and institutional lawns, and with patches of volunteer woods in "waste" lands, vacant lots, or along transportation corridors. Remnant forests of notable extent appear in some cities on old estate or plantation grounds, and new forests are often planted in parks or as amenities in new developments. Finally, of course, street trees, including stately arches of canopy, blocks in length, or widely spaced trees along sidewalks of busy commercial thoroughfares, are also part of the heterogeneous urban forest. Urban forestry, as a research and practical field, is concerned with all of these trees and conditions.

(3) Urban forests perform many services, and this is especially so because they interdigitate with the

spaces where people live, travel, and recreate. Well-being, calming, recreation, nature observation, and interaction with other than human organisms are key benefits that people often experience as individuals. Aggregate benefits emerge from heat mitigation and stormwater control, for example. Notable is that urban forest supports other kinds of biodiversity. Birds, mammals, and insects find shelter, nesting habitat, and food in urban trees. Other plants, ranging from mosses, ferns, epiphytes, and shrubs to spring ephemerals, benefit from the structures and conditions that trees create, either alone or in stands. The interactions between trees and the other plant, animal, and microbial residents in cities, especially those in which evolution may play a part, require additional research. The services differ based on the size and layering of particular parts of the urban forest.

(4) Urban forest and urban forestry are different from "biomic" forestry for another important reason. Understanding and managing the benefits—and hazards—of the urban forest is an integrated social-ecological process. The presence and condition of every tree in a city or town result from human decisions. Where trees are planted, removed, or pruned are all decisions and mean the social and the ecological are inextricably linked in any consideration of the urban forest. Ironically, it is human decisions that lead to neglect and lack of management of urban trees and stands as well.

(5) All forests are complex in their past and spatial relationships, with differences in habitat, herbivores, succession, and disturbance (and more) influencing the distribution, health, and success of individual trees and stands. However, there are additional layers of complexity that must be accounted for in the urban forest. How is the urban forest (including individual trees) apportioned across different parcels? Who manages or makes decisions about each parcel—recognizing that the decision maker may not be the property owner of record. Each parcel will likely have its own history of management of the urban forest, reflecting such things as aesthetic or other values of the occupants, or rules of a homeowners' association, or the life cycle of a family unit as it evolves from raising children to retirement, or a business as it moves from entrepreneurial vigor to a decline in its market. The complexities just outlined can be considered stacked layers that generate virtually unique patterns of heterogeneity that influence urban forests. One particular complication in urban forests is that what may look like a continuous canopy of a single stand may spread over several parcels. Conserving and managing such multi-owner/multi-manager stands requires formal arrangements across ownerships, and may be facilitated by municipal policy that recognizes the social value of contiguous forest canopy for biodiversity and environmental quality.

(6) One implication of the fact that the urban forest is a social-ecological entity, is that people in different social, economic, gendered, or racialized positions may interact with a single urban forest stand or even individual tree in very different ways. It is crucial to understand the variety of perspectives that are in play in a single place or forest. One-size-fits-all policies, or management strategies, or values about urban forest are likely to run afoul of the social diversity of real places where people care about, fear, or ignore trees.

(7) The great social heterogeneity affecting how urban forests are viewed also affects who benefits from them, and who suffers disproportionately from their presence, their condition, or their loss. This highlights the urban forest as a subject of environmental justice. What communities control where trees are located and who allocates resources to their care? What communities are located at convenient distances from large stands in the urban forest, or to tree-shaded sidewalks? What communities can afford the costs or time for pruning, cleaning of leaf litter, and removal of deadfall? What communities have members who participate in professional activities associated with trees, as compared to manual labor? And finally, what are the origins of contemporary patterns of how communities are associated with the urban forest? Much research has documented the role of past activities in determining the location

and composition of the current urban forest. Residential redlining, industrial development, transportation corridors, municipal investment (or lack thereof), power line location and maintenance, and past horticultural fashions are among the many legacies of past decisions and practices that may better explain the contemporary urban forest and how its benefits and burdens are spatially apportioned than contemporary actions. Contemporary actions can, however, restore just urban forest relationships, and can do so with respectful dialog with communities and other stakeholders.

(8) One of the most pervasive themes we discuss here is the productive complexity by which the understanding of the urban forest to date has been produced, and how that understanding is applied to the real world. Understanding the urban forest and ensuring that its benefits are equitably shared, and its burdens likewise are not concentrated on those with less power and fewer resources, combine three activities. The most fundamental component may be improving the data on the urban forest. Remote sensing was a major advance in the discovery of just how pervasive and spatially fragmented the urban forest was. Up until the advent of remote sensing, the term "forest" in urban contexts was associated mainly with large tracts in signature parks, such as the Ramble in New York's Central Park, or even larger stands in its Van Cortlandt or Pelham Bay Parks, Portland's Forest park, or Boston's Emerald Necklace, to name a few. But urbanists also recognized early on in the professional attention of arborists and foresters that the trees along streets were important to the structure, form, and life of cities. Unfortunately, remote sensing of the early generations of satellite technology actually was blind to many of the trees that contributed widely to the urban forest. This was corrected by high-resolution imagery and LiDAR. When these sensors were applied to cities around the world, a much larger urban forest appeared. When those remotely sensed images were combined with LiDAR, trees and canopy could be reliably differentiated within the shadows of tall buildings. Thus, advances in technology were key to understanding that the urban forest was much more extensive, and much more involved with private and small property parcels, than the prior attention to the public rights of way or large public and private institutional grounds suggested.

Theory was a second player in the new discoveries about the urban forest. For example, urban ecologists have shifted from an assumption that cities comprised distinct natural components, such as vegetation and streams, versus distinct socially derived components, such as buildings, infrastructure, and impervious surfaces, to a view that cities are constituted by hybrid social-natural patches. This is a radical reconceptualization of the urban, which interacted with the advances in data.

The third leg of the new urban platform is practice. Whereas ecologists in wild or rural places have, throughout most of the history of their science, largely ignored people, urban ecologists have had to become people-savvy. In wild places it was the assumed absence of contemporary human presence (along with ignorance of long histories of Indigenous presence and management) that permitted dismissal of the human from models and the search for data. Urban ecologists may have in their early days paid relatively scant attention to people as environmental agents, but as the study expanded to more cities, looked in greater historical depth at how the contemporary form and process of cities came about, and became more sensitive to the needs of all communities—not just the powerful economic or racialized elites, a new kind of urban ecology emerged. This sequence can be described as ecology IN the city, ecology OF the city, and ecology WITH the city. Along this continuum, ecologists first stretched their disciplinary boundaries to include other sciences, including the social. And secondly, they began to be sensitive to the diversity of communities, power relations in environmental decision making, and the deep history of inequity and injustice that afflicted cities in the US and indeed elsewhere in the world based on colonialism, slavery, and other extractive labor and resource systems. Environmental justice and environmental racism, areas that emerged first in social movements, have come to be important players in how ecologists understand and work in—and on behalf of—marginalized communities.

These eight insights show the significance of urban forestry understanding, data, and practices as core pillars of understanding urban biodiversity, and improving its benefit and linkages with humans.

References

1. Pataki DE, Alberti M, Cadenasso ML, Felson AJ. The benefits and limits of urban tree planting for environmental and human health. Frontiers in Ecology and Evolution. 2021;9:603757.
2. Ulmer JM, Wolf KL, Backman DR, Tretheway RL, Blain CJ, O'Neil-Dunne JP, et al. Multiple health benefits of urban tree canopy: the mounting evidence for a green prescription. Health and Place. 2016;42: 54–62.
3. Nesbitt L, Hotte N, Barron S, Cowan J, Sheppard SRJ. The social and economic value of cultural ecosystem services provided by urban forests in North America: a review and suggestions for future research. Urban Forestry & Urban Greening. 2017;25:103–11. Available from: https://doi.org/10.1016/j.ufug.2017.05.005
4. Wilson LJ. The Resource Enhancement and Protection Act - one method of meeting our public and resource needs. In: Adams LW, Leedy DL (eds.) Wildlife Conservation in Metropolitan Environments. Columbia, MD: National Institute for Urban Wildlife; 1991. p. 7–14.
5. Wood EM, Esaian S. The importance of street trees to urban avifauna. Ecological Applications. 2020;30(7):e02149.
6. Ossola A, Locke DH, Lin B, Minor ES. Yards increase forest connectivity in urban landscapes. Landscape Ecology. 2019;34(12):2935–48. Available from: https://doi.org/10.1007/s10980-019-00923-7
7. Schell CJ, Dyson K, Fuentes TL, Des Roches S, Harris NC, Miller DS, et al. The ecological and evolutionary consequences of systemic racism in urban environments. Science. 2020;369(6510):eaay4497.
8. Lambert MR, Brans KI, Des Roches S, Donihue CM, Diamond SE. Adaptive evolution in cities: progress and misconceptions. Trends in Ecology and Evolution. 2021;36(3):239–57.
9. Lambert MR, Donihue CM. Urban biodiversity management using evolutionary tools. Nature Ecology & Evolution. 2020;4(7):903–10. Available from: https://doi.org/10.1038/s41559-020-1193-7
10. Roman LA, Pearsall H, Eisenman TS, Conway TM, Fahey RT, Landry S, et al. Human and biophysical legacies shape contemporary urban forests: a literature synthesis. Urban Forestry & Urban Greening. 2018;31:157–68. Available from: http://linkinghub.elsevier.com/retrieve/pii/S1618866717307665
11. Gerrish E, Watkins SL. The relationship between urban forests and income: a meta-analysis. Landscape and Urban Planning. 2017;170:293–308. Available from: https://doi.org/10.1016/j.landurbplan.2017.09.005
12. Schwarz K, Fragkias M, Boone CG, Zhou W, McHale MR, Grove JM, et al. Trees grow on money: urban tree canopy cover and environmental justice. PLoS One. 2015;10(4):e0122051. Available from: https://doi.org/10.1371/journal.pone.0122051
13. Watkins SL, Gerrish E. The relationship between urban forests and race: a meta-analysis. Journal of Environmental Management. 2018;209:152–68. Available from: https://doi.org/10.1016/j.jenvman.2017.12.021
14. Grove JM, Ogden L, Pickett S, Boone CG, Buckley GL, Locke DH, et al. The legacy effect: understanding how segregation and environmental injustice unfold over time in Baltimore. Annals of the American Association of Geographers. 2018;108(2):524–37. Available from: https://doi.org/10.1080/24694452.2017.1365585
15. Nardone A, Rudolph KE, Morello-Frosch R, Casey JA. Redlines and greenspace: the relationship between historical redlining and 2010 greenspace across the United States. Environmental Health Perspectives. 2021;129(1):1–9.
16. Locke DH, Hall B, Grove JM, Pickett STA, Ogden LA, Aoki C, et al. Residential housing segregation and urban tree canopy in 37 US cities. npj Urban Sustainability. 2021; 1(1):15. Available from: https://doi.org/10.1038/s42949-021-00022-0
17. Nowak DJ, Ellis A, Greenfield EJ. The disparity in tree cover and ecosystem service values among redlining classes in the United States. Landscape and Urban Planning. 2022;221:104370. Available from: https://doi.org/10.1016/j.landurbplan.2022.104370
18. Namin S, Xu W, Zhou Y, Beyer K. The legacy of the Home Owners' Loan Corporation and the political ecology of urban trees and air pollution in the United States. Social Science & Medicine. 2020;246:112758.
19. Hoffman JS, Shandas V, Pendleton N. The effects of historical housing policies on resident exposure to intra-urban heat: a study of 108 US urban areas. Climate. 2020;8(1):12. Available from: https://www.mdpi.com/2225-1154/8/1/12
20. Hauer R, Peterson WD. Municipal Tree Care and Management in the United States: a 2014 urban & community forestry census of tree activities [Internet]. University of Wisconsin—Stevens Point;

2016. 78 p. Available from: https://www3.uwsp.edu/cnr/Documents/MTCUS%20-%20Forestry/Municipal%202014%20Final%20Report.pdf

21. Kimball LL, Wiseman PE, Day SD, Munsell JF. Use of urban tree canopy assessments by localities in the Chesapeake Bay watershed. Cities and the Environment. 2014;7(2):9. Available from: http://digitalcommons.lmu.edu/cate/vol7/iss2/9
22. Locke DH, Romolini M, Galvin MF, Strauss EG. Tree canopy change in coastal Los Angeles, 2009–2014. Cities and the Environment. 2017;10(2):3. Available from: http://digitalcommons.lmu.edu/cate/vol10/iss2/3
23. McPherson EG. Monitoring Million Trees LA: tree performance during the early years and future benefits. Journal of Arboriculture and Urban Forestry. 2014;40(5):285–300. Available from: https://www.fs.usda.gov/psw/publications/mcpherson/psw_2014_mcpherson003.pdf
24. Eisenman TS, Flanders T, Harper RW, Hauer RJ, Lieberknecht K. Traits of a bloom: a nationwide survey of U.S. urban tree planting initiatives (TPIs). Urban Forestry & Urban Greening. 2021;61:127006. Available from: https://doi.org/10.1016/j.ufug.2021.127006
25. Doroski DA, Ashton MP, Duguid MC. The future urban forest – a survey of tree planting programs in the Northeastern United States. Urban Forestry & Urban Greening. 2020;55:126816. Available from: https://doi.org/10.1016/j.ufug.2020.126816
26. Yao N, Konijnendijk van den Bosch CC, Yang J, Devisscher T, Wirtz Z, Jia L, et al. Beijing's 50 million new urban trees: strategic governance for large-scale urban afforestation. Urban Forestry & Urban Greening. 2019;44:126392. Available from: https://doi.org/10.1016/j.ufug.2019.126392
27. Berland A, Locke DH, Herrmann DL, Schwarz K. Residential land owner type mediates the connections among vacancy, overgrown vegetation, and equity. Urban Forestry & Urban Greening. 2023;80:127826. Available from: https://doi.org/10.1016/j.ufug.2022.127826
28. Grove JM, Locke DH, O'Neil-Dunne JPM. An ecology of prestige in New York City: examining the relationships among population density, socio-economic status, group identity, and residential canopy cover. Environmental Management. 2014;54(3):402–19. Available from: http://link.springer.com/10.1007/s00267-014-0310-2
29. Coleman AF, Harper RW, Eisenman TS, Warner SH, Wilkinson MA. Street tree structure, function, and value: a review of scholarly research (1997–2020). Forests. 2022;13(11):1779.
30. Burghardt KT, Avolio ML, Locke DH, Morgan GJ, Sonti NF, Swan CM. Current street tree communities reflect race-based housing policy and modern attempts to remedy environmental injustice. Ecology. 2023;104(2):e3881.
31. Anderson EC, Locke DH, Pickett ST, LaDeau SL. Just street trees? Street trees increase local biodiversity and biomass in higher income, denser neighborhoods. Ecosphere. 2023;14(2):e4389.
32. Ponte S, Sonti NF, Phillips TH, Pavao-Zuckerman MA. Transpiration rates of red maple (*Acer rubrum* L.) differ between management contexts in urban forests of Maryland, USA. Scientific Reports. 2021;11(1):22538. Available from: https://doi.org/10.1038/s41598-021-01804-3
33. Phillips TH, Baker ME, Lautar K, Yesilonis I, Pavao-Zuckerman MA. The capacity of urban forest patches to infiltrate stormwater is influenced by soil physical properties and soil moisture. Journal of Environmental Management. 2019;246:11–18. Available from: https://linkinghub.elsevier.com/retrieve/pii/S0301479719307558
34. Pregitzer CC, Charlop-Powers S, Bradford MA. Natural area forests in US cities: opportunities and challenges. Journal of Forestry. 2021;119(2):141–51.
35. Johnson LR, Johnson ML, Aronson MFJ, Campbell LK, Carr ME, Clarke M, et al. Conceptualizing social-ecological drivers of change in urban forest patches. Urban Ecosystems. 2021;24(4):633–48.
36. Doroski DA, Bradford MA, Duguid MC, Hallett RA, Pregitzer CC, Ashton MS. Diverging conditions of current and potential future urban forest patches. Ecosphere. 2022;13(3):e4001.
37. Ossola A, Locke DH, Lin B, Minor E. Greening in style: urban form, architecture and the structure of front and backyard vegetation. Landscape and Urban Planning. 2019;185:141–57. Available from: https://doi.org/10.1016/j.landurbplan.2019.02.014
38. Sonti NF, Campbell LK, Svendsen ES, Johnson ML, Novem Auyeung DS. Fear and fascination: use and perceptions of New York City's forests, wetlands, and landscaped park areas. Urban Forestry & Urban Greening. 2020;49:126601.
39. Sonti NF. Ambivalence in the Woods: Baltimore resident perceptions of local forest patches. Society and Natural Resources; 33(7):823–841. https://www.tandfonline.com/doi/abs/10.1080/08941920.2019.1701162?journalCode=usnr20
40. Perkins DJ. Blind Spots in Citizen Science Data: implications of volunteer bias in eBird data [Master's Thesis]. [Raleigh, NC]: North Carolina State University; 2020. Available from: https://www.proquest.com/docview/2478031545

41. Boone CG, Cadenasso ML, Grove JM, Schwarz K, Buckley GL. Landscape, vegetation characteristics, and group identity in an urban and suburban watershed: why the 60s matter. Urban Ecosystems. 2010;13(3):255–71. Available from: https://link.springer.com/article/10.1007/s11252-009-0118-7
42. Locke DH, Baine G. The good, the bad, and the interested: how historical demographics explain present-day tree canopy, vacant lot and tree request spatial variability in New Haven, CT. Urban Ecosystems. 2014;18(2):391–409. Available from: http://link.springer.com/10.1007/s11252-014-0409-5
43. Grant A, Millward AA, Edge S, Roman LA, Teelucksingh C. Where is environmental justice? A review of US urban forest management plans. Urban Forestry & Urban Greening. 2022;77:127737. Available from: https://doi.org/10.1016/j.ufug.2022.127737
44. Nesbitt L, Meitner MJ, Girling C, Sheppard SRJ, Lu Y. Who has access to urban vegetation? A spatial analysis of distributional green equity in 10 US cities. Landscape and Urban Planning. 2019;181:51–79. Available from: https://doi.org/10.1016/j.landurbplan.2018.08.007
45. Browning MHEM, Rigolon A, Ogletree S, Wang R, Klompmaker JO, Baile C, et al. The PAD-US-AR dataset: measuring accessible and recreational parks in the contiguous United States. Scientific Data. 2022;9(773):1–15.
46. Konijnendijk CC. Evidence-based guidelines for greener, healthier, more resilient neighbourhoods: introducing the 3–30–300 rule. Journal of Forestry Research. 2023;34(3):821–30. Available from: https://doi.org/10.1007/s11676-022-01523-z
47. Esperon-Rodriguez M, Tjoelker MG, Lenoir J, Baumgartner JB, Beaumont LJ, Nipperess DA, et al. Climate change increases global risk to urban forests. Nature Climate Change. 2022;12(10):950–5.
48. Esperon-Rodriguez M, Ordoñez C, van Doorn NS, Hirons A, Messier C. Using climate analogues and vulnerability metrics to inform urban tree species selection in a changing climate: the case for Canadian cities. Landscape and Urban Planning. 2022;228:104578.
49. Pallai C, Wesson K. Chesapeake Bay Program Partnership High-Resolution Land Cover Classification accuracy assessment methodology. 2017. Available from: https://www.chesapeakeconservancy.org/wp-content/uploads/2017/01/Chesapeake_Conservancy_Accuracy_Assessment_Methodology.pdf
50. Claggett P, Ahmed L, Buford E, Czawlytko J, MacFaden S, McCabe P, et al. Chesapeake Bay Program's one-meter resolution land use/land cover data: overview and production. 2014. Available from: https://cicwebresources.blob.core.windows.net/docs/LU_Classification_Methods_2017_2018.pdf
51. O'Neil-Dunne JPM, Macfaden SW, Royar A. A versatile, production-oriented approach to high-resolution tree-canopy mapping in urban and suburban landscapes using GEOBIA and data fusion. Remote Sensing. 2014;6(12):12837–65. Available from: http://www.mdpi.com/2072-4292/6/12/12837/
52. O'Neil-Dunne JPM, MacFaden SW, Royar A, Reis M, Dubayah R, Swatantran A. An object-based approach to statewide land cover mapping. In: ASPRS 2014 Annual Conference, Mar 23–28, 2014, Louisville, KY. Available from: https://www.asprs.org/a/publications/proceedings/Louisville2014/ONeilDunne.pdf
53. O'Neil-Dunne JPM, MacFaden SW, Royar AR, Pelletier KC. An object-based system for LiDAR data fusion and feature extraction. Geocarto Interational. 2013;28(3):227–42. Available from: http://www.tandfonline.com/doi/abs/10.1080/10106049.2012.689015
54. MacFaden SW, O'Neil-Dunne JPM, Royar AR, Lu JWT, Rundle AG. High-resolution tree canopy mapping for New York City using LIDAR and object-based image analysis. Journal of Applied Remote Sensing. 2012;6(1):063567. Available from: https://www.spiedigitallibrary.org/journals/journal-of-applied-remote-sensing/volume-6/issue-1/063567/High-resolution-tree-canopy-mapping-for-New-York-City-using/10.1117/1.JRS.6.063567.short
55. McGee JA III, Day SD, Wynne RH, White MB. Using geospatial tools to assess the urban tree canopy: decision support for local governments. Journal of Forestry. 2012;110:275–86. Available from: https://academic.oup.com/jof/article/110/5/275/4599516
56. Ossola A, Hopton ME. Measuring urban tree loss dynamics across residential landscapes. Science of the Total Environment. 2018;612:940–9. Available from: https://doi.org/10.1016/j.scitotenv.2017.08.103
57. Ossola A, Jenerette GD, McGrath A, Chow W, Hughes L, Leishman MR. Small vegetated patches greatly reduce urban surface temperature during a summer heatwave in Adelaide, Australia. Landscape and Urban Planning. 2021;209:104046.
58. Landry SM, Northrop RJ, Andreu MG, Beck K. City of Tampa urban forest management, monitoring and policy. Florida Geography. 2014;45:44–62.
59. Pu R, Landry SM. A comparative analysis of high spatial resolution IKONOS and WorldView-2 imagery for mapping urban tree species. Remote Sensing of the Environment. 2012;124:516–33. Available from: https://doi.org/10.1016/j.rse.2012.06.011

60. Pham T-T-H, Apparicio P, Séguin A, Landry SM, Gagnon M. Spatial distribution of vegetation in Montreal: an uneven distribution or environmental inequity? Landscape and Urban Planning. 2012;107(3):214–24. Available from: https://doi.org/10.1016/j.landurbplan.2012.06.002
61. Landry S. Connecting Pixels to People: management agents and social-ecological determinants of changes to street tree distributions [Ph.D. Thesis]. [Tampa, FL]: University of South Florida; 2013. Available from: http://scholarcommons.usf.edu/etd/4715/
62. Landry SM, Chakraborty J. Street trees and equity: evaluating the spatial distribution of an urban amenity. Environmental Planning. 2009;41(11):2651–70. Available from: https://journals.sagepub.com/doi/10.1068/a41236
63. Alonzo M, Bookhagen B, Roberts DA. Urban tree species mapping using hyperspectral and lidar data fusion. Remote Sensing of the Environment. 2014;148:70–83. Available from: https://doi.org/10.1016/j.rse.2014.03.018
64. Alonzo M, McFadden JP, Nowak DJ, Roberts DA. Mapping urban forest structure and function using hyperspectral imagery and lidar data. Urban Forestry & Urban Greening. 2016;17:135–47. Available from: https://doi.org/10.1016/j.ufug.2016.04.003
65. Alonzo M, Bookhagen B, McFadden JP, Sun A, Roberts DA. Mapping urban forest leaf area index with airborne lidar using penetration metrics and allometry. Remote Sensing of the Environment. 2015;162:141–53. Available from: https://doi.org/10.1016/j.rse.2015.02.025
66. Alonzo M, Baker ME, Gao Y, Shandas V. Spatial configuration and time of day impact the magnitude of urban tree canopy cooling. Environmental Research Letters. 2021;16(8):84028. Available from: https://doi.org/10.1088/1748-9326/ac12f2
67. Locke DH, Grove JM, Lu JWT, Troy AR, O'Neil-Dunne JPM, Beck BD. Prioritizing preferable locations for increasing urban tree canopy in New York City. Cities and the Environment. 2010;3(1):4. Available from: https://digitalcommons.lmu.edu/cgi/viewcontent.cgi?article=1066&context=cate
68. Locke DH, Grove JM, Galvin MF, O'Neil-Dunne JPM, Murphy C. Applications of urban tree canopy assessment and prioritization tools: supporting collaborative decision making to achieve urban sustainability goals. Cities and the Environment. 2013;6(1):7. Available from: https://digitalcommons.lmu.edu/cate/vol6/iss1/7/
69. Galvin MF, Grove JM, O'Neil-Dunne JPM. A Report on Baltimore City's present and potential urban tree canopy. Annapolis, MD: Maryland Forest Service; 2006. Available from: https://www.fs.usda.gov/nrs/utc/reports/2006_UTC_Report_Baltimore.pdf
70. Goetz SJ, Wright RK, Smith AJ, Zinecker E, Schaub E. IKONOS imagery for resource management: tree cover, impervious surfaces, and riparian buffer analyses in the mid-Atlantic region. Remote Sensing of the Environment. 2003;88(1–2):195–208. Available from: https://www.sciencedirect.com/science/article/abs/pii/S0034425703002414
71. Nguyen VD, Roman LA, Locke DH, Mincey SK, Sanders JR, Fichman ES, et al. Branching out to residential lands: missions and strategies of five tree distribution programs in the U.S. Urban Forestry & Urban Greening. 2017;22:24–35. Available from: https://doi.org/10.1016/j.ufug.2017.01.007
72. Locke DH, Morgan GJ. A market analysis of opt-in tree planting and rain barrel installation in Baltimore, MD, 2008—2012. 2015. Available from: https://dexterlocke.com/wp-content/uploads/2023/05/Locke-Morgan_2015_A-Market-Analysis-of-Opt-In-Tree-Planting-and-Rain-Barrel-Installation-in-Baltimore-MD-2008-%E2%80%94-2012.pdf
73. Locke DH, Mitchell M, Turner C, Douglas J. A market analysis of New York Restoration Project's Tree Giveaway Program, spring 2008—fall 2013. New York Restoration Project; 2014. Available from: http://dexterlocke.com/wp-content/uploads/2023/01/giveawaymarketanalysis_d_locke_m_mitchell_june2014.pdf
74. Carmichael CE, McDonough MH. Community stories: explaining resistance to street tree-planting programs in Detroit, Michigan, USA. Society and Natural Resources. 2019;32(5):588–605. Available from: https://doi.org/10.1080/08941920.2018.1550229
75. Battaglia M, Buckley GL, Galvin M. It's not easy going green: obstacles to tree-planting programs in East Baltimore. Cities and the Environment. 2014;7(2):6. Available from: http://digitalcommons.lmu.edu/cate/vol7/iss2/6/
76. Locke DH, Roman LA, Murphy-Dunning C. Why opt-in to a planting program? Long-term residents value street tree aesthetics. Arboriculture and Urban Forestry. 2015;41(6):324–33. Available from: https://www.fs.usda.gov/nrs/pubs/jrnl/2015/nrs_2015_locke_001.pdf
77. Locke DH, Smith-Fichman E, Blaustein J. A market analysis of TreePhilly's Yard Tree Program, spring 2012—spring 2014. TreePhilly; 2014. Available from: http://dexterlocke.com/wp-content/uploads/2023/01/MarketAnalysis_Report_TreePhillyGiveaways20150120.pdf

78. Roman LA, Nguyen VD, Locke DH, Mincey SK, Sanders JR, Fichman ES, et al. Best practices for yard tree distribution programs. Arborist News. 2017;(Oct);30–5. Available from: https://www.isa-arbor.com/Publications/Arborist-News#archived

79. Hand MS, Roman LA, Locke DH, Fichman ES. Phone-call reminders narrow the intention-action gap by increasing follow-through for a residential tree give-away program. Urban Forestry & Urban Greening. 2019;44:126425. Available from: https://linkinghub.elsevier.com/retrieve/pii/S161886671930247X

CHAPTER 7

Participatory Science for Equitable Urban Biodiversity Research and Practice

Deja J. Perkins, Lauren M. Nichols, and Robert R. Dunn‡

Introduction

The term ecology, originally written oecology, means the study ("ology") of the home ("oikos"). Ironically, however, as Sharon Kingsland argues in an excellent recent history of urban ecology (1), the field was founded not to study the home, nor, more generally, where people live, but instead regions far away from the daily lives of humans. Although there were early attempts to imagine alternatives, ecologists chose to focus on remote regions for simplicity. As Kingsland put it, "ecological writing focused on botanical and zoological problems, excluding humans" (p. 26). As a result of this history, early ecological theory developed largely without reference to humans and the ways in which human history, culture, and social systems interact with the rest of the living world. Once ecologists began to focus more on humans, the approach was often to consider the effects of humans on remote habitats. This was manifested in the context of hundreds and perhaps thousands of studies of edge effects in forests (2,3,4), the effects of hunting in forests (5, see 6), and the effects of roads on species moving among forests (7,8). Yet, even these studies were often more likely to be categorized as conservation biology rather than ecology per se, a reality that implicitly situated humans outside of (and hence as unnaturally impacting) remote ecosystems. The lens of ecology was still trained on the remote; it had just become harder to ignore that the remote was influenced (and had long been influenced) by humanity. Eventually, ecologists began to consider cities. However, as they did, the first generation of studies of animals (our remit here) in cities tended to focus on the urban environments that were like more remote environments. For example, building on MacArthur and Wilson's *Theory of Island Biogeography* (9), studies considered habitats in cities as smaller island-like versions of larger habitats elsewhere (10,11), but the habitats that were considered in such studies were only green habitats, not the "gray" habitats that constitute a much larger proportion of most cities (12). Only very recently have cities come to be studied as an entirely different sort of ecosystem, one in which the behavior, culture, sociopolitics, and economics of humans must be part of the theory. It is an exciting new moment in urban ecology, one in which ecologists are working with sociologists, psychologists, anthropologists, and geologists to see the city more holistically.[1]

Here, we begin by considering the variety of models participatory science can follow, and how these various models have been used to study the

‡ North Carolina State University, USA

[1] Lambert's and Schell's chapter discusses a brief history of urban ecology and conservation.

Deja J. Perkins, Lauren M. Nichols, and Robert R. Dunn, *Participatory Science for Equitable Urban Biodiversity Research and Practice*. In: *Urban Biodiversity and Equity*. Edited by: Max R. Lambert & Christopher J. Schell, Oxford University Press. © Oxford University Press (2023).
DOI: 10.1093/oso/9780198877271.003.0007

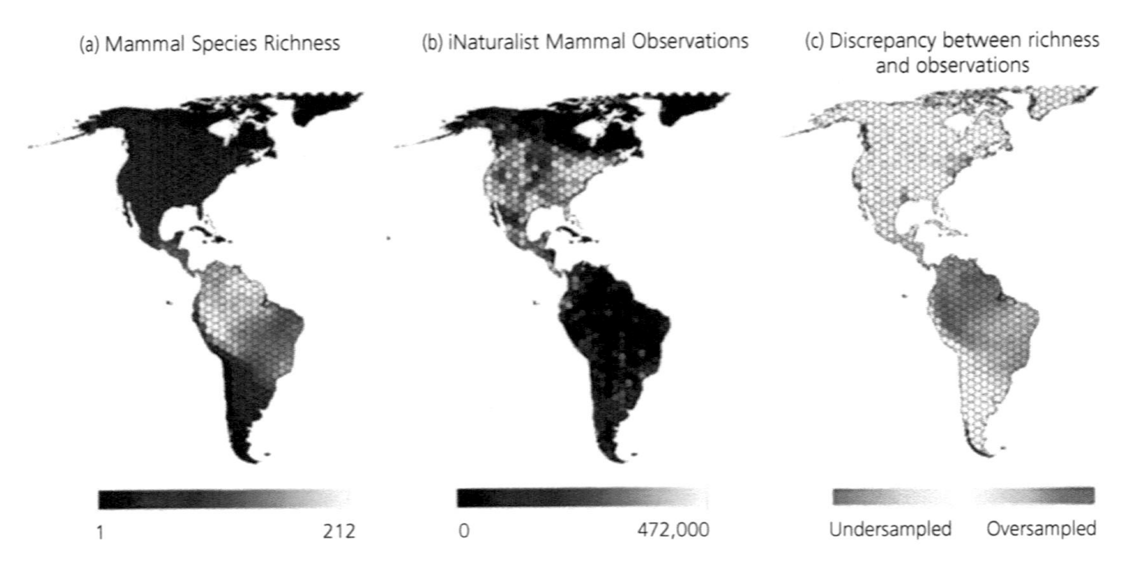

Figure 7.1 Although crowd-sourced data are invaluable, they can be biased, such as by over- or under-sampling geographies relative to how much biodiversity is actually present. The relationship between (a) regional mammal species richness (data derived from (80)) and (b) Research Grade mammal observations logged on the iNaturalist platform (data downloaded from GBIF.org (4 August 2022) GBIF Occurrence Download: https://doi.org/10.15468/dl.2qvkdm), highlighting the (c) regions that are relatively over- or under-sampled based on their regional richness. Discrepancy between richness and sampling effort for each 200-km hexagon was calculated as the difference between the relative species richness and the relative observation effort within each hexagon.

urban landscape. We point out the potential biases (Figure 7.1) and challenges that have been observed with participatory methods, before scaling out to examine the use of these tools globally. We conclude with special considerations for designing projects that aim to be inclusive, diverse, equitable, accessible, produce high data quality, and retain volunteer participation.

The utility of participatory science for urban biodiversity

Participatory science is an umbrella term that includes a range of practices united by engaging the public in scientific research based on the principles of Public Participation in Scientific Research (PPSR). (13). At one extreme, participatory science includes projects that are grassroots and community-led, often by marginalized and historically excluded communities. Such participatory projects are often described as community science (e.g., Louisiana Bucket Brigade; https://labucketbrigade.org/). Various forms of community science, such as community-based participatory research (14) and community-owned and -managed research (15), explicitly address equity and inclusion in their local, place-based structures (16). These methods bring the knowledge and expertise of communities together with formally trained researchers to, ideally, form an ethical, equitable, and transparent participatory partnership to answer community-articulated problems (17). Due to their local focus, community-driven projects are not often available to use in other geographic areas. One solution to this issue of scale is the use of large-scale participatory projects as tools to implement at local scales. For example, the City Nature Challenge (CNC) is a way for a large-scale project (iNaturalist) to connect at local scales through local ambassadors and organizers of the project. In 2022—while the global Covid-19 pandemic was still active—the CNC was still able to accumulate participation in over 450 cities across the globe, aiding in over a million observations of over 50,000 species and engaging over 60,000 people worldwide through the iNaturalist platform (https://www.inaturalist.

org/blog/65702-the-2022-city-nature-challenge-results-are-in).

At the other extreme, participatory science also includes projects driven by institutions in which public contribution is primarily through data collection, such as massive surveys of birds (e.g., Breeding Bird Survey; https://www.pwrc.usgs.gov/bbs/). Such projects are often described as citizen science, but here we call them contributory projects. Contributory projects are typically larger in scale and can cross geographic regions. Institution-driven contributory approaches are particularly well-suited to large-scale studies that cover many communities and even many regions and countries. For example, the eBird platform includes data from over 796,000 people around the world in 250 subregions, recording over 10,000 bird species, with a total of over 69 million checklists from around the world (https://ebird.org/home). Similarly, the iNaturalist platform also has a global reach with many projects focused on individual taxa or habitats. Such large-scale platforms are best suited for understanding questions related to biogeography and macroecology, and where time data are available, they can also be used to study population, community, taxonomic, and biomass change. In a subset of cases, data from these large-scale platforms are sufficiently densely sampled in space and time that they can be used to ask questions at smaller scales. Such large-scale projects are also amenable to relatively rapid expansions to consider new kinds of organisms. For example, one of the authors (Dunn) was able to start a new project on the iNaturalist platform called "Never Home Alone" to consider the animals living in homes around the world (https://www.inaturalist.org/projects/never-home-alone-the-wild-life-of-homes). Relatively rapidly and with little additional investment, this project was able to compile the biggest existing database of indoor animal distribution patterns (based on more than 30,000 records).

There are many different terms around the world used to describe participatory science and public participation in scientific research (see (18) for an expansive list of terms. There are also a number of rubrics that frame the types of participation. One such rubric frames projects with regard to their core characteristics and dimensions. Montanari et al. (19) elaborated a model for categorizing participatory science projects based on three dimensions: (1) their level of collaboration within project creation (i.e., level of public collaboration during project creation); (2) the types of tasks undertaken by participants; and (3) the project's underlying goal (Figure 7.2). With regard to the type of tasks (2), volunteer tasks can be categorized as those that have action-based activities (e.g., initiated by volunteers to encourage local action), conservation-based activities (e.g., raise awareness and foster stewardship by answering questions about the environment), investigation-based activities (e.g., provide large quantities of data through monitoring), virtual-based activities (e.g., online investigations), and education-based activities (e.g., increase scientific literacy) (20). As for goals (3), projects can and often will have multiple goals or outcomes. These goals and outcomes fall into three main categories: outcomes for researchers (e.g., scientific discovery), outcomes for individuals/participants (e.g., gaining new skills/knowledge), and outcomes for social-ecological systems (e.g., societal impact, building community capacity for action) (13). Projects rarely, if ever, achieve all goals equally because project goals don't always align well with project activities (13), but in theory, proper support systems (e.g., managing staff for each goal; funding) may aid in overcoming poor execution due to conflict between project goals and activities.

Here we use participatory science as the umbrella term (Figure 7.2) that encompasses projects that vary across all these dimensions. Participatory science encompasses a variety of approaches to engage the public(s) in scientific activities for research, monitoring, policy action, or other goals that scientists cannot typically undertake alone. Public engagement in scientific activities can help address some of the challenges of urban biodiversity research, particularly access to sampling sites. With most people living in cities, there is a highly diverse and dense potential pool of collaborators and therefore there are opportunities to expand research in metropolitan environments. Participatory science activities can include, but are not limited to, generating scientific data, mapping data, annotating images, and making or deploying low-cost sensors.

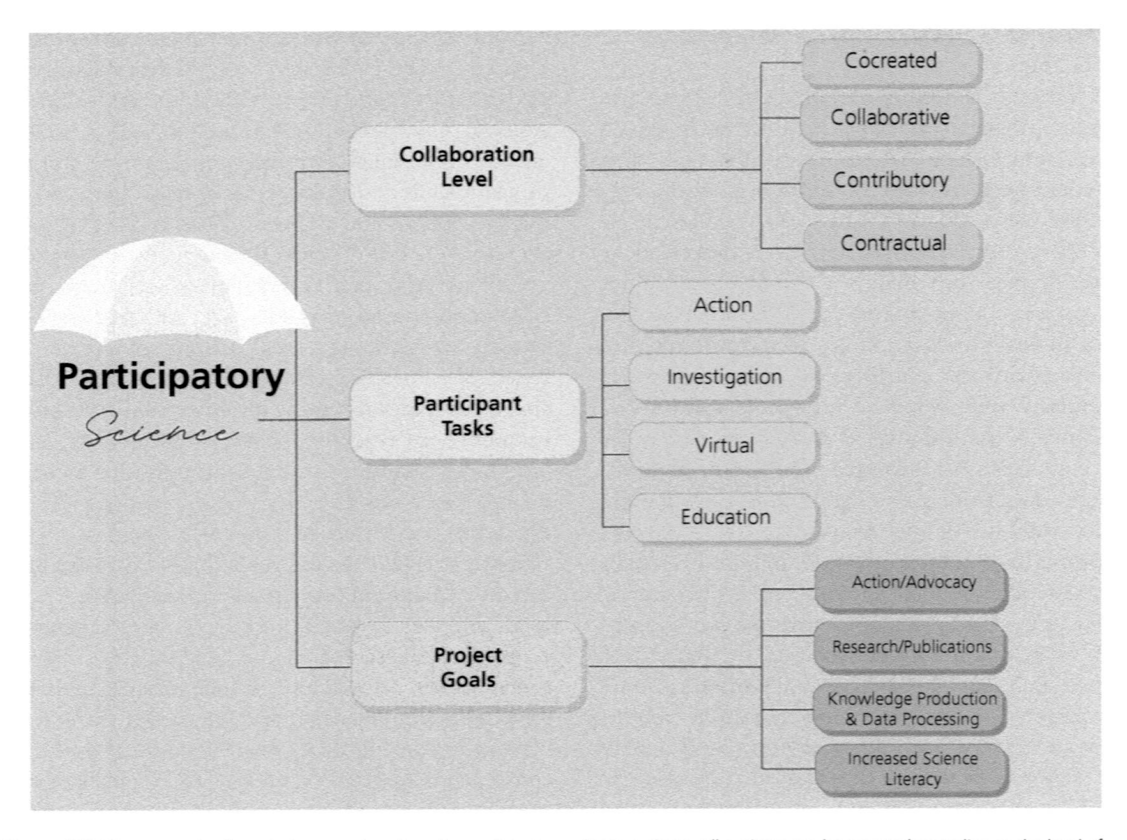

Figure 7.2 The non-mutually exclusive ways to categorize participatory science projects. All projects can be grouped according to the level of collaboration within their project (19), tasks undertaken by participants during the project (20), and the overarching project goals (13). Projects can have multiple goals and participants can undertake multiple actions within a project. A participatory project can be described as a cocreated, action-based project aiming to increase scientific literacy.

Participatory science has the advantage of generating observations in areas typically inaccessible to scientists, such as homes, backyards, schoolyards, and other private property. Participatory science data can occur over long timescales and across large geographic extents yet at fine resolutions (21). Participatory science has the potential to advance research in ways that are not possible with conventional methods.

The study of biodiversity and conservation is one of the areas of study in urban settings that benefit from participatory methods. For example, studies of urban populations of amphibians (22), bats (23), mammals (24), birds and butterflies (25), and even microbes (26) have relied on participatory sciences. Topically, participatory studies of biodiversity sciences have contributed to understanding the influence of climate change on urban ecosystems, particularly urban heat island effects. Some research on urban heat islands has examined biotic proxies such as snail shells (27) and camera traps (28). Other projects have directly measured temperatures at fine scales (29,30,31,32). Participatory sciences have also been useful for studies of environmental quality in urban settings, including air quality (33,34,35), light pollution (36), microplastics (37,38), and soundscapes (39,40). Some of these studies gathered data by setting up projects, recruiting participants, and managing volumes of data, whereas other studies tapped into participant-generated databases archived by projects at sites like the Global Biodiversity Infrastructure Facility

(GBIF). However, participatory science can contribute far beyond monitoring urban biodiversity. Various characteristics of urban ecosystems influence biodiversity, and fortunately, multiple disciplines have adopted, and adapted, methods of the participatory sciences for their research agendas.

Given the diversity of participatory projects in ecology, researchers who seek to start new projects would be well-advised to learn from existing projects or collaborate with them. This is true particularly for urban environments because of the variety of projects that exist, and the potential to create micro-sites and improve preexisting datasets. Candid Critters (mammals), eBird (birds), eMammal (mammals), iNaturalist (all taxa), Zooniverse (image-based projects), and Foxes in the City (mammals in Germany) are all partially or wholly focused on urban environments. These projects are examples of how even within a single domain of study (urban wildlife) participatory methods operate at local (Foxes in the City), regional, and global (eBird) scales, but they can also operate online (Zooniverse) or in person, and both individually (Candid Critters) or collectively. In addition, these projects have a range of scientific goals. In the context of birds, for example, participatory methods have been used to investigate migration, population, and distribution, as well as behavior (41). Despite the utility of participatory science for urban biodiversity research, it also has its challenges.

Spatial, racial, and other challenges

A core feature that makes participatory science data valuable is the geographic information submitted along with each observation or data point. The growing number of individuals with cell phones, digital cameras, and other handheld devices with GPS capabilities means that the quantity of data collected can scale with the size of populations, such that data should be richest where human populations are most dense. Such data might be "rich" but not representative if the individuals who participate in projects are a nonrandom subset of the entire population. For example, in the US when the participants in citizen science projects are predominantly White, and the distribution of humans in cities is nonrandom with regard to race and socioeconomics, citizen science projects will tend to reveal a racially biased perspective, a picture weighted toward White perspectives on the subject of study, whether it be biodiversity, pollution, or anything else. Such biases in data are particularly problematic where the subject of study is, itself, influenced by socioeconomic and cultural factors. For example, socioeconomics has been shown to often influence bird and plant communities (50) and biodiversity more generally (47). But our view of such influences is muddied by biases in who collects data and where those data are collected.

The White lens on biodiversity that emerges from projects driven by participant interest systematically obscures the ability of projects to study and understand the influence of race, racism, class, and caste systems on ecological realities of relevance to daily human life and well-being. One approach to remedying these biases is to employ randomized designs, where participants, houses, or geographic areas are chosen randomly, particularly with respect to variables of interest. However, randomized sampling designs assume participants can readily access or want to access those random points, which may not be true. Further, randomized sampling designs assume that the "patchiness" of cities is relatively equally distributed at the spatial scale of study. This assumption is almost never justified in US cities, which are nearly all segregated due to the history of racism in the US. For example, many cities across the US were subjected to the Home Owners' Loan Corporation (HOLC) redlining system which left lasting legacies on the geography[2] of those cities (51). These legacy patterns can also show up in our participatory science data. eBird, which allows its participants to randomly sample environments in space and time based on personal preference, significantly undersamples formerly D-grade or "redlined" areas in comparison to A-graded (pristine areas) by 69% (52). These areas being undersampled and underrepresented in the data allows the communities that reside there, and their environmental conditions, to be continuously overlooked, mirroring the other

[2] Hoover's and Scarlett's and Locke et al.'s chapters detail the diverse ways that systematic racism and classism—including redlining—have shaped urban nature.

Terminology Box	
Participatory Science	An umbrella term derived from the PPSR literature that includes a variety of approaches to engage the public in scientific activities for research, monitoring, policy action, or other goals that scientist cannot typically undertake alone.
Public Participation in Scientific Research (PPSR)	A framework created by Bonney et al. (42) and Shirk et al. (13) which covers a wide range of participatory approaches, including citizen science, crowdsourcing, participatory action research, community-based research, and volunteered geographic information
Citizen Science	"Scientific work undertaken by members of the general public, often in collaboration with or under the direction of professional scientists and scientific institutions." (18)
Community Science	Scientific research and monitoring that is driven and controlled by communities, characterized by place-based knowledge, collective action, and empowerment, with the aim to negotiate, improve, or transform governance for stewardship and socio-ecological sustainability. (43)
Contributory Projects	A participatory method generally designed by scientists where members of the public primarily contribute data; also includes studies in which scientists analyze citizens' observations, such as those in journals or other records, whether or not those citizens are still alive. (42)
Collaborative Projects	A participatory method generally designed by scientists and where members of the public contribute data but may also help to refine project design, analyze data, or disseminate findings. (42)
Cocreated Projects	A participatory method designed by scientists and members of the public working together where at least some of the public participants are actively involved in most or all steps of the scientific process; also includes research wholly conceived and implemented by amateur (nonprofessional) scientists. (42)
Crowd-derived Geographic Information (CdGI)	A general term referring to geographic information from the crowd/public that is and is not volunteered, that is and is not social media, and that is and is not user-generated content. (44). A broader term than VGI
Volunteered Geographic Information (VGI)	Digital spatial data produced by citizens/individuals who use tools with Global Positioning System (GPS) technology to gather and disseminate their observations and geographic knowledge. (45)
Critical GIS	Critical Geographic Information System (GIS) also includes GIS research and practice with an explicitly emancipatory agenda of engaging spatial technologies to disrupt socially and technologically mediated forms of exclusion and disempowerment (46). It examines relationships in a digital environment, and the consequences of these practices for social knowledge, representation, and power. (45)
Feminist GIS	Feminist GIS examines the implications of GIS for feminist research methodologies, exploring ways of working with knowledge as multiple and situated in a GIS, and challenging assumptions about inherent linkages between GIS and any specific epistemology. (45)
Biological Luxury Effect	A pattern of higher biodiversity in areas of higher affluence. (47)
Action-based activities	Activities initiated by volunteers to encourage local involvement. (48)
Conservation-based activities	Activities that address environmental management questions and help foster stewardship in participants by raising awareness via educational projects. (48)
Investigation-based activities	Activities that provide researchers with large quantities of data contributing to long-term monitoring and management goals. (48)
Virtual-based activities	Activities where the production and validation of scientific knowledge are primarily mediated by online participation. (48)
Education-based activities	Activities that aim at increasing scientific literacy and raising awareness. (48)
Datafication	Growing reliance on data-driven technologies across aspects of society. (49)
Data Justice	Analysis of data that involves highlighting structural inequality and the unevenness of the implications and experiences of data across different groups and communities in society. (49)

ways they have been overlooked by society. Beyond redlining, urban areas across the US and throughout the world are shaped by a diversity of racist and classist policies and practices (53,54). Citizen science, and science more broadly, must account for these inequities if they seek to collect meaningful data.

Another important bias to note is class bias. For example, Mahmoudi et al. (55) reveal a class bias in the Community Collaborative Rain, Hail, and Snow Network (CoCoRHaS), a large-scale, home-based precipitation project (explored further in Case study 1 below). This project has shown that although participation in large-scale projects is higher in urban areas, participation varies by income, even within rural areas where participation overall is lower. Mahmoudi and colleagues (55) attribute this bias, at least in part, to systemic racism and its influence on wealth, discretionary income, and, ultimately, the ability and willingness to participate in projects. While some spatial biases can be seen on a national scale, racial biases occur on a smaller scale within city boundaries. Depending on a project's methods (56), or its community norms and practices (44), these biases may be overlooked. Structured protocols (such as those used by atlases or systematic surveys) are better at overcoming this bias than unstructured or semi-structured protocols (such as eBird), which can not account for biases at local county scales (56). Therefore, these biases can either be perpetuated or overcome depending on the project's structure and protocols.

It is of note with the work of Mahmoudi and colleagues (55) that the impact of the biases present in the data depends on the question being asked. From the perspective of the impact of precipitation on people, the socioeconomic biases in data collection leave the project relatively oblivious to the impacts of precipitation on people who are poorer and/or people of color. It is important to note, however, that wealth/income/class are not synonymous with race despite how class and race are often correlated in the socio-ecological literature. However, it is the case that the over-representation of urban samples leads CoCoRHaS to undersample precipitation in rural areas. For some research questions, this kind of geographic nonrandomness might not be problematic. For example, if studying the ecology of a species that lives with humans, it is likely more important to sample human populations representatively rather than geography. However, for other questions, such as modeling the distribution of precipitation, under-represented regions prevent an accurate understanding of the topographic, climatic, land use, or other large-scale influences on the study subject. As a result, it is a key not only for studies to be aware of the potential biases in their participatory sampling, but also to understand which of those biases will have the largest impact on study conclusions.

Our knowledge of the causes of participation biases in participatory research is nascent. In general, it is thought that participatory methods can pose barriers to participation for lower-income individuals depending on the cost and availability of tools (sampling tools and cell phones), access to natural areas, and priority of other responsibilities (16). Just who such individuals are will vary among countries. In the US, a survey of eBird registrants (29,380 respondents) revealed that only 5.2% of respondents identified as an ethnic minority (Black, Asian, Pacific Islander, Multi-racial, Hispanic, or Indigenous/Native American) and 94.8% identified as non-Hispanic White (57). However, the most recent census indicates that the US White non-Hispanic population is now less than 60% of the total population, thus the Rutter et al. (57) analysis reveals a tremendous racial skew in eBird participation. The lack of participant diversity within bird-based projects suggests a lack of diversity within the recreational activity of bird-watching and the broader field of ornithology itself. We predict that as the demographics of participants are studied in other countries similar biases will come to light, in each case reflecting, at least in part, populations that have or do not have access to discretionary funds and time.

Social-ecological applications

We are learning that socioeconomic and cultural status both influence our perspective on urban biodiversity and independently influence the spatial structure of urban biodiversity and the patterns we

Case studies

Case study 1

Mahmoudi et al. (55) used data from the Community Collaborative Rain, Hail, and Snow Network (CoCoRHaS) to explore how sampling locations in citizen science projects might be connected to spatial patterns related to race and class. CoCoRHaS relies on volunteer project participants to monitor daily precipitation levels by installing and monitoring rain gauges at participants' homes. Mahmoudi et al. (55) examined rain gauge locations in Baltimore and Portland and used US Census Tract data to test the relationship between the number of rain gauges in a Census Tract and the race/ethnicity and income of a tract's inhabitant. Whiter and wealthier census tracts were more likely to have rain gauges. Race was a better predictor of rain gauge location in urban and suburban areas, whereas median income was a better predictor in rural areas. Generally, rain gauge data are more likely to be missing in poor, non-White places, resulting in data and knowledge gaps. Imagine if rainfall were non-random with regard to race or income, due to heat island effects (58), or differences in the large-scale geography of race and socioeconomics. In such a scenario, the effects of high rainfall on flooding or low rainfall on drought would be invisible in some poor, non-White places, which might exacerbate existing inequalities such as prioritizing aid funding. The uneven distribution of CoCoRHaS data based on patterns of race and wealth reflects a larger challenge shared amongst large-scale citizen science projects. Missing data that are biased by historical and ongoing racial and wealth inequalities will result in biased environmental projections, forecasts, and management decisions. Projects that fail to engage communities of color "run the risk of perpetuating social and environmental inequality through racialized and class-based knowledge production" (55, p. 383).

Case study 2

Perhaps the simplest way to partially account for sampling biases in urban biodiversity projects is through sampling designs that employ a mix of randomization and stratified sampling. For example, Turner (59) used a citywide biological monitoring tool, the Tucson Bird Count, to study ecology and conservation in Tucson, AZ. Superficially, the Tucson Bird Count is similar in structure to the Breeding Bird Survey, but with a specific focus on urban areas. It follows a structured sampling regime where project volunteers select from a series of points that are randomly distributed across a 1-km by 1-km grid within the boundaries of the city of Tucson. Points that fall on inaccessible areas (e.g., private property, tops of buildings) are moved to the nearest publicly accessible area, and GPS coordinate changes are detailed on subsequent maps and notes. As a result, the approach yields a relatively unbiased picture of urban systems with regard to space and hence has the potential to reduce spatial-racial biases. The Tucson Bird Count methods have been replicated for other bird counts in other cities such as the Fresno Bird Count in Fresno, CA, and the Triangle Bird Count based in Raleigh, NC.

Perkins (56) compared the protocol and project design of the structured atlas-style protocols of the Tucson Bird Count to the semi-structured protocols of eBird to gauge their ability to detect underlying patterns with regard to the influence of geographic patterns and affluence (wealth) on biodiversity. Perkins (56) found that the semi-structured protocols of eBird oversampled high-income areas and undersampled low-income areas resulting in an inability to detect patterns of biological luxury, whereas the structured protocols of the Tucson Bird Count reflected a more accurate depiction of the urban landscape, at least with regard to income (Figure 7.3). Using more structured protocols within sampling regimes, such as allowing volunteers to sample at predetermined locations randomly distributed across a gridded sampling plot, can help eliminate potential volunteer bias with regard to site selection even if the participants in the project are biased with regard to their race, ethnicity, income, or gender. It is worth noting, however, that stratified random sampling designs such as this one tend to be stratified by geography rather than population and so result in fewer samples per capita in high-density areas even though such areas are more reflective of the average lived experience (56). They are also likely to miss human populations that are very geographically concentrated and hence might not be sampled simply by chance. For example, in some cities, especially where segregation has persisted on the landscape, particular immigrant populations are concentrated in individual city blocks that might be missed by a stratified random sampling design. The outcome of this research is essential to urban conservation that relies on participatory science methods because spatial planning and prioritization for urban conservation efforts could be led astray if relying on such socially biased datasets.

Continued

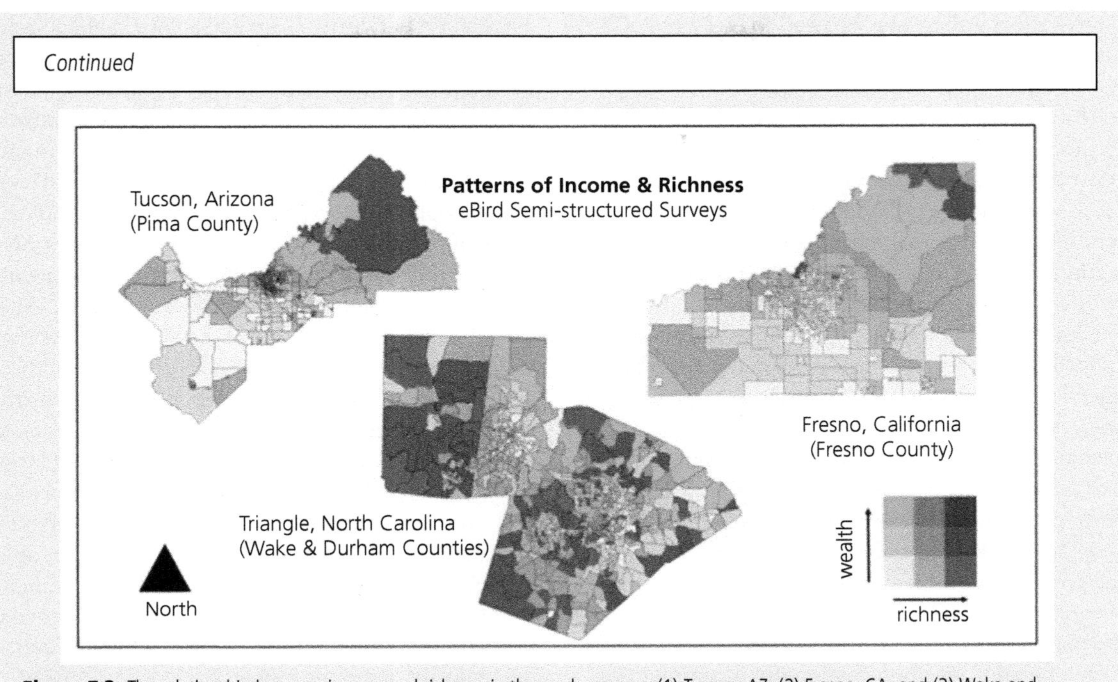

Figure 7.3 The relationship between income and richness in three urban areas: (1) Tucson, AZ; (2) Fresno, CA; and (3) Wake and Durham Counties of the Triangle Region of North Carolina. Species richness values were derived from April–May 2019 eBird checklist observations summarized to the census tract level. Income values were obtained from the 2016 US Census Median Household Income variable. This bivariate map displays income on a green scale with higher wealth values in dark green and bird diversity values on a purple scale with higher values in dark purple. Areas with low wealth and no bird diversity observations are in light gray, whereas dark areas have high wealth and bird diversity.

can observe with participatory methods (50). Yet, while this recognition is relatively new in the study of biodiversity, it is part of a larger history of scholarly study that documents the ways in which socioeconomics and cultural status influence the ways in which urban spaces are conceived and built (60). Some of these social-ecological applications include participatory science approaches. For example, participatory sciences are used in urban planning (61,62), the design of built infrastructure (63), improvements to water sanitation (64,65), and sustainable development (66).

Ultimately, designing biodiversity-focused participatory science projects that reflect and serve diverse populations will depend not only upon new approaches to designing projects, but also a cognizance of the many other disciplines that bear an understanding of the social systems that influence and are influenced by biodiversity. For example, a rich literature drawing upon theory and empirical studies from the fields of Geography, Critical GIS, Feminist GIS, and Participatory GIS shows the many ways in which identity, power, and spatial knowledge are intertwined in cities. As ecologists develop participatory science projects in urban settings, they should consider these literatures in which a key dimension is the intertwining ways in which intersectional identities shape experiences (45). This includes the influence of race and anti-Black racism on the design of the built environment (e.g., segregation and distribution of amenities/resources), our neighborhood characteristics (e.g., presence and type of parks/green spaces), and how environmental disamenities (e.g., landfills, factories, sources of pollutants) are distributed across the landscape. Biologists who engage in participatory urban science need to be aware of the deep insights from this large and growing body of work.

The spatial patterns created on the landscape through racist and classist policies are reflected in contributory participatory science data and thus have social-ecological implications for people and wildlife when management decisions are made using participatory science data. Not only is this an issue for our perceived results, but also for any conclusions derived from contributory datasets. Not collecting data within Black, Indigenous, and people of color (BIPOC) and low-income areas not only excludes those stories from our research but also practices erasure of the real people and the real environmental issues occurring in the neighborhoods that are undersampled/excluded. At our core, scientists/researchers are supposed to be in service to the public, helping to investigate and answer questions of interest and of benefit to broader society. We are not serving this broader societal goal if we are intentionally or unintentionally excluding these geographical areas (and people) from our work. In addition, there is a plethora of research demonstrating the benefits to participants when engaged in participatory projects including increased science literacy, knowledge of the scientific process, fun/enjoyment, contribution to science and scientific knowledge, appreciation for nature, etc. (67,16,68). However, these benefits are not experienced by BIPOC and low-income communities who are not engaged in these processes.

Global perspectives

So far, we have focused on the consideration of the role of participatory science in cities within the US (our own home region). However, the potential and complexity of participatory science in cities get both more interesting and more complex if we consider how these same dynamics play out on a global scale. Within cities in the US, we tend to be oblivious to biodiversity and its changes and consequences in underrepresented, marginalized, and lower-income communities. As we have noted, this lack of awareness is particularly problematic in that it prevents us from seeing the changes in the very places where change is most likely, for example, the association between extreme heat island effects and poverty. At the global scale, these data gaps may be even greater. Cities around the world are diverse in size, culture, history, land use, climate, infrastructure, as well as the provision of, and demand for, urban nature (69). We must be careful not to extrapolate knowledge gained from the interactions of large western cities to smaller cities and towns which can have different economic, sociocultural, climatic, and biodiversity factors (69). At a global scale, very few studies have used participatory methods to study urban ecology in the tropics, yet most species (and most people) live in the tropics. Much of the published literature on the effects of urban nature has a geographic bias toward large cities in Asia, North America, Europe, Australia, and New Zealand, yet almost half of the world's population lives in cities of less than 300,000 people (69). As a result, we are not prepared to document changes in the majority of urban species, nor are we prepared to understand the changes that influence the majority of people globally.

Practices to maximize the effectiveness of urban participatory projects

Moving forward, how do we manage contributory participatory science projects in the future that include underrepresented participants and spaces? Successful community science projects center marginalized populations (70), and share control of project governance (71) and related practices to overcome inherent limitations given the composition of a project team. These are the practices that most large-scale contributory projects lack. It is important to keep these practices in mind as we use large-scale contributory datasets (eBird, iNat, etc.), in order to avoid bias and maintain equitable practices as we work with the public.

Defining the project's scope and structure early in the planning process will help identify constraints and potential solutions. Based on time and budgetary constraints, reaching a more diverse audience might require reevaluating the geographic scope of the project and how study sites are selected or assigned.

Inclusive participation does not happen by chance, project organizers must purposefully work toward it. Projects should be intentional in creating marketing materials that are accessible, approachable, and that allow potential participants to see themselves reflected in the work. Marketing materials may also need to be translated into additional languages, particularly in linguistically diverse regions. In addition, projects should reach out to potential participants using a range of media, whether those be newspapers, TV, various forms of social media, or radio, with an awareness of how audiences differ among those media types. Information sessions should give participants an overview of what project participation might entail. Sessions should be held in places where potential participants feel at ease and welcomed. If the work is being conducted in a space where a volunteer feels "othered" and they do not see themselves represented in the media, diverse volunteers may be disinclined to participate (72).

The importance of participant engagement after the initial recruitment phase is often overlooked. In our experience, this often happens because initial funding for projects begins to wane. Incorporating costs for labor, materials, and occasionally participant stipends into the initial budget is important to ensure engagement can continue throughout all stages of a project. Such mismatches may actually increase costs if they lead to high participant turnover rates where volunteers need to be repeatedly recruited and trained (72). It is important to recognize that participants are not a homogeneous group and may have a variety of reasons for participating in a project (72). Volunteer surveys can provide useful information about participant value motivations, time constraints, accessibility needs, and demographics (age, ethnicity, gender, education, etc.), as well as serve as a tool to gauge the effectiveness of the current strategies for eliminating spatial and racial bias.

Retention can be an issue with participatory projects. Volunteers may drop out of projects because they perceive that the project is poorly managed or because of a lack of communication. Reportbacks and good communication are essential to a project's success. Consistent and personable communication throughout the project's timeframe may help with participant retention and interest. However, communication that is too frequent, such as in-depth weekly conversations about data quality, might also lead to a decrease in engagement (73). Communication frequency and format must therefore be tailored and reevaluated throughout the lifetime of the project to match the needs of the intended audience.

Volunteers also drop out of projects when they feel undervalued or overburdened with tasks (72). A seemingly obvious, yet often overlooked, reality is that volunteers are people and not merely tools for success. As such, people have varying needs, interests, and motivations. This reframing recognizes that participants are not just free labor, but people, and "individuals will keep coming if their needs are fulfilled" (74, p. 645). Project managers must acknowledge that a volunteer's time is valuable and limited. Providing updates on how the data they are collecting are being used and reiterating the importance of the work being done help volunteers see that their work is valued (72).

Another way to retain interest and support from a community of volunteers is by providing volunteers access to their data. Although it is not always possible to share all project data with volunteers (e.g., where internal review boards and privacy are involved), but where possible volunteers should have access to, and the ability to use, the data and results. Data accessibility allows volunteers to have ownership over their results and is an important step in ensuring that projects avoid perpetuating data injustice (75). As a large-scale project, eBird allows anyone to request data from its database. However, manipulating and using the data requires extensive knowledge of statistical programming languages. Without this very specialized knowledge, volunteers don't have independent access to the data they have collected. As a solution, eBird has a science team that summarizes eBird data into "species status and trends" visualizations that can be downloaded from the website. Although this "done for you" format does not allow for individual analysis, it is beneficial for presentations, teaching, etc. It is not currently a standard practice for participatory projects to have a data science team focused on data accessibility, in no small part because only a tiny minority of projects can afford such teams. However, sharing the data with volunteers will help uphold ideals of data justice. If we plan to take advantage of the datafication of the world, we must recognize how the collection and use of data impact society, as well as how data help frame how social issues can be understood and solved (49).

In addition to being aware of how data are collected and shared, projects must be thoughtful in how the resulting data are used. Data from projects may be used to make restoration decisions or designate management priority areas, or other outcomes that have direct impact on social and environmental justice. Brainstorming the potential uses, and most importantly the potential users, of the data, prior to collection, will help provide a better understanding of how the data should be stored and managed. Check the statistical literature for data management best practices to assist in sharing, storing, and using raw data. Upholding "tidy data" principles helps maintain a standard structure where each variable forms a column and each observation forms a row to simplify the data cleaning process and aid in statistical analysis (76).

Building on research and our experiences, we outline the following recommendations for project managers when building or managing a successful participatory research practice that advances urban biodiversity conservation and moves toward more diverse, inclusive, equitable, and accessible practices:

- Decide on the type of project structure (Figure 7.2) and desired level of collaboration with the public(s) that would make the most sense given time and budgetary constraints. Depending on time, budget, and staffing it may be more sensible to use an existing project or create a microproject on an established platform (e.g., iNaturalist, Zooniverse).
- Plan an intentional recruitment strategy and use a variety of recruitment methods (newspapers, social media, radio, or podcast ads) in a variety of locations (faith-based centers, libraries, visitor centers) centered on the target audience.
- Produce recruitment material that centers diversity in its media.
- After participants have been secured, conduct volunteer surveys to assess values, motivations, contact preferences, questions of interest, and group demographics to inform social needs. Use this information to inform decision-making.
- Provide volunteers with materials to help them understand the context and history of the study site where the work is being conducted.
- Evaluate survey protocol for biases. Use structured (atlas-style) or semi-structured protocols if possible.
- Have information sessions or workshops that allow participants to experience an example of what might occur while participating in the project. These sessions can serve as training for volunteers and an opportunity for projects to receive feedback.
- Maintain consistent communication throughout the project to help evaluate and assess if participant needs are being met, if data collection is going as intended, or if assistance is needed. Communication should occur throughout all project stages in a constant feedback loop.
- Follow tidy data structure principles of one observation per row and one variable per column for ease of post-collection analysis.
- Make project results and raw data available to the public via a project website. No matter the level of collaboration a project undertakes with the public, it is important for a project to be data just and provide access to the raw data (not just data summaries) when possible.

We recognize that establishing a new participatory science project takes money, time, and support (staff), and may not be feasible for many. If using existing large-scale datasets, consider what is and is not being captured in the project's scale. If designing a new project, the level of public involvement throughout a project's creation will influence how long it may take to establish, as establishing relationships with any specific community takes time (77). It may take a few years for projects to obtain adequate spatial coverage of the study area, depending on the project's scale and ability to recruit and retain volunteers. As such, long-term urban initiatives may benefit most from creating new participatory projects because their longevity is better suited for establishing meaningful relationships with the public that retain year-to-year participation. No matter the scale, participatory methodologies are of value to urban research.

Conclusion

Participatory science has come to be valued for its ability to help us gather data to better understand the complex interactions within the urban ecosystem, which can provide the information that communities and agencies need to make effective biodiversity management and policy decisions. It also provides direct benefits to those who participate as contributors, in the form of science education, increased scientific literacy (78,13), skill building, entertainment, and more. Particularly important from the perspective of conservation is democratizing: access to biodiversity; the benefits associated with access to nature; inclusion in the scientific process; and the stewardship of nature. However, embedded within their practice, participatory methods can be constrained by a series of hidden biases. Those biases are a function of who participates. To date, citizen science participants in the US and Europe are disproportionately affluent, older, White, and science-educated (79). The demographics of those who contribute data, though seemingly unimportant to the data being collected, directly impact the quality of the science as well as those who miss out on the benefits of participation. First, the demographics of the participants are directly tied to sociopolitics, socioeconomics, race, and geography, which directly impacts the data and resulting science. This means that our ability to understand the natural world—our world—is severely constrained by who participates in research. Second, those who participate and gain the direct benefits of participation are largely those who already have access to science education and similar benefits. In this way, the benefits of participatory science perpetuate existing injustices of access to opportunities. Third, data gaps based on biased participation will limit the ability of communities, governments, and agencies to make effective management plans and decisions. Management might either be overlooked in understudied locations or be applied inappropriately based on biased data. Ineffective management and policies can therefore perpetuate existing injustices in understudied communities. If participatory science is meant to benefit participants, reveal general truths, and share the benefits of those truths, its efforts will be limited by who is included.

Applied research scientists and conservation practitioners have a public responsibility to help conduct science that benefits all individuals in society, regardless of ability, race, gender, income, or sexual orientation. Utilizing participatory methods in urban ecology can help us better understand the complex interactions within the urban ecosystem. However, without an awareness of the potential racial-spatial biases and other challenges that occur when capturing crowd-derived data at varying scales, we will continue to perpetuate harm on historically excluded and low-income individuals. Neglecting to capture data from low-income or high-minority populated areas continues to erase the important stories of the people and ecological interactions that inhabit those data gaps. Participatory methods can be a valuable tool for understanding urban ecosystems and advancing urban conservation. But like any other approach, participatory methods can be arduous and require careful and intentional planning to reduce bias and accurately capture underlying social-ecological processes. Understand that participatory practices take time, money, support, collaboration, and a variety of skills, especially if designing a new project.

Most importantly, researchers and project managers must remember that while conservation is about biodiversity, it is also fundamentally about people, and participants are people, not tools for scientific advancement. Understanding participant motivations and needs; maintaining consistent communication; and providing a system/community of support where volunteers can see themselves reflected in the work can help projects retain year-to-year participation and provide robust insights about conservation where people live, work, and play.

References

1. Kingsland SE. Urban ecological science in America: the long march to cross-disciplinary research. In Pickett STA, Cadenasso ML, Grove MJ, Irwin EG, Rosi EJ, Swan CM (eds.) Science for the Sustainable City: Empirical Insights from the Baltimore School of Urban

Ecology. New Haven, CT: Yale University Press; 2019. p. 24–44.

2. Lidicker WZ. Responses of mammals to habitat edges: an overview. Landscape Ecology. 1999;14(Aug): 333–43.
3. Murcia C. Edge effects in fragmented forests: implications for conservation. Trends in Ecology & Evolution. 1995;10(2):58–62. Available from: https://doi.org/10.1016/S0169-5347(00)88977-6
4. Sisk T, Battin J. Habitat edges and avian ecology: geographic patterns and insights for western landscapes. Studies in Avian Biology. 2002;25(Jan):30–48.
5. Fa J, Brown D. Impacts of hunting on mammals in African tropical moist forests: a review and synthesis. Mammal Review. 2009;39(4):231–64. Available from: https://doi.org/10.1111/j.1365-2907.2009.00149.x
6. Stoner KE, Vulinec K, Wright SJ, Peres CA. Hunting and plant community dynamics in tropical forests: a synthesis and future directions. Biotropica. 2007;39(3):385–92. Available from: https://doi.org/10.1111/j.1744-7429.2007.00291.x
7. Cushman SA. Effects of habitat loss and fragmentation on amphibians: a review and prospectus. Biological Conservation. 2006;128(2):231–40. Available from: https://doi.org/10.1016/j.biocon.2005.09.031
8. Laurance WF, Goosem M, Laurance SGW. Impacts of roads and linear clearings on tropical forests. Trends in Ecology & Evolution. 2009;24(12):659–69. Available from: https://doi.org/10.1016/j.tree.2009.06.009
9. MacArthur RH, Wilson EO. The Theory of Island Biogeography. Princeton, NJ: Princeton University Press; 1967.
10. Davis AM, Glick TF. Urban ecosystems and island biogeography. Environmental Conservation. 1978;5(4):299–304. Available from: https://doi.org/10.1017/S037689290000638X
11. Fattorini S, Mantoni C, Simoni L, Galassi D. Island biogeography of insect conservation in urban green spaces. Environmental Conservation. 2017;45(1):1–10. Available from: https://doi.org/10.1017/S0376892917000121
12. Dunn RR, Burger JR, Carlen E, Koltz AM, Light JE, Martin RA, et al. A theory of city biogeography and the origin of urban species. Frontiers in Conservation Science. 2022;3:761449.
13. Shirk J, Ballard H, Wilderman C, Phillips T, Wiggins A, Jordan R, et al. Public participation in scientific research: a framework for deliberate design. Ecology and Society. 2012;17(2):29–48. Available from: https://doi.org/10.5751/ES-04705-170229
14. Israel BA, Parker EA, Rowe Z, Salvatore A, Minkler M, López J, et al. Community-based participatory research: lessons learned from the Centers for Children's Environmental Health and Disease Prevention Research. Environmental Health Perspectives. 2005;113(10):1463–71. Available from: https://doi.org/10.1289/ehp.7675
15. Heaney CD, Wing S, Campbell RL, Caldwell D, Hopkins B, Richardson D, et al. Relation between malodor, ambient hydrogen sulfide, and health in a community bordering a landfill. Environmental Research. 2011;111(6):847–52. Available from: https://doi.org/10.1016/j.envres.2011.05.021
16. Pandya RE. A framework for engaging diverse communities in citizen science in the US. Frontiers in Ecology and the Environment. 2012;10(6):314–17. Available from: https://doi.org/10.1890/120007
17. Tucker C, Taylor D. Good science: principles of community-based participatory research. Race, Poverty & the Environment. 2004;11(2):27–9.
18. Eitzel MV, Cappadonna JL, Santos-Lang C, Duerr RE, Virapongse A, West SE, et al. Citizen science terminology matters: exploring key terms. Citizen Science: Theory and Practice. 2017;2(1):1.
19. Montanari M, Jacobs L, Haklay M, Donkor F, Mondardini M. Agenda 2030's, "Leave no one behind," in citizen science? Journal of Science Communication. 2021;20(6):A07. Available from: https://doi.org/10.22323/2.20060207
20. Wiggins A, Crowston K. From conservation to crowdsourcing: a typology of citizen science. In: 2011 44th Hawaii International Conference on System Sciences, Kauai, HI, USA. New York: IEEE; 2011. p. 1–10.
21. Cooper C, Hochachka W, Dhondt A. The opportunities and challenges of citizen science as a tool for ecological research: public participation in environmental research. In: Dickinson JL, Bonney R (eds.) Citizen Science. Ithaca, NY: Cornell University Press; 2017. p. 97–113.
22. Lee T, Kahal N, Kinas H, Randall L, Baker T, Carney V, et al. Advancing amphibian conservation through citizen science in urban municipalities. Diversity. 2021;13(5):211. Available from: https://doi.org/10.3390/d13050211
23. Lewanzik D, Straka TM, Lorenz J, Marggraf L, Voigt-Heucke S, Schumann A, et al. Evaluating the potential of urban areas for bat conservation with citizen science data. Environmental Pollution. 2022;297(Mar):118785. Available from: https://doi.org/10.1016/j.envpol.2021.118785
24. Mueller MA, Drake D, Allen ML. Using citizen science to inform urban canid management. Landscape and Urban Planning. 2019;189(Sep):362–71. Available from: https://doi.org/10.1016/j.landurbplan.2019.04.023

25. Wei JW, Lee BPY-H, Wen LB. Citizen science and the urban ecology of birds and butterflies — a systematic review. PloS ONE. 2016;11(6):e0156425. Available from: https://doi.org/10.1371/journal.pone.0156425
26. McKenney E, Flythe T, Millis C, Stalls J, Urban JM, Dunn RR, et al. Symbiosis in the soil: citizen microbiology in middle and high school classrooms. Journal of Microbiology & Biology Education. 2016;17(1):60–2. Available from: https://doi.org/10.1128/jmbe.v17i1.1016
27. Kerstes N, Breeschoten T, Kalkman V, Schilthuizen M. Snail shell colour evolution in urban heat islands detected via citizen science. Communications Biology. 2019;2(Jul):264. Available from: https://doi.org/10.1038/s42003-019-0511-6
28. Herrera DJ, Cove MV. Camera trap serendipity and citizen science point to broader effects of urban heat islands on food webs. Food Webs. 2020;25(Dec):e00176. Available from: https://doi.org/10.1016/j.fooweb.2020.e00176
29. Czajkowski KP, Struble J. Engaging citizen scientists to study the urban heat island effect through the GLOBE program. American Geophysical Union, Fall Meeting 2018, abstract GH22A–01. Dec. 2018.
30. Liebowitz AW, Sebastian EA, Yanos C, Bilik M, Ginchereau J, Valentine M, et al. Engaging citizen scientists in characterizing urban heat island at the neighborhood scale using satellite and ground observations. American Geophysical Union, Fall Meeting 2019, abstract IN51E–0677. Dec. 2019.
31. Lucille A. The use of citizen science in the characterization of the Lyon's urban heat and cool islands. In: 2019 20th IEEE International Conference on Mobile Data Management (MDM), Jun 10–13, 2019, Hong Kong, China. New York: IEEE; 2019. p. 387–8.
32. Slack S, Blake R, Norouzi H, Deas AA, Foley E, Zhang M. Engaging citizen scientists in urban heat island investigations. American Geophysical Union, Fall Meeting 2021, held in New Orleans, LA, December 13–17, 2021, abstract ED55D–0311. Dec. 2021.
33. Huddart JEA, Thompson MSA, Woodward G, Brooks SJ. Citizen science: from detecting pollution to evaluating ecological restoration. WIREs Water. 2016;3(3):287–300. Available from: https://doi.org/10.1002/wat2.1138
34. Lu T, Liu Y, Garcia A, Wang M, Li Y, Bravo-Villasenor G, et al. Leveraging citizen science and low-cost sensors to characterize air pollution exposure of disadvantaged communities in southern California. International Journal of Environmental Research and Public Health. 2022;19(14):8777. Available from: https://doi.org/10.3390/ijerph19148777
35. Mahajan S, Kumar P, Pinto JA, Riccetti A, Schaaf K, Camprodon G, et al. A citizen science approach for enhancing public understanding of air pollution. Sustainable Cities and Society. 2020;52(Jan):101800. Available from: https://doi.org/10.1016/j.scs.2019.101800
36. Schroer S, Kyba CCM, van Grunsven R, Celino I, Corcho O, Hölker F. Citizen science to monitor light pollution – a useful tool for studying human impacts on the environment. In: Hecker S, Haklay M, Bowser A, Makuch Z, Vogel J, Bonn A (eds.) Citizen Science. London: UCL Press; 2018. p. 353–66.
37. Barrows APW, Christiansen KS, Bode ET, Hoellein TJ. A watershed-scale, citizen science approach to quantifying microplastic concentration in a mixed land-use river. Water Research. 2018;147(Dec):382–92. Available from: https://doi.org/10.1016/j.watres.2018.10.013
38. Syberg K, Palmqvist A, Khan FR, Strand J, Vollertsen J, Westergaard Clausen LP, et al. A nationwide assessment of plastic pollution in the Danish realm using citizen science. Scientific Reports. 2020;10(1):17773.
39. Carson B, Cooper C, Larson L, Rivers L. How can citizen science advance environmental justice? Exploring the noise paradox through sense of place. Cities & Health. 2021;5(1–2):1–13. Available from: https://doi.org/10.1080/23748834.2020.1721222
40. Maisonneuve N, Stevens M, Niessen ME, Hanappe P, Steels L. Citizen noise pollution monitoring. In: Chun SA, Sandoval R, Regan P (eds.) Proceedings of the 10th Annual International Conference on Digital Government Research: Social Networks: making connections between citizens, data and government (Puebla, Mexico, May 2009). Digital Government Society of North America; 2009. p. 96–103.
41. Greenwood JJD. Citizens, science and bird conservation. Journal of Ornithology. 2007;148(1):77–124. Available from: https://doi.org/10.1007/s10336-007-0239-9
42. Bonney R, Ballard H, Jordan R, McCallie E, Phillips T, Jennifer S, et al. Public participation in scientific research: defining the field and assessing its potential for informal science education. A CAISE Inquiry group report. Washington, DC: Center for Advancement of Informal Science Education; 2009. Available from: https://eric.ed.gov/?id=ED519688
43. Charles A, Loucks L, Berkes F, Armitage D. Community science: a typology and its implications for governance of social-ecological systems. Environmental Science & Policy. 2020;106:77–86.
44. Johnson I, Hecht B. Structural causes of bias in crowd-derived geographic information: towards a holistic understanding. Association for the Advancement of Artificial Intelligence (AAAI) Spring Symposium

on Observational Studies through Social Media and Other Human-Generated Content. 2015. Available from: https://isaacjoh.com/Publications/StructuralCausesBiases_aaaispring2016.pdf
45. Elwood S. Volunteered geographic information: future research directions motivated by critical, participatory, and feminist GIS. GeoJournal. 2008;72(3):173–83. Available from: https://doi.org/10.1007/s10708-008-9186-0
46. Harvey F, Kwan M-P, Pavloskaya M. Introduction: critical GIS. Cartograpica: The International Journal for Geographic Information and Geovisualization. 2005;40:1–4.
47. Leong M, Dunn RR, Trautwein MD. Biodiversity and socioeconomics in the city: a review of the luxury effect. Biology Letters. 2018;14(5):20180082. Available from: https://doi.org/10.1098/rsbl.2018.0082
48. Frigerio D, Pipek P, Kimmig S, Winter S, Melzheimer J, Diblikova L, et al. Citizen science and wildlife biology: synergies and challenges. Ethology. 2018;124(6):365–377.
49. Dencik L, Sanchez-Monedero J. Data justice. Internet Policy Review. 2022;11(1):1615.
50. Kinzig AP, Warren P, Martin C, Hope D, Katti M. The effects of human socioeconomic status and cultural characteristics on urban patterns of biodiversity. Ecology and Society. 2004;10(1):23. Available from: https://doi.org/10.5751/ES-01264-100123
51. Schell CJ, Dyson K, Fuentes TL, Des Roches S, Harris N, Miller DS, et al. The ecological and evolutionary consequences of systemic racism in urban environments. Science. 2020;369(6510):eaay4497.
52. Ellis Soto D, Chapman M, Locke D. Uneven biodiversity sampling across redlined urban areas in the United States. EcoEvoRxiv [Preprint]. 2022. Available from: https://ecoevorxiv.org/repository/view/3736/
53. Hamel P, Keil R. Suburban Governance: a global view. Toronto: University of Toronto Press; 2015.
54. Rothstein R. The Color of Law: a forgotten history of how our government segregated America. New York: Liveright; 2017.
55. Mahmoudi D, Hawn CL, Henry EH, Perkins DJ, Cooper CB, Wilson SM. Mapping for whom? Communities of color and the citizen science gap. ACME: An International Journal for Critical Geographies. 2022;21(4):372–88. Available from: https://doi.org/10.13016/m2oveu-gfbf
56. Perkins DJ. Blind Spots in Citizen Science Data: implications of volunteer biases in Ebird data [Master's Thesis]. [Raleigh, NC]: North Carolina State University; 2020. Available from: https://www.proquest.com/docview/2478031545
57. Rutter JD, Dayer AA, Harshaw HW, Cole NW, Duberstein JN, Fulton DC, et al. Racial, ethnic, and social patterns in the recreation specialization of birdwatchers: an analysis of United States eBird registrants. Journal of Outdoor Recreation and Tourism. 2021;35(Sep):100400. Available from: https://doi.org/10.1016/j.jort.2021.100400
58. Marelle L, Myhre G, Steensen BM, Hodnebrog Ø, Alterskjær K, Sillmann J. Urbanization in megacities increases the frequency of extreme precipitation events far more than their intensity. Environmental Research Letters. 2020;15(12):124072. Available from: https://doi.org/10.1088/1748-9326/abcc8f
59. Turner WR. Citywide biological monitoring as a tool for ecology and conservation in urban landscapes: the case of the Tucson Bird Count. Landscape and Urban Planning. 2003;65(3):149–66. Available from: https://doi.org/10.1016/S0169-2046(03)00012-4
60. Shelton T, Poorthuis A, Zook M. Social media and the city: rethinking urban socio-spatial inequality using user-generated geographic information. Landscape and Urban Planning. 2015;142(Oct):198–211. Available from: https://doi.org/10.1016/j.landurbplan.2015.02.020
61. Cooper C, Balakrishnan A. Citizen science perspectives on e-participation in urban planning. In: Silva CN (ed.) Citizen E-Participation in Urban Governance: crowdsourcing and collaborative creativity. Hershey, PA: IGI Global; 2013. p. 172–97.
62. Franco S, Cappa F. Citizen science: involving citizens in research projects and urban planning. TeMA—Journal of Land Use, Mobility and Environment. 2021;14(1):114–18. Available from: https://doi.org/10.6092/1970-9870/7892
63. Mueller J, Lu H, Chirkin A, Klein B, Schmitt G. Citizen Design Science: a strategy for crowd-creative urban design. Cities. 2018;72(Feb):181–8. Available from: https://doi.org/10.1016/j.cities.2017.08.018
64. Corburn J. Water and sanitation for all: citizen science, health equity, and urban climate justice. Environment and Planning B. 2022;49(8):2044–53.
65. McGoff E, Dunn F, Cachazo LM, Williams P, Biggs J, Nicolet P, et al. Finding clean water habitats in urban landscapes: professional researcher vs citizen science approaches. The Science of the Total Environment. 2017;581–582(Mar):105–16. Available from: https://doi.org/10.1016/j.scitotenv.2016.11.215
66. Cappa F, Franco S, Rosso F. Citizens and cities: leveraging citizen science and big data for sustainable

urban development. Business Strategy and the Environment. 2022;31(2):648–67. Available from: https://doi.org/10.1002/bse.2942

67. Dickinson JL, Shirk J, Bonter D, Bonney R, Crain RL, Martin J, et al. The current state of citizen science as a tool for ecological research and public engagement. Frontiers in Ecology and the Environment. 2012;10(6):291–7. Available from: https://doi.org/10.1890/110236
68. Wehn U, Almomani A. Incentives and barriers for participation in community-based environmental monitoring and information systems: a critical analysis and integration of the literature. Environmental Science & Policy. 2019;101(Nov):341–57. Available from: https://doi.org/10.1016/j.envsci.2019.09.002
69. Kendal D, Egerer M, Byrne JA, Jones PJ, Marsh P, Threlfall CG, et al. City-size bias in knowledge on the effects of urban nature on people and biodiversity. Environmental Research Letters. 2020;15(12):124035. Available from: https://doi.org/10.1088/1748-9326/abc5e4
70. Cooper CB, Hawn CL, Larson LR, Parrish JK, Bowser G, Cavalier D, et al. Inclusion in citizen science: the conundrum of rebranding. Science. 2021;372(6549):1386–8. Available from: https://doi.org/10.1126/science.abi6487
71. Cooper CB, Rasmussen LR, Jones ED. A toolkit for data ethics in the participatory sciences. Citizen Science Association; 2022. Available from: https://citizenscience.org/data-ethics/
72. West S, Pateman R. Recruiting and retaining participants in citizen science: what can be learned from the volunteering literature? Citizen Science: Theory and Practice. 2016;1(2):15. Available from: https://doi.org/10.5334/cstp.8
73. Grace-McCaskey C, Iatarola B, Manda A, Etheridge J. Eco-ethnography and citizen science: lessons from within. Society and Natural Resources. 2019;32(10):1123–38. Available from: https://doi.org/10.1080/08941920.2019.1584343
74. Ryan RL, Kaplan R, Grese RE. Predicting volunteer commitment in environmental stewardship programmes. Journal of Environmental Planning and Management. 2001;44(5):629–48. Available from: https://doi.org/10.1080/09640560120079948
75. Christine DI, Thinyane M. Citizen science as a data-based practice: a consideration of data justice. Patterns (New York, N.Y.). 2021;2(4):100224. Available from: https://doi.org/10.1016/j.patter.2021.100224
76. Wickham H. Tidy data. Journal of Statistical Software. 2014;59(10):1–23. Available from: https://doi.org/10.18637/jss.v059.i10
77. Cartwright LA, Cvetkovic M, Graham S, Tozer D, Chow-Fraser P. URBAN: development of a citizen science biomonitoring program based in Hamilton, Ontario, Canada. International Journal of Science Education, Part B. 2015;5(2):93–113. Available from: https://doi.org/10.1080/21548455.2013.855353
78. Pateman R, Dyke A, West S. The diversity of participants in environmental citizen science. Citizen Science: Theory and Practice. 2021;6(1):9.
79. Allf BC, Cooper CB, Larson LR, Dunn RR, Futch SE, Sharova M, et al. Citizen science as an ecosystem of engagement: implications for learning and broadening participation. BioScience. 2022;72(7):651–63. Available from: https://doi.org/10.1093/biosci/biac035
80. Jenkins CN, Pimm SL, Joppa LN. Global patterns of terrestrial vertebrate diversity and conservation. Proceedings of the National Academy of Sciences of the United States of America. 2013;110(28):E2602–10.

CHAPTER 8

Multicity Ecological Networks for Addressing Urban Biodiversity Conservation

Seth Magle[‡], Mason Fidino[‡], Elizabeth W. Lehrer[‡], Tobin Magle[§], and Myla F.J. Aronson[¶]

Introduction

Different cities have different wildlife species. We find pumas (*Puma concolor*) in Los Angeles, long-tailed macaques (*Macaca fascicularis*) in Singapore, and grey herons (*Ardea cinerea*) in Amsterdam, partly because cities are embedded in different ecoregions of the world with their own unique biodiversity (Figure 8.1). But the structure of the built environment within a city also creates "winners" and "losers" and thus shapes which animals persist or are extirpated from the city. In the US, for example, Chicago, IL and Madison, WI are a little over 200 km apart and located in similar ecoregions. However, Madison has a much greater proportion of green space and a much lower human population density than Chicago. As a result, red fox (*Vulpes vulpes*) are widely distributed and highly abundant in Madison but are quite rare in Chicago (1).

To most people, however, biological differences pale in comparison to the cultural, economic, architectural, and other anthropocentric factors that characterize each urban region. When one compares New York City to Tokyo, a biologist might imagine the surrounding landforms, the differing bird and plant communities, and so on. But the average city resident is more likely to envision the buildings, the people, the highways, and the downtown nightlife. Each city also has its own local laws and ordinances, some of which relate to the management and maintenance of green spaces like parks or street trees. If cities are so different from one another, what value can conservation gain from comparing cities? Can policy makers, city planners, park managers, and backyard conservationists conserve biodiversity using research from another city? In fact, multicity comparisons allow the practice of conservation to better use science to inform actions. Multicity comparisons teach us which patterns and processes are generalizable across cities and which are specific to particular cities or city attributes.

Of course, the perceptive biologist knows that all of these factors impact urban biodiversity as well and can have as strong an influence as the underlying habitat (2–4). Roads, buildings, people, and their pets act as both threats and opportunities that shape the urban communities of plants and animals worldwide (5).

As both ecological and human factors influence urban biodiversity, a full understanding of the distribution, natural history, and ecology of these species requires data collection across multiple cities, regions, and continents (6). The history of urban wildlife research is, however, largely restricted to single-city examples (7). This has created an amalgamation of case studies but has limited general urban conservation insight and an

‡ Urban Wildlife Institute, Lincoln Park Zoo, USA
§ Research Computing Services, Northwestern University, USA
¶ Rutgers University, USA

Seth Magle et al., *Multicity Ecological Networks for Addressing Urban Biodiversity Conservation.* In: *Urban Biodiversity and Equity.* Edited by: Max R. Lambert & Christopher J. Schell, Oxford University Press. © Oxford University Press (2023). DOI: 10.1093/oso/9780198877271.003.0008

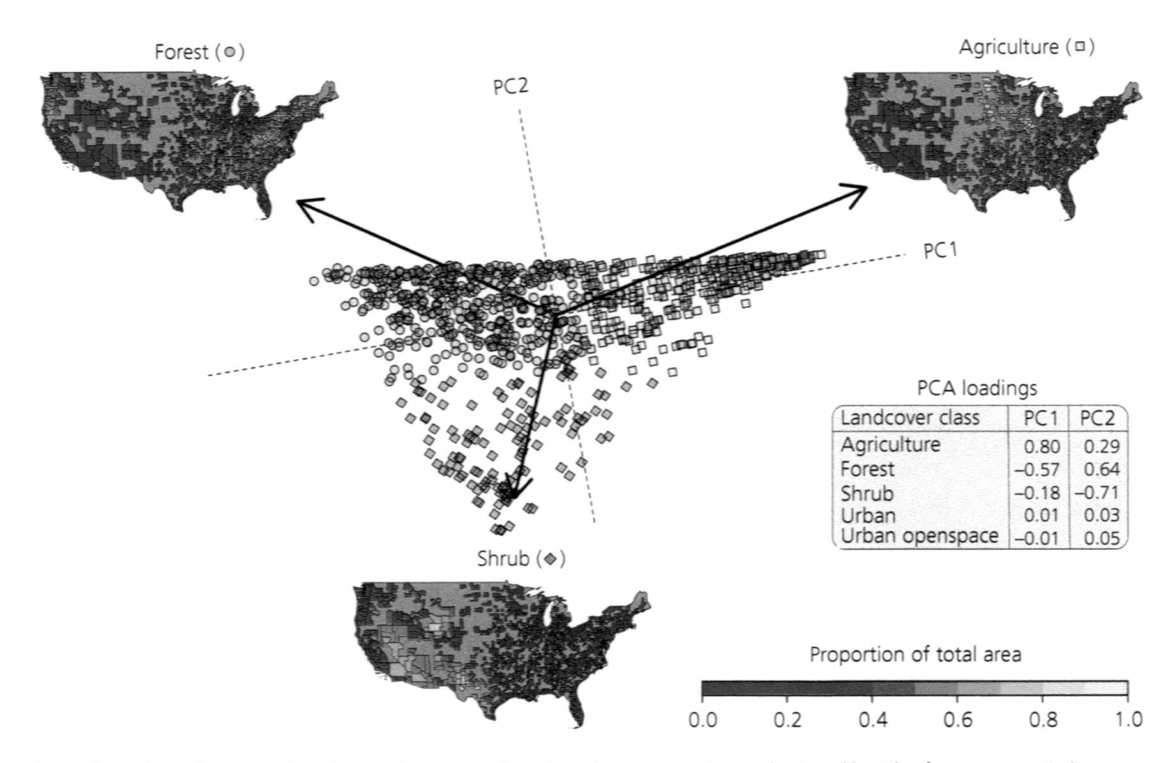

Landcover class	PC1	PC2
Agriculture	0.80	0.29
Forest	−0.57	0.64
Shrub	−0.18	−0.71
Urban	0.01	0.03
Urban openspace	−0.01	0.05

Figure 8.1 Metropolitan and micropolitan statistical areas throughout the contiguous US are dominated by either forest cover, agriculture, or shrub cover. However, there is substantial variation among these 925 areas. Land-use land cover data were compiled from the 2019 National Landcover Dataset and class values were simplified to create five landcover classes (Agriculture: 81, 82; Forest: 41, 42, 43; Shrub: 51, 52; Urban: 22, 23, 24; Urban openspace: 21). Following dimension reduction by principal component analysis (PCA), a k-means clustering algorithm was used in conjunction with a silhouette analysis which classified three dominant land cover classes among statistical areas. The principal components plot in the center of this figure demonstrates the clustering of land cover classes among cities, where dots represent individual statistical areas and arrows represent the dominant PCA loadings. The three maps illustrate the proportion of these dominant land cover classes across the contiguous US.

understanding of what lessons can be translated among cities versus which are largely relevant to the city a study occurred in. While these single-city studies have certainly advanced the field, they fundamentally do not allow us to address how ecological and human factors in diverse and varied cities shape ecological communities at the scale of a region, a country, a continent, or our planet. Fortunately, numerous multicity networks and projects that employ a variety of methods to identify patterns of urban biodiversity have emerged to close this gap.

Case study The diversity of multicity urban wildlife research networks

Multicity research networks span local, regional, continental, and global scales and sample a variety of taxa. This section is not intended as a census of all existing multicity research networks, but we hope that by outlining some illustrative examples we can give a sense of their current scope and scale.

Urban Wildlife Information Network (UWIN)

UWIN was founded in 2015 and is currently composed of 47 partnering cities, 44 of which are in the US and Canada (6). UWIN uses systematic sampling protocols across cities (1,3), and started as a camera trapping project for terrestrial mammals. Since its creation, UWIN has added protocols to sample birds, bats, small mammals, and more. UWIN also connects researchers with architects and planners to incorporate their research into city planning and management (8; https://www.urbanwildlifeinfo.org). As one example, UWIN members in Chicago worked with the city to create a new Wildlife Management and Coexistence Plan, based partially on data collected in their study.

Urban Biodiversity Research Coordination Network (UrBioNet)

UrBioNet was founded in 2015 and is a US National Science Foundation (NSF)-funded global network for urban biodiversity research and practice. UrBioNet provides a forum for discussion and data sharing on topics relevant to urban biodiversity research as well as the management, design, and planning for urban biodiversity. With over 400 members from 40 countries, UrBioNet engages a global audience of researchers, landscape architects, urban planners, and students.

Urban Long-Term Ecological Research Program (LTER)

The LTER program is an NSF-funded network of 28 research sites (https://lternet.edu) that was intended to be a catalyst for ecological research in general and for documenting generalizable patterns. Recognizing the importance of long-term data to capture stochastic ecological events, the program began in 1980 with an emphasis on transdisciplinary ecological research across extended timescales. Within the broader LTER program, three sites focus on the urban biome: Baltimore (BES; 1997–2020), Central Arizona—Phoenix (CAP; 1997), and Minneapolis-Saint Paul (MSP; 2021). The urban LTERs examine land use and land cover change, human–environment interactions, and often leverage social science and community engagement in their research.

The "Ecological Homogenization of Urban America" project (EHUA)

This NSF-funded project is a multi-scale, multidisciplinary research network that studies residential yards across six cities in the contiguous US (9). All three urban LTER cities are part of EHUA, as well as Miami, FL; Boston, MA; and Los Angeles, CA. Research topics in EHUA were initially focused on the ecological homogenization of soil and plants across residential yards but have expanded as this network continues to grow (10).

Global Urban Evolution Project (GLUE)

GLUE is a consortium of researchers examining how urbanization drives evolution (11). Following standardized protocols, researchers from over 160 cities are sampling and assaying the same genetic loci from white clover (*Trifolium repens*) populations to understand if urbanization leads to similar parallel evolutionary changes in the antipredator defense chemical produced by the plant and what factors influence this relationship (https://www.globalurbanevolution.com/).

Global perspectives

Single-city studies—while important and useful—only provide information about patterns of biodiversity within the city they are conducted, and moreover are biased to certain parts of the world (12). As a result, disagreements about patterns of urban biodiversity abound in the literature. For example, the distribution of Virginia opossum (*Didelphis virginiana*) throughout Amherst, MA is not strongly associated with open water sources (13) but opossum throughout Chicago, IL are (14). Urban bird species richness throughout various cities has been observed to either increase (15), decrease (16), or not change over time (17) depending on study location. Plant species richness throughout cities can positively or negatively covary with socioeconomics (18). If the goal of science is to seek generality, then such disagreements in the literature may seem troubling, as they collectively suggest that generality may not be possible. Yet, we contend that this is not the case.

In fact, the amount of potentially conflicting information among urban ecological studies is welcome. Why? Because, as in other fields of ecology, these differences indicate that hypotheses are likely supported or rejected depending on the system—in our case the city—where they are studied (19). The fact that context matters should not be surprising. Spatial or temporal variation in the relative fitness of species, and therefore patterns of biodiversity, is nearly ubiquitous in nature (19). As such, the extent to which ecological patterns generalize from one location to the next depends on how similar two locations are. When we recognize that cities are not carbon copies of one another it becomes clear that multicity research is imperative to add a regional, continental, or global perspective (depending on the scale of the cities involved) to urban ecology (20).

Urban ecologists, ourselves included, have recognized that future research should broaden to include cross-city comparisons across multiple scales, such as from local parks, to neighborhoods, to entire regions (6,20–25). In the last decade, significant progress has been made on this front, mostly through meta-analyses (analyses of other published studies (26–30)), though most are not the product of research networks (but see (31)). Meta-analyses are often used because they make it possible to synthesize past research to assess general patterns in the literature which could—in turn—help identify ways to conserve or increase biodiversity within cities. For example, through an analysis of 87 different publications across 75 cities worldwide, Beninde et al. (27) found that large (>50 ha) habitat patches with corridors have greater biodiversity, and therefore biodiversity-friendly management in cities should focus on increasing the area and connectivity of green space (Figure 8.2). After all, as cities typically contain only roughly 8% of nearby native regional birds and 25% of their regional plants (31), there are substantial opportunities to make cities more welcome to native flora and fauna through biodiversity-friendly management.

While meta-analyses do provide a way to synthesize past literature, they also come with caveats. For example, meta-analyses may have some bias because there is always uncertainty in how comparable published studies are given differences in methodology. Likewise, publication bias may mean some findings are overrepresented in the literature, which could distort results (32). These issues do not mean that these analyses are not valid, but additional tools are needed. Coordinated distributed experiments (33) and observational research networks (6,34) institute standardized methodologies across studies to address these shortcomings. Any one city is unlikely to be able to do all the research necessary to fully understand urban wildlife, and so it is essential to know what patterns can be applied to other cities and what information is likely to be context or city dependent. By using a common

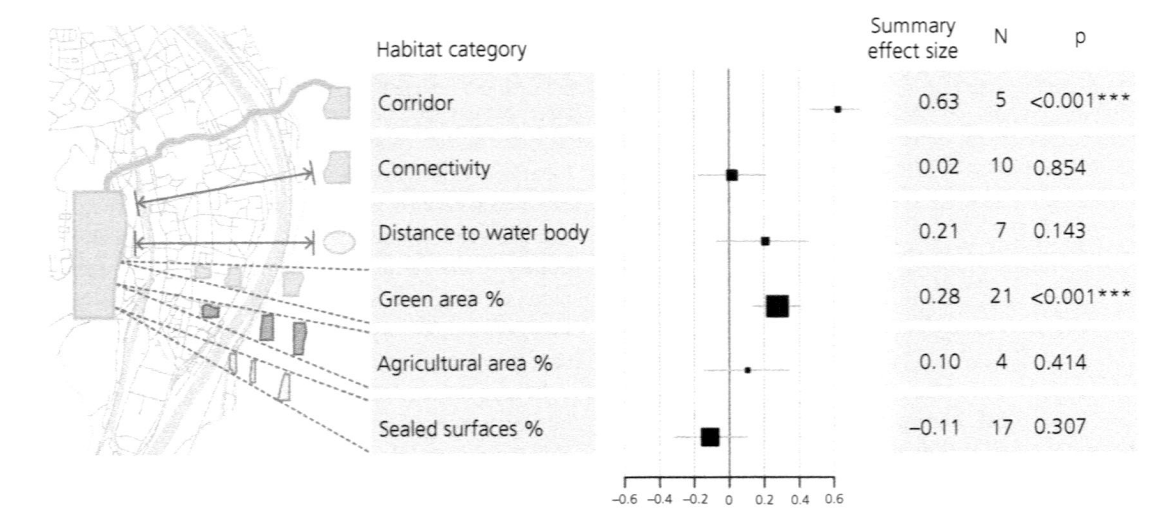

Figure 8.2 Illustrating the importance of area and connectivity for urban biodiversity via a meta-analysis. Reprinted with permissions from Beninde et al. (27).

study design, differences observed across a research network are far more likely to be ecological than methodological, which makes it possible to investigate how factors at various scales shape urban biodiversity (Figure 8.3).

While we have much to discover about regional and global urban ecological patterns, multicity research networks have started to provide the data necessary to understand how among-city variability relates to differences in urban biodiversity. A study from UWIN (see the Case study) showed that the distribution and relative scarcity of mammals across the US and Canada vary as a function of a city's average housing density and green space availability (1). As mammals can generate conflict with humans, findings like these can help target efforts toward outreach, coexistence, and management at scales previously impossible. Perhaps the largest example comes from the Global Urban Evolution Project (GLUE, see the Case study). By sampling over 6000 white clover populations (*Trifolium repens*) across urban gradients in 169 cities, Santangelo et al. (11) found that among-city variability in vegetation, impervious cover, and aridity was strongly associated with the production of hydrogen cyanide (HCN) in white clover, which is used as an anti-herbivore defense. These findings demonstrate not only that the magnitude of evolutionary or ecological phenomena can vary among cities, but also the direction of the relationship. Had these datasets been analyzed as single-city studies, generality would be hard to grasp given the dissimilar responses among cities. But through these collaborations, we see that such inconsistencies are the result of large-scale differences among cities. As such, multicity research networks can provide

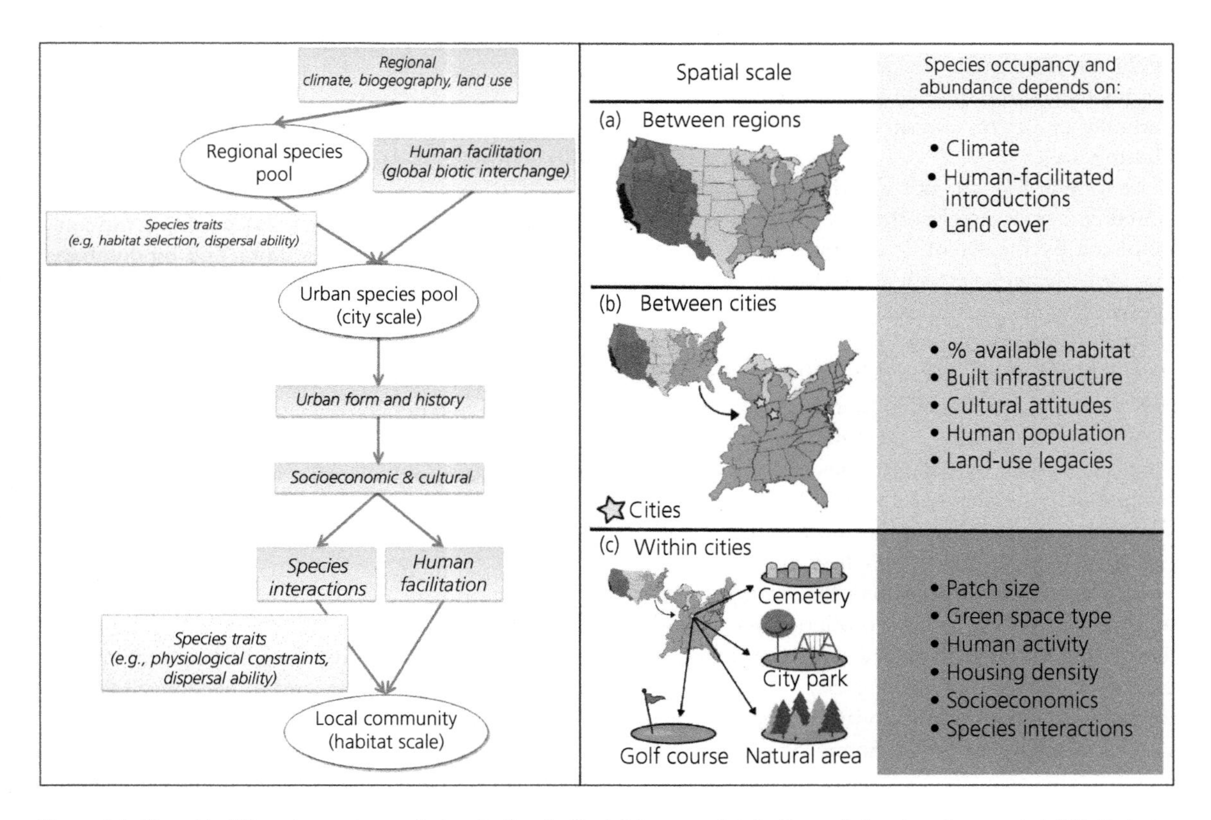

Figure 8.3 Hierarchical filters shape patterns of urban biodiversity. The left figure, reprinted with permissions from Aronson et al. (20), illustrates how community assembly of urban species pools is determined by a series of hierarchical filters. Green boxes represent filters hypothesized to be important determinants of species distributions at different scales. White circles represent species pools. Species life history and functional trait filters are represented in blue boxes. The right figure, reprinted with permissions from Magle et al. (6), provides examples of factors at different spatial scales that may determine patterns of species occupancy or abundance.

mechanistic understanding of global patterns of urban biodiversity and how to manage it (35).

Social-ecological applications

In urban areas, wildlife and people are inextricably linked, a phenomenon observed across disciplines such as human dimensions of wildlife, human–wildlife conflict mitigation, and environmental justice. Just as multicity networks are well-positioned to explore ecological questions, they have enormous power to explore linked social-ecological systems. Cities undoubtedly have as much variance in sociological factors as in ecological ones. As such, sampling an array of cities provides greater power and insight to understand how human communities influence other species. As an example, one study evaluated how mammalian wildlife were distributed with respect to wealth in a set of 20 North American cities (3). While species richness positively covaried with socioeconomic gradients in roughly half of the cities, it negatively covaried with gradients of urban intensity in nearly every city (Figure 8.4).

Urban wildlife are not distributed equally, or equitably, across neighborhoods, a fact that went unnoticed by the scientific community for far too long (36).[1] Systemic racism drives the structure and layout of cities, including where green space is situated. These decisions influence what habitats are colonized and inhabited by wildlife. As such, both positive and negative effects of living with wildlife are unequally distributed, with conflicts more often borne by communities of color and the rewards of living near nature, including health benefits and ecosystem services, more often coming to White neighborhoods. Schell et al. (36) outline several key research questions that integrate systemic racism, ecology, and evolution. These include "How does biodiversity vary with the degree of residential segregation within a city?" and "Is functional or structural connectivity reduced in cities with more pronounced economic or racial segregation?" The answers to these questions will be much more satisfying if they are asked across multiple cities. Quite often, wealth and racism are used synonymously in ecological studies (e.g., (3)) but they are not the same (36). With data from multiple cities, the effects of these separate (but potentially correlated) factors can be teased apart. Multicity networks also have the potential to evaluate the effects of systemic racism on wildlife across cities with varying levels and configurations of segregation and inequity.

Cities have enormous potential to support biodiversity, especially if properly managed (37). Decisions regarding urban planning and land use across multiple scales of governance, from residents up to municipal governments, can influence both the biodiversity present and the benefits to residents gained by exposure to such biodiversity (38–41). Multicity networks are ideal for studying such questions, identifying patterns, and informing management of urban areas to maximize biodiversity. For example, by examining the role of land management practices on bird species richness in six US metropolitan cities, Lerman et al. (10) identified key strategies for land management to support bird species diversity that were consistent across regions, with natural areas and residential yards playing an especially important role. Through UrBioNet (see the Case study), a meta-analysis examining the convergence of socioeconomic status and biodiversity and the role of human decision-making in 84 case studies across 34 cities identified a strong relationship between socioeconomic status and biodiversity in most cases. However, in cases where there was a negative relationship, social policy and human decision-making were able to mitigate this inequality, demonstrating the important role of governance and associated institutions in shaping urban biodiversity (40).[2]

Education and outreach

Urban-based biodiversity research has tremendous opportunity for outreach and education given the close proximity of city residents to their wild neighbors. Many urban residents may be unaware of the diversity of species that are present in their cities, or the wealth of nature that surrounds them, even in the most densely populated cities. Residents

[1] Hoover's and Scarlett's chapter discusses biodiversity and environmental inequities in cities.

[2] Larson's and Brown's chapter outlines the role of human motivations and governance in urban wildlife conservation.

may also be surprised to learn of the enormous benefits that connecting to nature can provide (42), including mental and physical health, recreation, and relaxation. Public outreach and education is a key component to reducing negative interactions between humans and wildlife (43). As with other advantages of multicity research networks, the impact of public outreach and education is amplified because programs can be deployed across multiple cities and reach more and more people. Between shared lesson plans, knowledge of existing programs, and repositories of materials, researchers have access to a wealth of resources that have been developed and deployed by others successfully. For researchers or managers who have little experience with public outreach and little time to develop a program from scratch, plugging in to successful programs saves a lot of time and effort, and has a much stronger potential for success. For example, UWIN has an education committee that shares resources among partners in the network. As part of this committee, UWIN partners in St. Louis replicated a program originally developed in Chicago by Lincoln Park Zoo called Partners in Fieldwork, an award-winning and free year-long program implemented in local grade-schools. The program provides the opportunity for middle and high school students to become "student field researchers." Students collect data from a field station they set up on their school campus. Teachers are provided with ongoing professional training to implement the curriculum in their classroom and support students in designing their research projects.

Fundraising opportunities

Another advantage of a multicity network is that they can open doors to more opportunities for funding, which can enhance the capacity of organizations to engage in conservation actions. Although

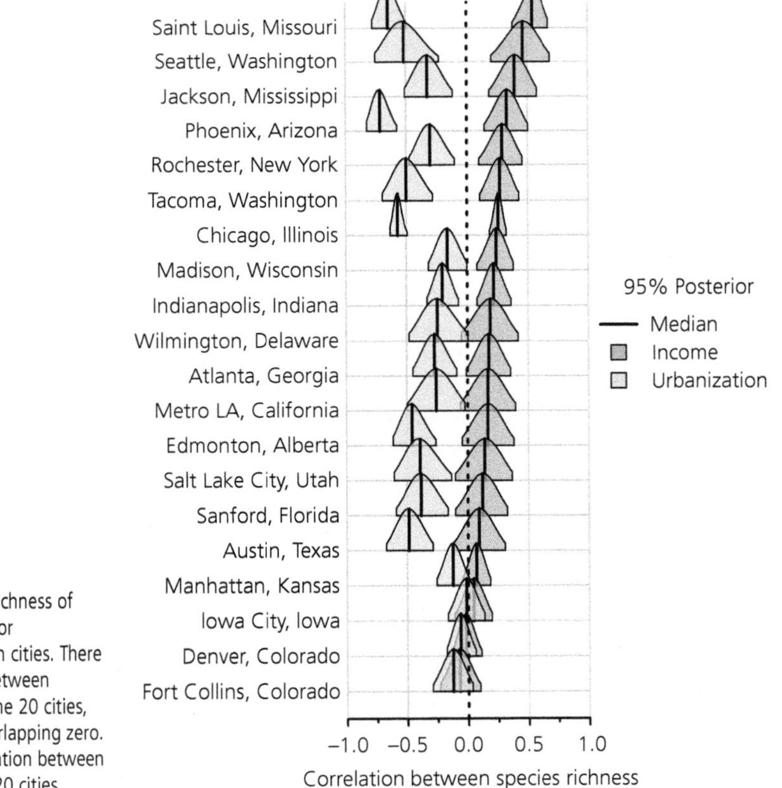

Figure 8.4 The correlation between species richness of medium to large mammals and a city's income or urbanization gradient across 20 North American cities. There was strong evidence of a positive correlation between income gradients and species richness in 9 of the 20 cities, as evidenced by 95% credible intervals not overlapping zero. There was strong evidence of a negative correlation between species richness and urbanization in 16 of the 20 cities. Modified from Magle et al. (3).

some funding agencies may be focused on local research, many prioritize multidisciplinary, collaborative research with large impacts (e.g., NSF; (8)). As we have described earlier, the impact of research conducted by a multicity network is immense and amplified by its numerous members and vast geographic scope. Even if a funding agency is limited to supporting research that is conducted in a specific geographic area, that locally conducted research can contribute to a much broader dataset as a member of a multicity network. Furthermore, as part of a network, researchers can learn about foundations or grants that they may not be aware of, and apply as a group rather than independently, while emphasizing the multidisciplinary, collaborative, and broad-scale impact of the research.

Increased temporal scale

Multicity research networks can also increase temporal scale, if data are collected multiple times per year—for example, during separate seasons—or across years. Speaking from experience, starting and maintaining multicity research networks is a tremendous amount of work, but adding a temporal component to such research has many advantages. Ongoing sampling means new interested partners have the opportunity to collaborate. Furthermore, increasing temporal scale makes it possible to study local or global change over time if data are collected before, during, and after relevant events. For example, the Covid-19 global pandemic led to dramatic changes in human activity patterns, which urban wildlife no doubt responded to (44,45). At a more local scale, urban development patterns no doubt change over time, and long-term datasets are an ideal way to determine how biodiversity may respond to changing patterns of urban development.

Increased taxonomic breadth

Urban wildlife research is highly biased by taxa, with most work being conducted on birds and mammals (7,12). Arthropods, insects, fish, reptiles, amphibians, and other taxa remain relatively unexplored. This is a tremendous missed opportunity for urban ecology. Multicity networks have the potential to expand the number of species sampled, most obviously for the simple reason that sampling across wider regions expands the pool of available species based on their geographic ranges. Access to broader species pools increases our ability to model interactions and energy flows between individuals, populations, and differing trophic levels.

We have also observed a less obvious benefit to multicity networks with respect to diversity of sampling. Researchers who only have experience in sampling one taxon (e.g., mammals or birds) are often reluctant to launch research into a new one, largely because the methodologies are different, and learning new techniques takes a great deal of time and effort. Multicity networks bring together teams of researchers with different experiences, who can share expertise and draft flexible, straightforward data collection protocols, thus making the process of designing new studies less daunting. UWIN (see the Case study) began as a consortium dedicated solely to monitoring midsize terrestrial mammals using camera traps, but now is developing protocols for birds, bats, ticks, reptiles, and small mammals. Likely none of the members would have launched a large-scale study on these taxa on their own, but the resources of the network enabled them to reach beyond their normal comfort zone and capture data on understudied species.

Providing community

Arguably the greatest benefit of a coordinated research network is the extensive professional community that they can provide. Inherent in that community is a diversity of viewpoints, knowledge, skills, and experience of researchers from around the world with a variety of backgrounds that can act as a resource for its members. For example, a challenge faced by practitioners and researchers in one city may have been faced and solved by those in another, one network member may have developed software that would be useful for answering a question posed by another city, or another member may have a compelling idea for a dimension to add to a collaborative paper which others had not considered. Combining efforts across many cities nationally and even globally enriches and elevates the quality of the research by bringing together people from diverse backgrounds with a variety of skills and ideas.

Current challenges for multicity research

While the advantages of multicity research are extensive, there are reasons why single-city studies are more common—it is difficult to manage large research alliances. Coordinating researchers and their associated projects is challenging. Partners can have different goals and objectives, as well as their own unique resources and limitations. Academics, for example, are most often motivated by publications and grants, whereas researchers with nonprofit institutions may be more invested in outreach or educational gains. Managing a network means managing all of the associated personalities and motivations of collaborators, a task that is as much social as intellectual.

Some of these coordination tasks are straightforward, if not actually simple. When multicity research projects are proposed within the network, someone needs to lead them. Assigning research projects to individual principal investigators (PIs) can be challenging if multiple people are interested in similar questions. These issues, and those of determining authorship on collaborative projects, can often be best handled by a representative body composed of members from across the network. For example, UWIN handles these issues with a research committee with one vote from each participating city.

Research prioritization at a network scale is its own challenge, of course. Some networks have central leadership, a PI in charge of the project as a whole. These roles are challenging, requiring not only specific topic knowledge of the methods for the urban wildlife study in question, but also the ability to step back from the day-to-day field collection and analysis to focus on administration and logistics—not why most scientists went to graduate school. Others use a more democratic approach, with an advisory board or group of coequal PIs. This can be effective in sharing the work and ensuring effective outcomes, but it can also add time to the process of decision making. Whatever the approach to project management, it is critical to be upfront about the structure of the organization so that all parties are aware of their rights and responsibilities as members.

If a feature of the network is shared research protocols, a careful approach to the design of those protocols is absolutely essential. It is bad enough to design a local study poorly and find one cannot answer the intended research questions, but catastrophic to make the same mistake across multiple studies at once. Consultation with field ecologists and statisticians is critical to ensure the study design can address a given research question and can reasonably be executed. There is, however, also a risk of making the design too rigid. Each city is designed differently, and a spatial design that might work in an expansive, open metropolitan area, for example, Houston, TX, may not fit whatsoever for an island like Manhattan. Finding the proper compromise between scientific rigor and flexibility will be a task for a team that includes statisticians as well as people with deep experience in on-the-ground data collection.

As networks grow, logistical issues of scale follow. A network with three or four collaborators could be managed informally, with e-mails and spreadsheets perhaps the only needed tools. When that same network reaches 40 collaborators, disaster will ensue if new tools are not adopted. For example, with a small number of collaborators, distributed databases that individual partners manage, such as duplicate copies of the same local database, may be the best and cheapest option. As the number of partners increases, however, maintaining data integrity becomes difficult due to small differences in how data are entered among partners, such as the name or spelling of a species. These differences can snowball into a significant amount of data cleaning when the time comes to compile data across a network.

From the logistical side of coordinating networks, the larger and more complex the project, the more effort will be required to keep the partners moving in the same direction. If possible, a full-time coordinator is an invaluable asset, and perhaps essential for large networks. Funding these types of positions can be a challenge as they may not seem as flashy or productive to a granting agency, but large-scale research is all but impossible without them. The same types of productivity tools that work for non-research applications—cloud-based spreadsheets, calendar coordination, social productivity tools such as Slack—are also beneficial, though researchers are often hesitant to adopt them.

Future needs for multicity research

While tremendous progress has been made, multicity urban wildlife research networks in any form are still in their infancy, and huge gaps remain, both in our knowledge, and in projects designed to acquire that knowledge. Our review of existing broad-scale research has revealed several promising areas for future work.

Many existing networks (e.g., Urban LTER) are restricted to a certain region or country, and even those that have a global scope (e.g., UrBioNet, UWIN) have patchy distributions of partners that leave significant gaps. In particular, existing urban wildlife research is heavily focused in North America, Europe, and Oceania, with poor representation in South America, Africa, and Asia (7,12). Given the huge variation in cities, a universal understanding of urban wildlife research will be impossible without networks that sample cities all around the world. In addition to spatial gaps, there are other limitations in existing research networks. Mammals and birds are fairly well studied, with arthropods, fish, reptiles, amphibians, and other taxa still mostly unknown (7,12).

It is a tired refrain, but true: to address these limitations and expand on the power of existing urban wildlife networks, additional funding will be needed. These large-scale projects are expensive, and while funding sources exist to initiate this type of work through the NSF's grants (LTER, infrastructure) and other sources, sustaining and growing them is another matter. Data managers, research software engineers, coordinators, graduate students, and postdoctoral researchers are all needed to make these multicity projects a reality. Funding is desperately needed to take these fledgling networks to the next level.

We have described several existing urban wildlife research networks, and others that include urban wildlife as part of a larger portfolio. Each has its own focus, strengths, and limitations. Most are focused on collecting ecological data, and an added focus on collecting and interfacing with social, economic, and cultural data is likely to be useful given the interdisciplinary nature of cities themselves. An interesting next step could be for urban data networks, whether focused on wildlife ecology or not, to start communicating and sharing data with one another, in essence creating metanetworks.

Within these proposed metanetworks, the process of conceptualizing and formatting data such that they can interface in a useful way will be a challenge, and in the case of transdisciplinary metanetworks, even differing language and terminology will need to be navigated. Although difficult, this process will be deeply rewarding. Real progress toward understanding, managing, and conserving wildlife in cities will require deep collaboration between ecologists, urban planners, architects, and urban residents at large (8). Multicity networks represent a critical first step toward this end goal, and a tantalizing glimpse at what we can achieve as the scope of our efforts keeps expanding.

Conclusion

Urban wildlife research has become more prominent over the past few decades (12) and is now beginning to fill in knowledge gaps that have existed since the birth of modern wildlife ecology. The field has grown to encompass new spaces (e.g., suburban, exurban), new topics (e.g., justice, human dimensions), and new taxa beyond birds and mammals. All of these advances are important and useful, but to move the discipline beyond examination of local patterns, and toward an analysis of urban wildlife as global phenomena, multicity networks are essential.

If we are to conserve biodiversity and manage wildlife on an urban planet, we cannot do it alone. Not only must we work together within our discipline, but researchers must also work with urban planners, landscape architects, sociologists, economists, community organizers, residents, and everyone else who is a part of designing, creating, and maintaining our urban environments. These conversations will not get far if our ecological understanding is restricted to each researcher's own field sites. The only way to move beyond a patchwork of local studies and toward these global principles is multicity networks.

We have outlined many of the advantages of these networks, given examples of several growing networks, and described some of the difficulties of creating and maintaining them. We urge in the strongest possible terms that if you are starting up an urban wildlife study, or even if you have one ongoing, take a little time to research available networks that might be relevant to your project. Joining them may not require much additional work from you and could open up your research questions to all-new scales and in entirely new directions.

References

1. Fidino M, Gallo T, Lehrer EW, Murray MH, Kay CA, Sander HA, et al. Landscape-scale differences among cities alter common species' responses to urbanization. Ecological Applications. 2021;31(2):e02253.
2. Magle SB, Lehrer EW, Fidino M. Urban mesopredator distribution: examining the relative effects of landscape and socioeconomic factors. Animal Conservation. 2016;19(2):163–75.
3. Magle SB, Fidino M, Sander HA, Rohnke AT, Larson KL, Gallo T, et al. Wealth and urbanization shape medium and large terrestrial mammal communities. Global Change Biology. 2021;27(21):5446–59.
4. Hassell JM, Ward MJ, Muloi D, Bettridge JM, Phan H, Robinson TP, et al. Deterministic processes structure bacterial genetic communities across an urban landscape. Nature Communications. 2019;10(1):2643.
5. Beninde J, Feldmeier S, Werner M, Peroverde D, Schulte U, Hochkirch A, et al. Cityscape genetics: structural vs. functional connectivity of an urban lizard population. Molecular Ecology. 2016;25(20):4984–5000.
6. Magle SB, Fidino M, Lehrer EW, Gallo T, Mulligan MP, Ríos MJ, et al. Advancing urban wildlife research through a multi-city collaboration. Frontiers in Ecology and the Environment. 2019;17(4):232–9.
7. Magle SB, Hunt VM, Vernon M, Crooks KR. Urban wildlife research: past, present, and future. Biological Conservation. 2012;155(Oct):23–32.
8. Kay CAM, Rohnke AT, Sander HA, Stankowich T, Fidino M, Murray MH, et al. Barriers to building wildlife-inclusive cities: insights from the deliberations of urban ecologists, urban planners and landscape designers. People and Nature. 2021;4(1):62–70. Available from: https://doi.org/10.1002/pan3.10283
9. Groffman PM, Cavender-Bares J, Bettez ND, Grove JM, Hall SJ, Heffernan JB, et al. Ecological homogenization of urban USA. Frontiers in Ecology and the Environment. 2014;12(1):74–81.
10. Lerman SB, Narango DL, Avolio ML, Bratt AR, Engebretson JM, Groffman PM, et al. Residential yard management and landscape cover affect urban bird community diversity across the continental USA. Ecological Applications. 2021;31(8):e02455.
11. Santangelo JS, Ness RW, Cohan B, Fitzpatrick CR, Innes SG, Koch S, et al. Global urban environmental change drives adaptation in white clover. Science. 2022;375(6586):1275–81.
12. Collins MK, Magle SB, Gallo T. Global trends in urban wildlife ecology and conservation. Biological Conservation. 2021;261(Sep):109236.
13. Kanda LL, Fuller TK, Sievert PR. Landscape associations of road-killed Virginia opossums (Didelphis virginiana) in central Massachusetts. The American Midland Naturalist. 2006;156(1):128–34.
14. Fidino MA, Lehrer EW, Magle SB. Habitat dynamics of the Virginia opossum in a highly urban landscape. The American Midland Naturalist. 2016;175(2):155–67. Available from: https://doi.org/10.1674/0003-0031-175.2.155
15. Fidino M, Limbrick K, Bender J, Gallo T, Magle SB. Strolling through a century: replicating historical bird surveys to explore 100 years of change in an urban bird community. The American Naturalist. 2022;199(1):159–67.
16. Strohbach MW, Hrycyna A, Warren PS. 150 years of changes in bird life in Cambridge, Massachusetts from 1860 to 2012. The Wilson Journal of Ornithology. 2014;126(2):192–206.
17. Shultz AJ, Tingley MW, Bowie RC. A century of avian community turnover in an urban green space in northern California. Condor. 2012;114(2):258–67.
18. Leong M, Dunn RR, Trautwein MD. Biodiversity and socioeconomics in the city: a review of the luxury effect. Biology Letters. 2018;14(5):20180082. Available from: https://doi.org/10.1098/rsbl.2018.0082
19. Vellend M. The Theory of Ecological Communities (MPB-57). Princeton, NJ: Princeton University Press; 2016.
20. Aronson MF, Nilon CH, Lepczyk CA, Parker TS, Warren PS, Cilliers SS, et al. Hierarchical filters determine community assembly of urban species pools. Ecology. 2016;97(11):2952–63.
21. Childers DL, Cadenasso ML, Grove JM, Marshall V, McGrath B, Pickett ST. An ecology for cities: a transformational nexus of design and ecology to advance climate change resilience and urban sustainability. Sustainability. 2015;7(4):3774–91.

22. Johnson MT, Munshi-South J. Evolution of life in urban environments. Science. 2017;358(6363):eaam8327.
23. McDonnell MJ, Hahs AK. The future of urban biodiversity research: moving beyond the "low-hanging fruit". Urban Ecosystems. 2013;16(3):397–409.
24. McPhearson T, Pickett ST, Grimm NB, Niemelä J, Alberti M, Elmqvist T, et al. Advancing urban ecology toward a science of cities. BioScience. 2016;66(3): 198–212.
25. Niemelä J. Ecology of urban green spaces: the way forward in answering major research questions. Landscape and Urban Planning. 2014;125(May):298–303.
26. Batáry P, Kurucz K, Suarez-Rubio M, Chamberlain DE. Non-linearities in bird responses across urbanization gradients: a meta-analysis. Global Change Biology. 2018;24(3):1046–54.
27. Beninde J, Veith M, Hochkirch A. Biodiversity in cities needs space: a meta-analysis of factors determining intra-urban biodiversity variation. Ecology Letters. 2015;18(6):581–92.
28. Gámez S, Potts A, Mills KL, Allen AA, Holman A, Randon PM, et al. Downtown diet: a global meta-analysis of increased urbanization on the diets of vertebrate predators. Proceedings of the Royal Society B. 2022;289(1970):20212487.
29. Jung K, Threlfall CG. Trait-dependent tolerance of bats to urbanization: a global meta-analysis. Proceedings of the Royal Society B. 2018;285(1885):20181222.
30. Murray MH, Sánchez CA, Becker DJ, Byers KA, Worsley-Tonks KE, Craft ME. City sicker? A meta-analysis of wildlife health and urbanization. Frontiers in Ecology and the Environment. 2019;17(10): 575–83.
31. Aronson MF, La Sorte FA, Nilon CH, Katti M, Goddard MA, Lepczyk CA, et al. A global analysis of the impacts of urbanization on bird and plant diversity reveals key anthropogenic drivers. Proceedings of the Royal Society B: Biological Sciences. 2014;281(1780):20133330.
32. Nakagawa S, Lagisz M, Jennions MD, Koricheva J, Noble DW, Parker TH, et al. Methods for testing publication bias in ecological and evolutionary meta-analyses. Methods in Ecology and Evolution. 2022;13(1):4–21.
33. Fraser LH, Henry HA, Carlyle CN, White SR, Beierkuhnlein C, Cahill Jr JF, et al. Coordinated distributed experiments: an emerging tool for testing global hypotheses in ecology and environmental science. Frontiers in Ecology and the Environment. 2013;11(3):147–55.
34. Cove MV, Kays R, Bontrager H, Bresnan C, Lasky M, Frerichs T, et al. SNAPSHOT USA 2019: a coordinated national camera trap survey of the United States. Ecology. 2021;102(6):e03353.
35. Knapp S, Aronson MF, Carpenter E, Herrera-Montes A, Jung K, Kotze DJ, et al. A research agenda for urban biodiversity in the global extinction crisis. BioScience. 2021;71(3):268–79.
36. Schell CJ, Dyson K, Fuentes TL, Des Roches S, Harris NC, Miller DS, et al. The ecological and evolutionary consequences of systemic racism in urban environments. Science. 2020;369(6510):eaay4497.
37. Lehrer EW, Gallo T, Fidino M, Kilgour RJ, Wolff PJ, Magle SB. Urban bat occupancy is highly influenced by noise and the location of water: considerations for nature-based urban planning. Landscape and Urban Planning. 2021;210(Jun):104063.
38. Belaire JA, Whelan CJ, Minor ES. Having our yards and sharing them too: the collective effects of yards on native bird species in an urban landscape. Ecological Applications. 2014;24(8):2132–43.
39. Kinzig A, Warren P, Martin C, Hope D, Katti M. The effects of human socioeconomic status and cultural characteristics on urban patterns of biodiversity. Ecology and Society. 2004;10(1):10.5751/ES-01264-100123.
40. Kuras ER, Warren PS, Zinda JA, Aronson MFJ, Cilliers S, Goddard MA, et al. Urban socioeconomic inequality and biodiversity often converge, but not always: a global meta-analysis. Landscape and Urban Planning. 2020;198(Jun):103799.
41. Pickett STA, Cadenasso ML, Rosi-Marshall EJ, Belt KT, Groffman PM, Grove JM, et al. Dynamic heterogeneity: a framework to promote ecological integration and hypothesis generation in urban systems. Urban Ecosystems. 2017;20(1):1–14.
42. Sandifer PA, Sutton-Grier AE, Ward BP. Exploring connections among nature, biodiversity, ecosystem services, and human health and well-being: opportunities to enhance health and biodiversity conservation. Ecosystem Services. 2015;12(Apr):1–15.
43. Espinosa S, Jacobson SK. Human-wildlife conflict and environmental education: evaluating a community program to protect the Andean bear in Ecuador. The Journal of Environmental Education. 2012;43(1):55–65. Available from: https://doi.org/10.1080/00958964.2011.579642
44. Kays R, Cove MV, Diaz J, Todd K, Bresnan C, Snider M, et al. SNAPSHOT USA 2020: A second coordinated national camera trap survey of the United States during the COVID-19 pandemic. Ecology. 2022;103(10):e3775.
45. Zellmer AJ, Wood EM, Surasinghe T, Putman BJ, Pauly GB, Magle SB, et al. What can we learn from wildlife sightings during the COVID-19 global shutdown? Ecosphere. 2020;11(8):e03215.

CHAPTER 9

Integrating Molecular Methods with a Social-Ecological Focus to Advance Urban Biodiversity Management

Kevin Avilés-Rodríguez[‡], Kim Hughes[‡], Jonathan L. Richardson[§], and Jason Munshi-South[‡]

Introduction

Approaches from molecular genetics and related methods have emerged as highly effective tools for biodiversity management (1–4). One advantage of molecular approaches is the use of DNA sequencing to study rare or elusive species (5). For example, the use of urban areas by newts in Switzerland was confirmed with techniques of environmental DNA (eDNA), which amplifies and analyzes DNA fragments shed by species into the environment (6). Similarly, DNA sequence analysis from scat samples has facilitated finding elusive animals that are difficult to observe or capture (7). Another advantage is evaluating the diet of urban adaptors (8). Further, molecular approaches using thousands of genome-wide markers have dramatically increased the resolution at which we can characterize patterns of genetic diversity (9,10).

The advantages of molecular approaches provide an important tool for managing urban biodiversity. Urban areas and populations are rapidly expanding. This increasing human density and altered urban habitat require explicit consideration of how urban environments and their related social-ecological factors impact biodiversity management (11–13). One advantage of molecular tools in cities is that they can accurately characterize patterns of genetic diversity even with patchy or low-density sampling (1,2), which often occurs in cities where not all property parcels are available for sampling. Molecular methods also have challenges, including the high costs of equipment and sequencing, as well as necessary expertise in working with nucleic acids and associated sequence data. Yet, here, we will outline why these genetic-based tools are particularly valuable for understanding biodiversity in cities. Additionally, we will discuss how molecular approaches can be applied to understand the intersections of human social conditions and nature, which is essential for urban conservation.

[‡] Louis Calder Biological Field Station, Fordham University, USA
[§] University of Richmond, USA

Molecular tools for characterizing and managing urban species

One of the management goals in any environment is evaluating patterns of biodiversity. Urbanization can reduce biodiversity primarily due to habitat fragmentation, high densities of inhospitable landscapes (e.g., highways, sidewalks, mown lawns with insecticides), and increased mortality rates (14–17). Quantifying diversity patterns across cities remains challenging due to the high level of landscape and environmental heterogeneity (18,19) and difficulties detecting species across cities.

Kevin Avilés-Rodríguez et al., *Integrating Molecular Methods with a Social-Ecological Focus to Advance Urban Biodiversity Management.*
In: *Urban Biodiversity and Equity.* Edited by: Max R. Lambert & Christopher J. Schell, Oxford University Press.
DOI: 10.1093/oso/9780198877271.003.0009

Conservation genetics to define urban management priorities

One challenge in managing urban biodiversity is defining conservation priorities and the management actions required (4). DNA-sequencing approaches allow managers to assess population viability inferred from the genetic variation and genetic connectivity throughout the city. Characterizing patterns of genetic diversity and gene flow helps practitioners define when intervention or management is appropriate (4).

Genetic diversity and population structure

Managing genetic diversity is essential to maintaining species and ecosystem viability and preserving species' capacity to adapt to rapidly changing environments (20). Genetic diversity is a measure of the genetic variability within a population, where low genetic diversity decreases fitness due to increased relatedness of individuals. High levels of fragmentation in cities may limit individuals' ability to disperse to and reproduce within the various habitat patches within range, which may reduce the genetic diversity of urban species (21). Recent reviews have documented that urbanization resulted in inconsistent increases, decreases, or no changes in the genetic diversity of urban carnivores and other wildlife populations depending on the city or species. The complex relationship between urban environmental heterogeneity and genetic diversity illustrates that the impact of urbanization on genetic diversity is inconsistent and highly dependent on a given city's conditions and the biology of the focal species (20,22). If genetic diversity is low within cities, managers may seek to increase connectivity between patchy populations using habitat corridors or human-mediated translocations to increase genetic diversity. Evaluating gene flow between urban and nonurban populations can define whether managers need to apply the intervention to preserve dispersal between these populations or if the best course of action is to allow these populations to diverge genetically. Additionally, socioeconomic stratification of resources can shape patterns of genetic diversity in cities (see Social-ecological applications below).

Characterizing relationships between urban environments and genetic diversity can be challenging when there is a lack of clarity regarding the ancestral or pre-urban genetic variation (19). Understanding the dynamics of population genetic structure in a species through time can inform management goals and reveal which ecological trends can lead to evolutionarily significant declines in diversity. Sequencing ancient DNA (aDNA) from fossils or museum specimens can provide insights into ancestral patterns of genetic diversity. For instance, aDNA sequencing provided important clarity regarding the current and ancestral genetic diversity of giant pandas (23) (*Ailuropoda melanoleuca*). While extant giant pandas have relatively high genetic diversity, sequencing ancient mitochondrial DNA from Holocene sub-fossil specimens revealed that ancestral genetic diversity was even higher (23). This finding opened up new management questions, such as trying to reconstruct the timescale of genetic diversity loss and the factors that contributed to this. Presently, little work has applied aDNA sequencing to frame contemporary adaptations and genetic diversity in urban environments within a species historical context. Thus, when appropriate, managers should consider the historical context when formulating conservation priorities within urban ecosystems.

A conservation challenge in any environment is defining the conservation or management units (24). Conservation units can be inferred by demonstrating evidence that landscapes and habitats limit dispersal or completely isolate populations. For instance, managers might define a conservation unit for species' urban and nonurban ranges. However, given the landscape heterogeneity in cities, there could be significant sub-city genetic variation which might result in multiple conservation units *within* a city. Additionally, urban populations could face similar selective pressures, urging managers to consolidate populations spread across multiple cities as a single conservation unit. Molecular approaches can define conservation units that consider the genetic diversity of units and the evolutionary mechanisms associated with urbanization. The advantage of defining evolutionarily significant units for conservation stems from identifying the species' potential for adaptation and managing the influx of genetic diversity to maintain this potential (25,26). For instance, if urban populations face strong natural selection, then translocating nonurban individuals might decrease fitness due to outbreeding depression (i.e., reduced fitness due to an influx of nonadaptive genes into a population).

Additionally, it is important to sample populations at large spatial scales to understand the mechanism contributing to the observed genetic diversity. For example, urban populations of black widow spiders were shown to have higher relatedness to the closest urban population rather than to adjacent nonurban populations (27). Similarly, urban pigeons in the northeastern US share high levels of genetic relatedness across multiple cities, with only the northernmost cities showing some genetic differentiation (28). Evolutionary dynamics can vary as a function of the environmental and temporal contexts of cities which interact with a species' population size and generation time. For example, following the installation of canal locks in 1827 at the Rideau Canal in Canada, painted turtles (*Chrysemys picta*) showed very little genetic differentiation (29). Authors referenced the large population size of painted turtles within the canal (over 10,000 individuals) and the long generation time (30–40 years) as explaining a potential lag between installation of dispersal barriers and a detectable signal of genetic isolation (29). In contrast, the recent colonization of the yellow-necked mouse (*Apodemus flavicollis*) into Warsaw, Poland, resulted in weak genetic differentiation; however, the strongest barrier to gene flow was the Vistula River, which crosses through the center of the city, rather than built urban features (30). Identifying conservation units within and between cities will require evaluating the genetic diversity and carefully considering spatial, temporal, and ecological factors.

Spatial evaluations of gene flow

Urban landscapes can influence the movement of individuals and their alleles (genetic variants), resulting in correlations between gene flow—the movement of alleles across populations and landscapes—and features of urbanization (31). Characterizing patterns of gene flow and genetic diversity can help clarify the mechanisms shaping the population genetics of species and characterize suitable and unsuitable urban habitat patches. For example, following urban colonization, a species can exhibit lower genetic diversity arising from a bottleneck or founder's effect, where only a small fraction of the genetic diversity in nonurban habitats is present (32). In contrast, genetic diversity can be greater in cities when dispersal is facilitated by human transportation or high natural mobility. For instance, the genetic connectivity of a species of land snail (*Cornu aspersum*) increases with transportation infrastructure due to human-facilitated dispersal (33). Some urban or semi-urban features may also function as patches that allow species persistence even in the most developed areas. For example, urban development (e.g., roads) reduced the genetic diversity of stream salamanders (*Eurycea bislineata*) however constructed water features in parks and golf courses functioned as corridors for this species (34).

Carefully managing gene flow is important as excessive gene flow can result in outbreeding depression, which can decrease fitness if gene flow dilutes adaptive genes (35). In contrast, a lack of gene flow can result in isolated populations subject to inbreeding depression, which decreases biological fitness by promulgating harmful mutations. Some urban taxa experience strong selection (36,37), so increases in gene flow might decrease fitness if populations are moved away from their adaptive peak. Understanding how human structures impact genetic diversity can inform management decisions such as constructing green spaces and corridors or facilitating translocation to combat genetic drift (changes in allele frequencies due to random chance) in isolated habitat patches (4).

Moreover, spatial modeling advances have facilitated inferring correlations between geographic distance, spatial and socioeconomic landscape heterogeneity (i.e., roads, buildings, median income, etc.), and genetic differentiation (38,39). Assessing patterns of gene flow spatially allows managers to test explicit hypotheses of which landscape features may be correlated with high and low genetic connectivity. For example, high mortality rates on roads and behavioral avoidance of roads by indigo snakes (40) (*Drymarchon couperi*) led to the suggestion that roads served as strong barriers that limited gene flow. However, landscape genomics approaches did not confirm a strong negative effect on population connectivity associated with roads and highways (41). Instead, the model identified the importance of undeveloped upland habitats as a strong dispersal corridor and open water as a strong barrier (42). A challenge of this approach is defining how humans and urbanization influence population evolutionary processes (12,14,42). For example, "development" can be subdivided into various relevant categories in urban environments (industrial, municipal, commercial, residential, etc.) and further subdivided by building age (12). Additionally, the ability to detect these interactions can vary as a function of the temporal and spatial scale analyzed (33,43). Incorporating socioeconomic factors appears to improve these models (discussed in Social-ecological applications below).

eDNA of cryptic species

Recently, eDNA-based methods have been proposed as a cheaper, easier, and faster alternative to survey biodiversity (5,44,45). In general, eDNA biomonitoring is achieved through amplifying DNA shed by organisms and found within water, soil, etc. Studies using eDNA can target single species or detect a wide variety of taxa using generic polymerase chain reaction (PCR) primers (short DNA sequences used for amplifying target DNA). eDNA is also used for the simple detection of species and quantifying the abundance of organisms (via amplification curves calibrated on biomass; 5). Moreover, eDNA approaches can be used to detect biodiversity occupancy in habitats that are hard to sample or to detect elusive species expanding their niches into cities. For instance, researchers in Australia used eDNA analyses within urban and nonurban tree hollows and found that tree hollows were used by both a cryptic marsupial and an invasive bird (46). Similarly, a species' diet can be inferred based on the sequences of DNA found in their stomach contents or feces (8,46).

A limitation of eDNA approaches is ensuring that species occupancy models constructed from eDNA data perform similarly to models based on traditional field count/observation approaches (44). Recently, invertebrate-derived DNA (iDNA) approaches, an offshoot of eDNA based on detecting DNA sequences ingested by invertebrates (e.g., blood meal in leeches), generated occupancy probabilities comparable to a model derived using camera trap observations (48).

Another limitation of the application of eDNA approaches in cities is whether the high density of impervious surfaces will limit the ability to capture and preserve the DNA material of urban organisms. Airborne eDNA analyses may be able to detect urban organisms even in patches dominated by impervious surfaces. For example, airborne eDNA approaches deployed at two zoos could detect many, but not all, zoo residents and other non-zoo species (49,50). Similarly, a field contrast between traditional insect sampling and airborne eDNA showed promising results for invertebrates and also detected vertebrate species within the study area (51). Thus, eDNA approaches can help urban conservation managers monitor biodiversity. Still, the resolution of this information will improve when using complementary eDNA survey strategies (sample eDNA in soil, air, water, blood meal of invertebrates) and/or when combined with traditional field approaches.

The application of eDNA approaches for urban conservation is still rare. However, two advantages of this approach could make this desirable for urban managers. Firstly, eDNA sequencing allows managers to avoid invasive sampling, such as tissue collection or blood draws. These can be particularly challenging when managing large animals or large mammals, which may be difficult and expensive to restrain and sedate for sample collection, particularly near the public. Moreover, recent advances in eDNA sequencing have facilitated the amplification of sufficient DNA markers to infer population genetics (52). By using a methylation-based DNA enrichment technique (53), researchers generated enough genome-wide markers from jungle cat (*Felis chaus*) scat to distinguish landscape genetic effects in a population where previous studies failed to find such patterns (52). Lastly, eDNA approaches can be applied to empirically test correlations between neighborhood social characteristics (e.g., wealth) and increased biodiversity. We anticipate that eDNA will become a more common tool in monitoring and managing urban biodiversity, particularly where the focus is to manage larger animals that readily produce scat samples and/or for monitoring urban aquatic biodiversity.

Managing invasive species

A pressing conservation issue in cities is detecting and managing non-native species. Urban ecosystems are vulnerable to species introduction due to the high levels of human traffic and trade (54,55). Understanding invasion pathways facilitates the rapid detection and eradication of new introductions (56). Introduction pathways in cities can vary due to climate, city location (i.e., coastal versus inland), and other factors, complicating the early detection and management of non-native species (55). Molecular approaches provide tools to infer invasion pathways and characterize population dynamics associated with the successful establishment and spread of invasive species (56). Generally, genetic approaches are used to infer the source populations and the genetic connectivity of invasive species (i.e., dispersal routes). For example, one study used genetic data to map the likely routes of recolonization after a campaign to eradicate invasive Norway rats from a socially vulnerable section of Salvador, Brazil (57).

Sequencing of mitochondria or chloroplast markers from samples collected across the native and invasive ranges can be used to infer the geographic origin and spread of invasive species (58,59). For instance, molecular approaches have shed new light into the origin and adaptability of invasive jumping earthworms (family Megascolecidae) in North America (60). Pheretimoid earthworms have been spreading globally due to human activity and occur in high abundance in urban parks, residential yards, greenhouses, and compost piles (61). Due to their impacts on hardwood forests, jumping earthworms are monitored as species of concern or pests in New York, Vermont, Wisconsin, and California (60). Mitochondrial sequencing identified three cryptic lineages of *Amynthas* in New York, Wisconsin, and Alabama, which were later confirmed to correspond to *Amynthas arrestis*, *Amynthas tokioensis*, and *Metaphire hilgendorfi* (61,62). Additionally, shared haplotype (DNA variants that tend to be inherited together) diversity across sampled states suggested multiple introductions and/or facilitated dispersal (61). A limitation of using only organellar markers or a few microsatellites is an over- or underestimation of the genetic variation and uncertainty of the evolutionary mechanism associated with the introduction of invasive species (63).

Another critical management strategy for invasive species is identifying the invasion front to enable early mitigation in key habitats (56). Recently, eDNA approaches have been validated as a cost-effective means to monitor invasive species (64,65). Yet, researchers have noted potential flaws in an overreliance on eDNA approaches to infer the impact of biological invasions (66). In particular, eDNA approaches may be constrained in differentiating between dead or alive organisms and between established or transient populations (66). Thus, while molecular approaches can reduce the cost and time associated with surveying invasive species, conservation efforts will likely require complementary strategies to mitigate the impact of invasive species.

Pathogens and transmission dynamics of urban wildlife

Another management consideration is understanding how landscape structures can mediate disease transmission in wild populations (67). In particular, disease transmission can increase due to higher contact rates, either due to fragmentation consolidating spatial corridor use or congregation on supplemental feeding sites (68,69).[1] Molecular methods can determine how the landscape shapes the movement and congregation of species and confirm the transmission of zoonotic pathogens between multiple urban hosts at suspected congregation sites. Managers thus seek to monitor the potential spillover/spillback of pathogens between urban and nonurban populations and between humans and other species (70). In general, urbanization can influence disease dynamics through increased contact rates of individuals that are at higher density, or by altering the presence or density of hosts that are less competent reservoirs/vectors (i.e., dilution effects due to biodiversity losses) (71,72).

[1] Byers et al.'s chapter discusses in detail the pathogen and disease dynamics of urban wildlife.

For example, in California, contact with cats and dogs has resulted in a spillover of canine and feline calicivirus into urban populations of gray foxes (*Urocyon cinereoargenteus*) and bobcats (*Lynx rufus*), (73). Genomic DNA and sequence polymorphism analyses are commonly used to infer the relationship between pathogen strains and the directionality of transmission (74). For example, the environmental changes in cities may favor the persistence of pathogens in the environment, as genetic evidence has suggested for Leptospira bacteria that persist in puddles and moist soil in cities (75).

eDNA/RNA approaches can also be implemented to monitor the presence of pathogens, with particular utility in aquatic habitats (76–78). For example, following the spread of Covid-19, the amplification of environmental RNA from wastewater was a valuable tool for detecting the emergence of new variants of Covid (79). Similarly, the presence of amphibian pathogens (e.g., *Batrachochytrium dendrobatidis* and ranavirus) in urban lakes has been monitored using eDNA sequencing (80). Rabies is another disease prevalent in cities with significant concerns for human health and wildlife management (81). Diagnosis and monitoring of rabies in cities have improved because of recent advances in RNA amplification, such as real-time quantitative PCRs (RT-qPCR) where probes are used to amplify known RNA isolates of rabies (81,82).

Microbiome analyses of urban species have also been implemented to infer the health of populations, community interactions, transmission, and to detect potential bacterial diseases (83–85). For example, a recent study used microbiome analyses to document the potential transmission of human microbiota to urbanized populations of coyotes (*Canis latrans*), anole lizards (*Anolis cristatellus* and *A.* spp.), and white-crowned sparrows (*Zonotrichia leucophrys*). The study documented a homogenization of gut microbiota in these species sampled across different urban environments in North America and Puerto Rico (86), where urban people and urban wildlife shared similar gut microbiomes. These results indicate the need for conservation to consider how urban metacommunities of microbes might impact the fitness of wildlife. Laboratory trials show that replacing mouse microbiomes with non-native microbes stunted their growth rates and immunological development. Thus, shifts in the microbiome community can impact urban wildlife (87). Conversely, soil samples collected from Central Park in New York City were shown to host microbial communities as diverse as nonurban biomes across the globe, suggesting that even highly urbanized environments host a high diversity of microbes (88). As the field of metagenomic approaches for conservation continues to expand, it will be critical to understand the function of microbiome communities as they relate to particular hosts' fitness, rather than draw inferences based on the role these microbiome communities play in relation to human health.

Evaluating gene expression to characterize responses to urban environments

Transcriptomics is a genomic approach where gene expression is quantified using transcribed mRNA sequences. Transcriptomics can be used to identify genes that are expressed differently between various environments (e.g., city patches, urban vs nonurban). Gene ontology (GO) annotations are used to identify the biological processes and molecular functions overrepresented among the differentially expressed genes. An advantage of transcriptomes over DNA-based sequencing is their ability to detect a genetic signal in response to environmental stimuli even when high-quality reference genomes are lacking, or when controlled experiments are not feasible (89). Thus, transcriptomes provide a robust approach to test hypotheses of how urban taxa respond to urban conditions. For example, urban populations of anole lizards (*Anolis cristatellus*) show increased expression of genes related to heat tolerance, such that plasticity and natural selection contribute to urban lizards' ability to survive in urban heat island environments (36,90). Similarly, urban populations of great tits (*Parus major*) have differentiated gene expression associated with immune and inflammatory responses, detoxification, protection against oxidative stress, lipid metabolism, and regulation of gene expression (91). By using transcriptomes to contrast urban and nonurban populations, practitioners can test how specific facets of urbanization or urban management actions affect species and, if necessary, define a mitigation strategy.

Molecular tools can be used to test the efficacy of urban management outcomes in laboratory and natural settings. For example, there is ample evidence that exposure to artificial lighting in great tits (*P. major*) can lead to changes in gene expression which directly impact reproductive behavior (92). Exposure to varying levels of artificial light at night (ALAN) can impact the level of gene transcripts associated with gonadal growth and, in the extreme, spermatogenesis (92). This dose-dependent triggering of reproductive cycles by human activity is significant to the birds, as improperly timed reproduction can increase energy costs at an inopportune time, possibly when food resources are scarce. To test the response to mitigating ALAN, a useful approach would be to evaluate *P. major*'s response to ALAN before and after ALAN mitigation, particularly if replicated in different parts of a city, between cities, or by different mitigation methods. Light intensity, color, and timing could all be factors to test. For example, laboratory trials of alternative light colors might reveal a light hue and intensity that mitigates the effect of ALAN. In turn, this evidence can inform conservation actions that modify lighting infrastructure implemented in cities. A similar application of this can already be seen with the uptake of sea turtle-friendly lighting to avoid the negative impacts of ALAN on sea turtle hatchlings (93). However, some studies have found that the moon cycle and incorrect installation of these lightbulbs can disrupt the efficacy of turtle-friendly lighting (93). Thus, combining laboratory and field assessments with molecular tools will provide the most useful and robust inferences for urban biodiversity management.

Molecular approaches to inform urban reintroduction and recovery

There has been considerable interest in translocating individuals to restore degraded ecosystems.[2] However, translocations can be challenging because many transplanted individuals often do not survive. For example, an ambitious conservation program in the early 2000s saw the reintroduction of 10 species to various parks in New York City (94). However, the survival rates and success of this project were very low. For example, rehabilitated Eastern Screech Owls (*Megascops asio*) sourced from populations outside of New York had a survival rate of 22.6% estimated throughout 13 months (95). Similarly, the survival of other species reintroduced to New York parks was low. In Pelham Bay Park 100 bobwhite quails (*Colinus virginianus*), 75 gray treefrogs (*Hyla versicolor*), and 100 spring peeper frogs (*Pseudacris crucifer*) were reintroduced with zero survival by the end of the year (94). These findings highlight the importance of understanding potential interactions between local environments and newly introduced transplants. A common conservation goal is often to increase local genetic diversity. Thus, genetic analyses should be used to inform which source populations might help diversify the local gene pool.

Recent nonurban translocation projects often use genetic tools to define genetic diversity when optimizing translocation projects. For instance, it was discovered in the 1990s that the 20–25 remaining Florida panthers (*Puma concolor coryi*) were significantly inbred, and that their population would likely go extinct. Eight panthers from Texas were translocated to fortify the genetic variation in the Florida population because of a known history of gene flow between the two populations. The result was an increase in average heterozygosity and a tripling of population size (96). Similarly, population genetics can be used to optimize translocations by minimizing the genetic similarity of sources and then assessing that offspring are the result of reproduction with translocated individuals (97,98). Post-release assessments of Stephen's kangaroo rats (*Dipodomys stephensi*) showed increased genetic diversity resulting from breeding with translocated individuals (99). Additionally, population monitoring confirmed that interpopulation breeding did not decrease fitness (i.e., comparable litter sizes) (99).

The success and monitoring of urban translocations and/or reintroductions will benefit greatly from using molecular approaches. We anticipate that as urban restoration and habitat enhancement efforts proceed, translocations to bolster genetic diversity or diversify the biodiversity in urban areas

[2] Spotswood et al.'s chapter discusses the various ways that urban translocations have occurred and could be used in the future.

will become increasingly necessary and common; molecular tools will play a pivotal role in the success of these efforts.

Assessing urban inter- and intra-species hybridization

Understanding patterns of unintended human-facilitated translocation and hybridization has become an important aspect of conservation. In general, managers balance the trade-offs between inbreeding depression (i.e., a drastic loss in fitness in wild populations due to lack of genetic variation), and outbreeding depression (i.e., a loss in fitness due to hybridization). Hybridization can increase genetic variation which might be important in aiding population recovery. For example, hybridization between coyotes (*Canis latrans*) and red wolves (*Canis rufus*) has resulted in some coyote individuals with over 50% red wolf DNA (99). Coyote hybrids represent an important relic of the genetic diversity of red wolves. This genetic vestige in coyotes is of conservation concern due to the low genetic diversity of wild red wolves, which are descended from 14 individuals (100,101). Thus, facilitated hybridization between captively bred red wolves and coyote hybrids could increase the genetic diversity of red wolves.

A growing conservation concern is understanding how human intervention and urbanization influence hybridization rates in nonhuman species. Theoretically, the hybridization rate is higher due to the increased contact rate of previously isolated species, environmentally driven shifts in mating patterns, and/or due to elevated fitness of hybrids in cities (102). For example, contact with domestic dogs has resulted in dog-coyote hybrids in New York City (103) (Figure 9.1; see Case study). Further, research has found support for increasing hybridization rates between black capped and mountain chickadees (*Poecile atricapillus* and *P. gambeli*, respectively) within cities and other human-disturbed habitats (104). Although

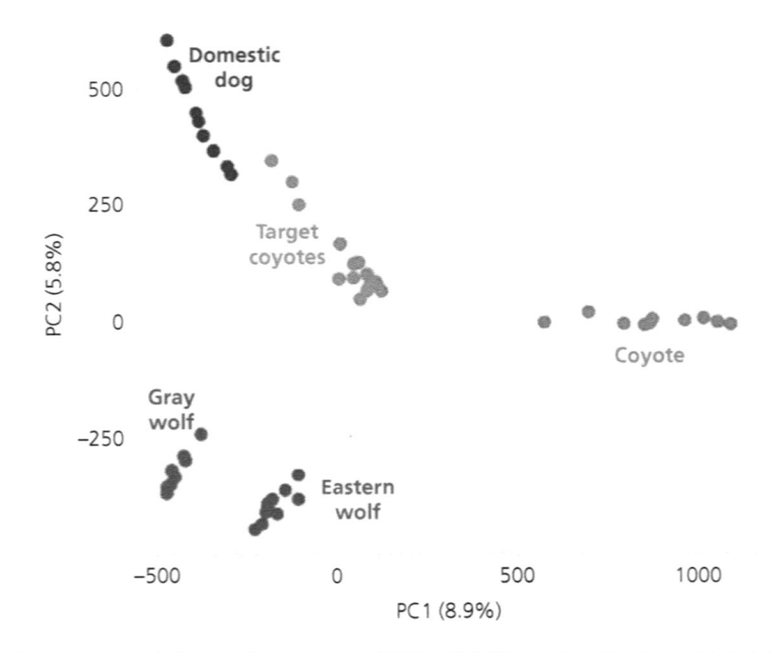

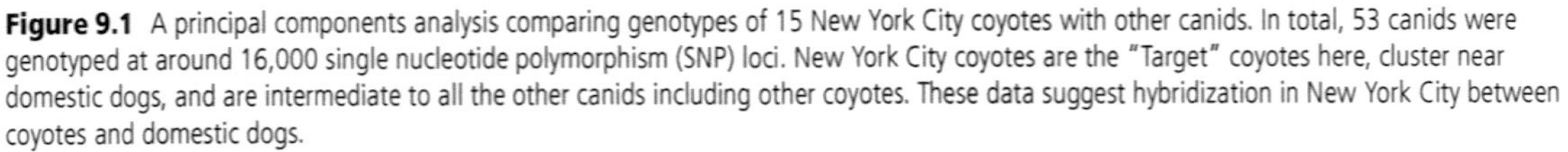

Figure 9.1 A principal components analysis comparing genotypes of 15 New York City coyotes with other canids. In total, 53 canids were genotyped at around 16,000 single nucleotide polymorphism (SNP) loci. New York City coyotes are the "Target" coyotes here, cluster near domestic dogs, and are intermediate to all the other canids including other coyotes. These data suggest hybridization in New York City between coyotes and domestic dogs.

This figure is abstracted from Caragiulo et al. (102) with permission.

examples connecting anthropogenically facilitated hybridization to management concerns are relatively rare (but see (105,106)), it is important to consider how interspecies breeding might impact genetic diversity in urban areas, which typically have a higher potential of contact between native and non-native species.

Global perspectives

An exciting application of molecular approaches is the ability to integrate samples across large spatial scales. A global scale allows managers to test global patterns of evolutionary change in urban environments and/or to consider how transboundary populations are affected by local anthropogenic barriers. Evidence supporting parallel evolutionary responses to urbanization might help practitioners identify the most common selective forces in cities. For example, patterns of adaptation to higher urban temperatures from urban heat islands have been demonstrated to be replicated across multiple urban environments in acorn ants (107) (*Temnothorax curvispinosus*) and white clover (37) (*Trifolium repens* L., Fabaceae). Urban populations of acorn ants have adapted to warmer urban temperatures but, as a result, are now maladapted to their ancestral nonurban environment due to losses in cold tolerance (107,108). Characterizing patterns of local urban adaptation and/or maladaptation to the nonurban environment can be critical for defining management strategies. For instance, the translocation of acorn ants into nonlocal environments could be deadly for the translocated individuals if their thermal tolerance doesn't match the new environment. Multicity contrasts can also be used to understand how trait–environment interactions differ across large spatial scales. For example, the selection of genes associated with hydrogen cyanide (HCN) production in white clover varies in direction and strength across the globe (37). Selection for HCN production is likely driven primarily by herbivory and drought and varies globally depending on local urban conditions. Thus, the local environmental context of where cities occur geographically matters when designing management plans for urban species, particularly non-native species (Figure 9.2).

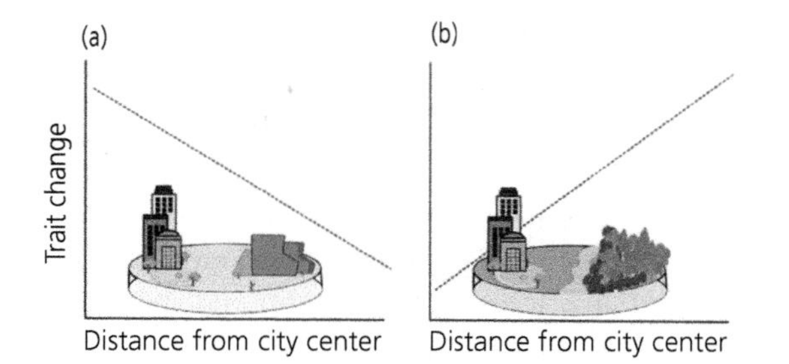

Figure 9.2 Theoretical relationship between trait changes within urban to rural gradients that differ in their local environment (arid versus temperate forest). (a) Here, a hypothetical city with more vegetation and more evaporation than a nearby arid nonurban area has a negative relationship between the trait/genotype and urbanization. (b) In contrast, a city in a region with more abundant vegetation in nonurban areas leads to a positive relationship. Molecular studies that compare cities around the world offer opportunities to understand how geographic contexts alter the impacts of urbanization on biodiversity management.

Similarly, it is important to consider the spatial context of how multiple cities within a region may impact species with large distributions. For instance, species with home ranges spanning many countries may be affected by numerous urban zones and physical barriers such as fences and walls which impact a species' dispersal and/or result in mortality (109). For example, 66 conservation units for lions (*Panthera leo*) were identified throughout Sub-Saharan Africa, with populations spanning multiple countries and anthropogenically fragmented habitats (110). Sequences from 20 microsatellite loci identified four primary populations and suggested that facilitated translocations and/or corridors were needed to mitigate the impact of cities and other anthropogenic habitats on lions. Molecular approaches will be valuable to city managers in coordinating conservation measures for species that occur in multiple cities, particularly across national boundaries.

Multicountry and global comparisons are challenging. Currently, molecular studies in urban areas are not evenly distributed worldwide. For instance, urban landscape genetics is a field that desperately needs better global representation and the areas experiencing the most rapid urbanization (Africa, Asia, and South America) have the fewest urban landscape genetic studies (12).

Social-ecological applications

Urban biodiversity research has been increasingly recognizing that socioeconomic conditions like race and class shape biodiversity and how we measure biodiversity.[3] For instance, structural racism and economic inequality directly shape the distribution of and access to the natural and the urban environment, and marginalized communities often also face more extreme environmental disservices (e.g., heat island, pollution, noise) from urbanization (111). Recently molecular approaches have been used to illustrate how systemic racism and inequality can directly impact the genetic health of wild populations (13). Specifically, an analysis of publicly available genetic data was used to demonstrate that wildlife populations living in majority non-White neighborhoods impacted by the legacy of racial discrimination have smaller effective population sizes, lower genetic diversity, and more genetic differentiation.

Molecular approaches have been used to highlight links between human social conditions and biodiversity in diverse ways. For instance, populations in New York City show decreased genetic diversity in districts with lower human density and less access to human food waste (112). In particular, Midtown New York City is a resource desert and may functionally differentiate uptown and downtown rats in New York (113). The spread of another urban commensal species—German cockroaches (*Blattella germanica*)—among cities and between continents is driven by human-mediated dispersal. Genetic analyses of microsatellite loci and single nucleotide polymorphisms found that cockroach populations structure along individual buildings within a city (114,115). Additionally, two cities connected by trains, and which have more movement of people between them, showed greater relatedness in cockroaches than would be predicted by distance (115). Such studies demonstrate how social, cultural, political, and economic forces can have measurable impacts on urban biodiversity that can be readily detected using molecular approaches. The interactions between these forces are complex and highlight the power of integrating molecular tools with analyses of sociopolitical contexts.

Beyond characterizing how human social conditions structure biodiversity, molecular tools can help urban managers assess disease transmission between wildlife and humans. Emerging diseases in urban areas can harm humans by impacting their health and/or their pets, livestock, and/or crops.[4] The most well-known examples are rodent diseases linked to extensive human mortality and economic loss. For instance, rodents can harbor plague (caused by the bacterium *Yersinia pestis*), leptospirosis, and hantavirus and can infect humans through bites, by contaminating food and water sources, or via ectoparasite transmissions (116,117). Moreover, socioeconomic and landscape features of urban development can influence the abundance of important disease vectors. For example, in Baltimore (Maryland, USA), DNA sequencing of mosquito blood meals found a lower prevalence of human hosts in *Aedes albopictus* mosquitoes from lower-income neighborhoods (118). The study suggested that the lack of outdoor recreational space in lower-income neighborhoods likely accounted for the reduced number of human hosts for mosquitos. Furthermore, brown rats (*Rattus norvegicus*) were the most common host detected in blood meals for two mosquito species. Given that brown rat populations are more abundant in lower-income neighborhoods in Baltimore and many other North American cities, these communities might have a higher risk of disease transmission from rats (119). Metabarcoding approaches targeting DNA and RNA sequences can be used to monitor zoonotic pathogens in urban populations (120). For instance, brown rats (*R. norvegicus*) in New York were shown to harbor approximately 20 distinct mammalian viruses, some of which have been associated with gastrointestinal diseases in humans (121). Genetic approaches can also be used to evaluate the success of lethal control campaigns against pest reservoirs. For example, genetic sequencing

[3] Hoover's and Scarlett's chapter illustrates the close relationship between urban inequities and injustices and biodiversity conservation. Perkins et al.'s chapter demonstrates how citizen science approaches—which are commonly used to monitor biodiversity—are heavily skewed to inequities in volunteer participation.

[4] Byers et al.'s chapter discusses pathogen and disease dynamics between wildlife and people in cities.

demonstrated a reduction in the effective population size and increased relatedness of rat populations following an eradication campaign, in a Brazilian community that has experienced high levels of human leptospirosis infection (57). Additionally, a genetic assessment confirmed that the rat population rebound following the eradication campaign came from local rats. Thus, molecular tools can be broadly applied to characterize transmission routes, pathogen loads, and the effectiveness of lethal control campaigns to decrease infection risks between people and wildlife in cities.

Conclusion and recommendations

Molecular tools provide powerful means to identify species' health, viability, and management needs in any environment. Urban areas in particular can benefit from these tools given the challenges with sampling across privately owned land, increased patchiness, and broader social-ecological complexity. For example, noninvasive sampling of eDNA can provide cost-effective means to monitor ecosystems and species. Similarly, genome-wide markers can help elucidate the mechanism shaping the genetic diversity of urban taxa. Managers can use this information to decide when and how to use an intervention.

DNA sequencing within and between cities allows managers to link spatial and environmental features (e.g., roads, cemeteries) to genetic diversity. The advantage of this approach is that candidate management actions can be identified when, for instance, genetic diversity is lower than expected, given repeated gene flow from nearby patches or populations. Managers could intervene by translocating individuals from other parts of the city or different environments to increase local genetic diversity. Additionally, installing suitable corridors may facilitate natural dispersal. However, a limitation of this approach is that spatial and temporal conditions can influence the ability to detect when a population is being affected by urbanization. For example, when genetic diversity is high and generation times of species are long, there could be a considerable lag between urban development and an observed decrease in genetic diversity.

Facilitated translocations to increase genetic diversity are controversial due to potential outbreeding depression (20,132). Managers can use "omics" approaches to source individuals from other populations suitable for transplantation based on the local environmental conditions. For instance, scans of expressed genome-wide sequences in *Eucalyptus gomphocephala* identified climate-driven diversifying selection (133). This information led to informed sourcing decisions for the transplantation of seeds adapted to current and/or predicted future climatic conditions (133). This approach could be fruitful when considering translocation into cities or between areas of a city. For example, an area thought to have a particularly strong heat island effect can be evaluated by applying molecular approaches to detect whether species are expressing known proteins or compounds associated with heat stress and contrasting genomic diversity in relation to other populations (36,37,90).

The combination of molecular approaches with increasing computational power can produce more precise models to predict how changing spatial, social, political, and economic features influence biodiversity. Molecular tools have played and will continue to play an essential role in providing high-resolution data linking social and environmental processes that underlie urban biodiversity management. Moving forward, we believe that higher access to publicly available genetic data and increases in modeling power could be used to predict how future habitat modifications might impact genetic diversity. We can liken this preemptive approach to triage, where molecular methods are used to predict the impact of habitat modification. For instance, models based on genetic diversity could be used to predict the implications of new constructions or ecosystem restorations. Understanding the molecular mechanisms underlying and constraining these changes necessitates understanding *all* aspects of the environment driving them. If scientists, managers, and community leaders avail themselves of the rapid advancements in this field, then equitable and informed urban conservation can be the new standard.

Case study: Molecularly untangling urban coyotes

Northeastern coyotes are a flagship species for understanding urban socio-eco-evolutionary dynamics (122). The very existence of the "northeastern coyote," a hybrid of multiple canid species (12), is tied to human actions. Specifically, the regional extirpation of wolves and pumas facilitated coyote range expansion across the entire continent of North America. Consequently, contact between coyotes and other canids in the Great Lakes region resulted in the hybridization of coyotes with wolves and domestic dogs (123–125). Hybrid traits, such as larger body size facilitated the hunting of larger prey species ranging from white-tailed deer to domestic pets (123,126). Consequently, wolf like coyotes are now the top predator in many areas of the northeast (125). Coyotes exemplify how the complex interactions between humans and nature can lead to novel eco-evolutionary dynamics.

A pressing issue regarding coyote conservation is defining conservation or management units, especially given high levels of hybridization (123). Molecular tools are critical for this work. Characterizing regional variation in genetic mixing can reveal the ecological role of hybridized coyotes in various ecosystems and their recent evolutionary past (99,100,123). Several loci linked to the exploration of novel environments like cities (potentially important for urban-tolerant species) may be under selection in eastern coyotes (100), but the role of hybridization is unclear. Indeed, New York City coyotes show alleles derived from hybridization with domestic dogs, which could be associated with notable human tolerance and sociability, another trait that can facilitate urban persistence (101). Increasing coyote tolerance to humans can have consequences for future range expansion and human conflict. Management strategies can be defined by evaluating historical and contemporary admixture patterns to determine conservation units where ecology may differ *within* species.

Evaluating urban coyotes has also provided key examples of how gray infrastructure can impact urban wildlife (127). Genetic analyses revealed that the southern California freeway functions as a hard barrier resulting in genetic differentiation between coyote populations on either side of it despite some evidence of dispersal across the freeway (127). In highly urbanized, recently colonized areas of New York City, populations have become genetically diverged from nonurban populations and show decreased genetic diversity due to a bottleneck effect (32).

Coyotes have also been a model species for understanding links between urban diet and health. For instance, anthropogenic food consumption is associated with gut microbiome shifts and larger spleens in urban coyotes (128). Additionally, urban coyotes had microbiomes that reflected a higher intake of carbohydrates. Scat metabarcoding to assess diet in New York City coyotes found that urban coyotes have a more diverse diet than nonurban coyotes; however, both urban and nonurban coyotes use anthropogenic food in addition to natural food sources. Both groups consumed deer, though less so for urban coyotes, and neither group significantly consumed housecats, with only <5% of urban coyote scat samples containing housecat DNA (8) (Figure 9.3). Understanding coyote diet can have implications for how humans and coyotes coexist—for instance, human residents might have a different perception of coyotes if they knew how rarely they consume their pets. In urban environments, coyotes can help alleviate deer overpopulation. Thus, coyote diet inference through molecular approaches can monitor physiological changes and track their impact on human pets and livestock.

Molecular tools also help assess the consequences of coyote–dog interactions. For instance, high contact rates with dogs can increase transmission of canine distemper virus and rabies (129). Further, increased urban exploitation can also impact the foraging behavior of urban species. Coyotes appear to have increased boldness behavior with urbanization, increasing the potential for human–wildlife conflict (130). Molecular work has identified genetic markers associated with boldness/aggression in coyotes (130,131). Two of these markers are also found in dogs and vary based on urbanization, suggesting that boldness is under selection in urban coyotes and possibly that hybridization with domestic dogs may have allowed coyotes to better tolerate people in cities (130,131). These types of mechanistic and genetic studies, while complex, could be important areas of study for predicting and minimizing wildlife conflict in an urbanizing world.

Case study: *Continued*

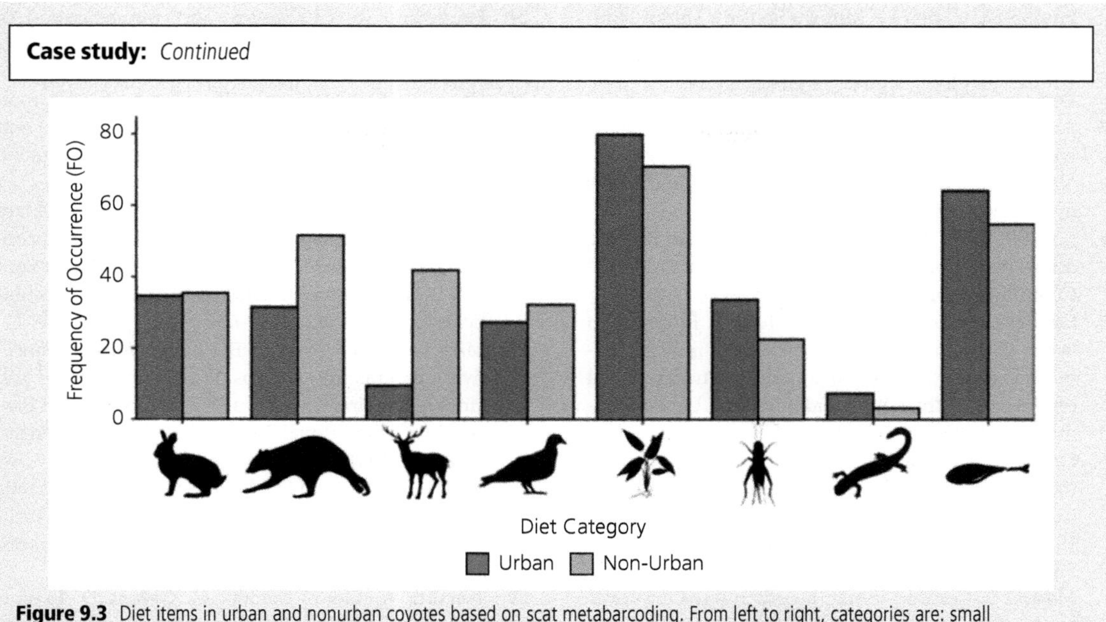

Figure 9.3 Diet items in urban and nonurban coyotes based on scat metabarcoding. From left to right, categories are: small mammals, other mammals, deer, birds, plants, insects, aquatic, and anthropogenic items (including cats).
This figure is reproduced with permissions without modification from Henger et al. (8).

References

1. Amos W, Balmford A. When does conservation genetics matter? Heredity. 2001;87(3):257–65.
2. Gibbs HL, Weatherhead P. Insights into population ecology and sexual selection in snakes through the application of DNA-based genetic markers. Journal of Heredity. 2001;92(2):173–9.
3. Trevelline BK, Fontaine SS, Hartup BK, Kohl KD. Conservation biology needs a microbial renaissance: a call for the consideration of host-associated microbiota in wildlife management practices. Proceedings of the Royal Society B: Biological Sciences. 2019;286(1895):20182448.
4. Lambert MR, Donihue CM. Urban biodiversity management using evolutionary tools. Nature Ecology & Evolution. 2020;4(7):903–10.
5. Thomsen PF, Willerslev E. Environmental DNA – an emerging tool in conservation for monitoring past and present biodiversity. Biological Conservation. 2015;183(Mar):4–18.
6. Charvoz L, Apothéloz-Perret-Gentil L, Reo E, Thiébaud J, Pawlowski J. Monitoring newt communities in urban area using eDNA metabarcoding. PeerJ. 2021;9:e12357.
7. Rodríguez-Castro KG, Saranholi BH, Bataglia L, Blanck DV, Galetti PM. Molecular species identification of scat samples of South American felids and canids. Conservation Genetics Resources. 2020;12(1):61–6.
8. Henger CS, Hargous E, Nagy CM, Weckel M, Wultsch C, Krampis K, et al. DNA metabarcoding reveals that coyotes in New York City consume wide variety of native prey species and human food. PeerJ. 2022;10:e13788.
9. Eaton DA, Ree RH. Inferring phylogeny and introgression using RADseq data: an example from flowering plants (*Pedicularis*: Orobanchaceae). Systematic Biology. 2013;62(5):689–706.
10. Janjua S, Peters JL, Weckworth B, Abbas FI, Bahn V, Johansson O, et al. Improving our conservation genetic toolkit: ddRAD-seq for SNPs in snow leopards. Conservation Genetics Resources. 2020;12(2):257–61.
11. Richardson JL, Burak MK, Hernandez C, Shirvell JM, Mariani C, Carvalho-Pereira TSA, et al. Using fine-scale spatial genetics of Norway rats to improve control efforts and reduce leptospirosis risk in urban slum environments. Evolutionary Applications. 2017;10(4):323–37.

12. Fusco NA, Carlen EJ, Munshi-South J. Urban landscape genetics: are biologists keeping up with the pace of urbanization? Current Landscape Ecology Reports. 2021;6:35–45.
13. Schmidt C, Garroway CJ. Systemic racism alters wildlife genetic diversity. Proceedings of the National Academy of Sciences of the United States of America. 2022;119(43):e2102860119.
14. Loss SR, Will T, Marra PP. The impact of free-ranging domestic cats on wildlife of the United States. Nature Communications. 2013;4:1396.
15. Chollet S, Brabant C, Tessier S, Jung V. From urban lawns to urban meadows: reduction of mowing frequency increases plant taxonomic, functional and phylogenetic diversity. Landscape and Urban Planning. 2018;180(Dec):121–4.
16. Garcês A, Queiroga F, Prada J, Pires I. A review of the mortality of wild fauna in Europe in the last century, the consequences of human activity. Journal of Wildlife and Biodiversity. 2020;4(2):34–55.
17. Francoeur XW, Dagenais D, Paquette A, Dupras J, Messier C. Complexifying the urban lawn improves heat mitigation and arthropod biodiversity. Urban Forestry & Urban Greening. 2021;60(May): 127007.
18. Szulkin M, Garroway CJ, Corsini M, Kotarba AZ, Dominoni D. How to quantify urbanization when testing for urban evolution? In: Szulkin M (ed.) Urban Evolutionary Biology. Oxford: Oxford University Press; 2020. p. 13–35.
19. Winchell KM, Aviles-Rodríguez KJ, Carlen EJ, Miles LS, Charmantier A, De León LF, et al. Moving past the challenges and misconceptions in urban adaptation research. Ecology and Evolution. 2022;12(11):e9552.
20. Weeks AR, Sgro CM, Young AG, Frankham R, Mitchell NJ, Miller KA, et al. Assessing the benefits and risks of translocations in changing environments: a genetic perspective. Evolutionary Applications. 2011;4(6):709–25.
21. Magura T, Kiss E, Lövei GL. No consistent diversity patterns in terrestrial mammal assemblages along rural-urban forest gradients. Basic and Applied Ecology. 2021;52:38–45.
22. Bateman PW, Fleming PA. Big city life: carnivores in urban environments. Journal of Zoology. 2012;287(1):1–23.
23. Sheng G-L, Barlow A, Cooper A, Hou X-D, Ji X-P, Jablonski NG, et al. Ancient DNA from giant panda (*Ailuropoda melanoleuca*) of south-western China reveals genetic diversity loss during the Holocene. Genes. 2018;9(4):198.
24. Frankham R. Challenges and opportunities of genetic approaches to biological conservation. Biological Conservation. 2010;143(9):1919–27.
25. DeYoung RW, Honeycutt RL. The molecular toolbox: genetic techniques in wildlife ecology and management. The Journal of Wildlife Management. 2005;69(4):1362–84.
26. Barbosa S, Mestre F, White TA, Paupério J, Alves PC, Searle JB. Integrative approaches to guide conservation decisions: using genomics to define conservation units and functional corridors. Molecular Ecology. 2018;27(17):3452–65.
27. Miles LS, Dyer RJ, Verrelli BC. Urban hubs of connectivity: contrasting patterns of gene flow within and among cities in the western black widow spider. Proceedings of the Royal Society B: Biological Sciences. 2018;285(1884):20181224.
28. Carlen E, Munshi-South J. Widespread genetic connectivity of feral pigeons across the Northeastern megacity. Evolutionary Applications. 2021;14(1):150–62.
29. Turcotte A, Blouin-Demers G, Garant D. Exploring the effect of 195 years-old locks on species movement: landscape genetics of painted turtles in the Rideau Canal, Canada. Conservation Genetics. 2022;23(3):467–79.
30. Gortat T, Rutkowski R, Gryczynska A, Kozakiewicz A, Kozakiewicz M. The spatial genetic structure of the yellow-necked mouse in an urban environment – a recent invader vs. a closely related permanent inhabitant. Urban Ecosystems. 2017;20(3): 581–94.
31. Munshi-South J, Richardson JL. Landscape genetic approaches to understanding movement. In: Szulkin M (ed.) Urban Evolutionary Biology. Oxford: Oxford University Press; 2020. p. 54–73.
32. DeCandia AL, Henger CS, Krause A, Gormezano LJ, Weckel M, Nagy C, et al. Genetics of urban colonization: neutral and adaptive variation in coyotes (*Canis latrans*) inhabiting the New York metropolitan area. Journal of Urban Ecology. 2019;5(1):juz002.
33. Balbi M, Ernoult A, Poli P, Madec L, Guiller A, Martin M-C, et al. Functional connectivity in replicated urban landscapes in the land snail (*Cornu aspersum*). Molecular Ecology. 2018;27(6):1357–70.
34. Fusco NA, Pehek E, Munshi-South J. Urbanization reduces gene flow but not genetic diversity of stream salamander populations in the New York City metropolitan area. Evolutionary Applications. 2021;14(1):99–116.
35. Richardson JL, Brady SP, Wang IJ, Spear SF. Navigating the pitfalls and promise of landscape genetics. Molecular Ecology. 2016;25(4):849–63.

36. Campbell-Staton SC, Winchell KM, Rochette NC, Fredette J, Maayan I, Schweizer RM, et al. Parallel selection on thermal physiology facilitates repeated adaptation of city lizards to urban heat islands. Nature Ecology & Evolution. 2020;4(4):652–8.
37. Santangelo JS, Ness RW, Cohan B, Fitzpatrick CR, Innes SG, Koch S, et al. Global urban environmental change drives adaptation in white clover. Science. 2022;375(6586):1275–81.
38. Peterman WE. ResistanceGA: an R package for the optimization of resistance surfaces using genetic algorithms. Methods in Ecology and Evolution. 2018;9(6):1638–47.
39. Winiarski KJ, Peterman WE, McGarigal K. Evaluation of the R package "resistancega": a promising approach towards the accurate optimization of landscape resistance surfaces. Molecular Ecology Resources. 2020;20(6):1583–96.
40. Bauder JM, Breininger DR, Bolt MR, Legare ML, Jenkins CL, Rothermel BB, et al. Multi-level, multi-scale habitat selection by a wide-ranging, federally threatened snake. Landscape Ecology. 2018;33(5):743–63.
41. Bauder JM, Peterman WE, Spear SF, Jenkins CL, Whiteley AR, McGarigal K. Multiscale assessment of functional connectivity: landscape genetics of eastern indigo snakes in an anthropogenically fragmented landscape in central Florida. Molecular Ecology. 2021;30(14):3422–38.
42. Lambert MR, Brans KI, Des Roches S, Donihue CM, Diamond SE. Adaptive evolution in cities: progress and misconceptions. Trends in Ecology & Evolution. 2021;36(3):239–57.
43. Manel S, Holderegger R. Ten years of landscape genetics. Trends in Ecology & Evolution. 2013;28(10):614–21.
44. Kelly RP. Making environmental DNA count. Molecular Ecology Resources. 2016;16(1):10–12.
45. Kelly RP, O'Donnell JL, Lowell NC, Shelton AO, Samhouri JF, Hennessey SM, et al. Genetic signatures of ecological diversity along an urbanization gradient. PeerJ. 2016;4:e2444.
46. Newton JP, Bateman PW, Heydenrych MJ, Mousavi-Derazmahalleh M, Nevill P. Home is where the hollow is: revealing vertebrate tree hollow user biodiversity with eDNA metabarcoding. Environmental DNA. 2022;4(5):1078–91.
47. Thuo D, Furlan E, Broekhuis F, Kamau J, Macdonald K, Gleeson DM. Food from faeces: evaluating the efficacy of scat DNA metabarcoding in dietary analyses. PLoS One. 2019;14(12):e0225805.
48. Abrams JF, Hörig LA, Brozovic R, Axtner J, Crampton-Platt A, Mohamed A, et al. Shifting up a gear with iDNA: from mammal detection events to standardised surveys. Journal of Applied Ecology. 2019;56(7):1637–48.
49. Clare EL, Economou CK, Bennett FJ, Dyer CE, Adams K, McRobie B, et al. Measuring biodiversity from DNA in the air. Current Biology. 2022;32(3):693–700.
50. Lynggaard C, Bertelsen MF, Jensen CV, Johnson MS, Frøslev TG, Olsen MT, et al. Airborne environmental DNA for terrestrial vertebrate community monitoring. Current Biology. 2022;32(3):701–7.
51. Roger F, Ghanavi HR, Danielsson N, Wahlberg N, Löndahl J, Pettersson LB, et al. Airborne environmental DNA metabarcoding for the monitoring of terrestrial insects—a proof of concept from the field. Environmental DNA. 2022;4(4):790–807.
52. Tyagi A, Khan A, Thatte P, Ramakrishnan U. Genome-wide single nucleotide polymorphism (SNP) markers from fecal samples reveal anthropogenic impacts on connectivity: case of a small carnivore in the central Indian landscape. Animal Conservation. 2022;25(5):648–59.
53. Chiou KL, Bergey CM. Methylation-based enrichment facilitates low-cost, noninvasive genomic scale sequencing of populations from feces. Scientific Reports. 2018;8(1):1975.
54. Hansen MJ, Clevenger AP. The influence of disturbance and habitat on the presence of non-native plant species along transport corridors. Biological Conservation. 2005;125(2):249–59.
55. Padayachee AL, Irlich UM, Faulkner KT, Gaertner M, Proches S, Wilson JRU, et al. How do invasive species travel to and through urban environments? Biological Invasions. 2017;19(12):3557–70.
56. Le Roux J, Wieczorek A. Molecular systematics and population genetics of biological invasions: towards a better understanding of invasive species management. Annals of Applied Biology. 2009;154(1): 1–17.
57. Richardson JL, Silveira G, Medrano IS, Arietta AZ, Mariani C, Pertile AC, et al. Significant genetic impacts accompany an urban rat control campaign in Salvador, Brazil. Frontiers in Ecology and Evolution. 2019;7:115.
58. Puckett EE. Variability in total project and per sample genotyping costs under varying study designs including with microsatellites or SNPs to answer conservation genetic questions. Conservation Genetics Resources. 2017;9:289–304.
59. Sleith RS, Karol KG. Global high-throughput genotyping of organellar genomes reveals insights into the origin and spread of invasive starry

stonewort (*Nitellopsis obtusa*). Biological Invasions. 2021;23(15):3471–82.
60. Chang C-H, Bartz MLC, Brown G, Callaham MA, Cameron EK, Dávalos A, et al. The second wave of earthworm invasions in North America: biology, environmental impacts, management and control of invasive jumping worms. Biological Invasions. 2021;23(11):3291–322.
61. Schult N, Pittenger K, Davalos S, McHugh D. Phylogeographic analysis of invasive Asian earthworms (*Amynthas*) in the northeast United States. Invertebrate Biology. 2016;135(4):314–27.
62. Chang C-H, Johnston MR, Görres JH, Dávalos A, McHugh D, Szlavecz K. Co-invasion of three Asian earthworms, *Metaphire hilgendorfi*, *Amynthas agrestis* and *Amynthas tokioensis* in the USA. Biological Invasions. 2018;20(4):843–8.
63. Puckett EE, Orton D, Munshi-South J. Commensal rats and humans: integrating rodent phylogeography and zooarchaeology to highlight connections between human societies. Bioessays. 2020;42(5):1900160.
64. Piaggio AJ, Engeman RM, Hopken MW, Humphrey JS, Keacher KL, Bruce WE, et al. Detecting an elusive invasive species: a diagnostic PCR to detect Burmese python in Florida waters and an assessment of persistence of environmental DNA. Molecular Ecology Resources. 2014;14(2):374–80.
65. Geerts AN, Boets P, Van den Heede S, Goethals P, Van der Heyden C. A search for standardized protocols to detect alien invasive crayfish based on environmental DNA (eDNA): a lab and field evaluation. Ecological Indicators. 2018;84(Jan):564–72.
66. Hewitt CL, Campbell ML, Dafforn K, Davis J, Deveney MR, McDonald JI, et al. RE: Environmental DNA is one tool among many in the biosecurity toolbox. Science (eLetters). 2018. Available from: https://www.science.org/doi/10.1126/science.aao3787
67. Gras P, Knuth S, Börner K, Marescot L, Benhaiem S, Aue A, et al. Landscape structures affect risk of canine distemper in urban wildlife. Frontiers in Ecology and Evolution. 2018;6:136.
68. Campbell TA, Long DB, Shriner SA. Wildlife contact rates at artificial feeding sites in Texas. Environmental Management. 2013;51(6):1187–93.
69. Meentemeyer RK, Haas SE, Václavík T. Landscape epidemiology of emerging infectious diseases in natural and human-altered ecosystems. Annual Review of Phytopathology. 2012;50:379–402.
70. Collins MK, Magle SB, Gallo T. Global trends in urban wildlife ecology and conservation. Biological Conservation. 2021;261(Sep):109236.
71. Bradley CA, Altizer S. Urbanization and the ecology of wildlife diseases. Trends in Ecology & Evolution. 2007;22(2):95–102.
72. LaDeau SL, Allan BF, Leisnham PT, Levy MZ. The ecological foundations of transmission potential and vector-borne disease in urban landscapes. Functional Ecology. 2015;29(Jul):889–901.
73. Riley SP, Foley J, Chomel B. Exposure to feline and canine pathogens in bobcats and gray foxes in urban and rural zones of a national park in California. Journal of Wildlife Diseases. 2004;40(1):11–22.
74. Robertson LJ, Clark CG, Debenham JJ, Dubey JP, Kváč M, Li J, et al. Are molecular tools clarifying or confusing our understanding of the public health threat from zoonotic enteric protozoa in wildlife? International Journal for Parasitology: Parasites and Wildlife. 2019;9(Aug):323–41.
75. Casanovas-Massana A, de Oliveira D, Schneider AG, Begon M, Childs JE, Costa F, et al. Genetic evidence for a potential environmental pathway to spillover infection of rat-borne leptospirosis. The Journal of Infectious Diseases. 2022;225(1):130–4.
76. Strand DA, Johnsen SI, Rusch JC, Agersnap S, Larsen WB, Knudsen SW, et al. Monitoring a Norwegian freshwater crayfish tragedy: eDNA snapshots of invasion, infection and extinction. Journal of Applied Ecology. 2019;56(7):1661–73.
77. Amarasiri M, Furukawa T, Nakajima F, Sei K. Pathogens and disease vectors/hosts monitoring in aquatic environments: potential of using eDNA/eRNA based approach. Science of the Total Environment. 2021;796(Nov):148810.
78. Farrell JA, Whitmore L, Duffy DJ. The promise and pitfalls of environmental DNA and RNA approaches for the monitoring of human and animal pathogens from aquatic sources. BioScience. 2021;71(6):609–25.
79. Randazzo W, Cuevas-Ferrando E, Sanjuán R, Domingo-Calap P, Sánchez G. Metropolitan wastewater analysis for COVID-19 epidemiological surveillance. International Journal of Hygiene and Environmental Health. 2020;230(Sep):113621.
80. Barnes MA, Brown AD, Daum MN, de la Garza KA, Driskill J, Garrett K, et al. Detection of the amphibian pathogens chytrid fungus (*Batrachochytrium dendrobatidis*) and ranavirus in West Texas, USA, using environmental DNA. Journal of Wildlife Diseases. 2020;56(3):702–6.
81. Sterner RT, Smith GC. Modelling wildlife rabies: transmission, economics, and conservation. Biological Conservation. 2006;131(2):163–79.
82. Faye M, Dacheux L, Weidmann M, Diop SA, Loucoubar C, Bourhy H, et al. Development and validation of sensitive real-time RT-PCR assay for

broad detection of rabies virus. Journal of Virological Methods. 2017;243(May):120–30.

83. Stumpf RM, Gomez A, Amato KR, Yeoman CJ, Polk JD, Wilson BA, et al. Microbiomes, metagenomics, and primate conservation: new strategies, tools, and applications. Biological Conservation. 2016;199(Jul):56–66.
84. Soares-Castro P, Araújo-Rodrigues H, Godoy-Vitorino F, Ferreira M, Covelo P, López A, et al. Microbiota fingerprints within the oral cavity of cetaceans as indicators for population biomonitoring. Scientific Reports. 2019;9:13679.
85. Bhagat N, Sharma S, Ambardar S, Raj S, Trakroo D, Horacek M, et al. Microbiome fingerprint as biomarker for geographical origin and heredity in *Crocus sativus*: a feasibility study. Frontiers in Sustainable Food Systems. 2021;5:688393.
86. Dillard BA, Chung AK, Gunderson AR, Campbell-Staton SC, Moeller AH. Humanization of wildlife gut microbiota in urban environments. eLife. 2022;11:e76381.
87. Moeller AH, Gomes-Neto JC, Mantz S, Kittana H, Segura Munoz RR, Schmaltz RJ, et al. Experimental evidence for adaptation to species-specific gut microbiota in house mice. mSphere. 2019;4(4):e00387–19.
88. Ramirez KS, Leff JW, Barberán A, Bates ST, Betley J, Crowther TW, et al. Biogeographic patterns in below-ground diversity in New York City's Central Park are similar to those observed globally. Proceedings of the Royal Society B: Biological Sciences. 2014;281(1795):20141988.
89. Alvarez M, Schrey AW, Richards CL. Ten years of transcriptomics in wild populations: what have we learned about their ecology and evolution? Molecular Ecology. 2015;24(4):710–25.
90. Campbell-Staton SC, Velotta JP, Winchell KM. Selection on adaptive and maladaptive gene expression plasticity during thermal adaptation to urban heat islands. Nature Communications. 2021;12(Oct):6195.
91. Watson H, Videvall E, Andersson MN, Isaksson C. Transcriptome analysis of a wild bird reveals physiological responses to the urban environment. Scientific Reports. 2017;7(Mar):44180.
92. Dominoni DM, de Jong M, Bellingham M, O'Shaughnessy P, van Oers K, Robinson J, et al. Dose-response effects of light at night on the reproductive physiology of great tits (*Parus major*): integrating morphological analyses with candidate gene expression. Journal of Experimental Zoology Part A: Ecological and Integrative Physiology. 2018;329(8–9):473–87.
93. Robertson K, Booth DT, Limpus CJ. An assessment of "turtle-friendly" lights on the sea-finding behaviour of loggerhead turtle hatchlings (*Caretta caretta*). Wildlife Research. 2016;43(1):27–37.
94. MacDonald J. Animal Species Reintroductions in 2001. Urban Park Rangers Wildlife Management Program. End of Year Report. New York: New York City Parks; 2001.
95. Christopher N. The Eastern Screech Owl Reintroduction Program in Central Park, New York City: habitat, survival, and reproduction [Dissertation]. [Bronx, NY]: Fordham University; 2004.
96. Johnson WE, Onorato DP, Roelke ME, Land ED, Cunningham M, Belden RC, et al. Genetic restoration of the Florida panther. Science. 2010;329(5999): 1641–5.
97. Bragg JG, Cuneo P, Sherieff A, Rossetto M. Optimizing the genetic composition of a translocation population: incorporating constraints and conflicting objectives. Molecular Ecology Resources. 2020;20(1):54–65.
98. Shier DM, Navarro AY, Tobler M, Thomas SM, King SND, Mullaney CB, et al. Genetic and ecological evidence of long-term translocation success of the federally endangered Stephens' kangaroo rat. Conservation Science and Practice. 2021;3(9):e478.
99. VonHoldt BM, Hinton JW, Shutt AC, Murphy SM, Karlin ML, Adams JR, et al. Reviving ghost alleles: genetically admixed coyotes along the American Gulf Coast are critical for saving the endangered red wolf. Science Advances. 2022;8(26):eabn7731.
100. Heppenheimer E, Brzeski JW, Wooten R, Waddell W, Rutledge LY, Chamberlain MJ, et al. Rediscovery of red wolf ghost alleles in a canid population along the American Gulf Coast. Genes (Basel). 2018;9(12):618.
101. Agan SW, Treves A, Willey LL. Estimating poaching risk for the critically endangered wild red wolf (*Canis rufus*). PLoS One. 2021;16(5):e0244261.
102. McFarlane SE, Mandeville EG. Diverse data sources and new statistical models offer prospects for improving the predictability of anthropogenic hybridization. Global Change Biology. 2023;29(4):923–5.
103. Caragiulo A, Gaughran SJ, Duncan N, Nagy C, Weckel M, vonHoldt BM. Coyotes in New York City carry variable genomic dog ancestry and influence their interactions with humans. Genes. 2022;13(9):1661.
104. Grabenstein KC, Otter KA, Burg TM, Taylor SA. Hybridization between closely related songbirds is related to human habitat disturbance. Global Change Biology. 2023;29(4):955–68.

105. Osório H, Zé-zé L, Amaro F, Nunes A, Alves M. Sympatric occurrence of *Culex pipiens* (Diptera, Culicidae) biotypes *pipiens*, *molestus* and their hybrids in Portugal, Western Europe: feeding patterns and habitat determinants. Medical and Veterinary Entomology. 2014;28(1):103–9.
106. Haba Y, McBride L. Origin and status of *Culex pipiens* mosquito ecotypes. Current Biology. 2022;32: R237–46.
107. Diamond SE, Chick LD, Perez A, Strickler SA, Zhao C. Evolution of plasticity in the city: urban acorn ants can better tolerate more rapid increases in environmental temperature. Conservation Physiology. 2018;6(1):coy030.
108. Martin RA, Chick LD, Garvin ML, Diamond SE. In a nutshell, a reciprocal transplant experiment reveals local adaptation and fitness trade-offs in response to urban evolution in an acorn-dwelling ant. Evolution. 2021;75(4):876–87.
109. Pokorny B, Flajšman K, Centore L, Krope FS, Šprem N. Border fence: a new ecological obstacle for wildlife in Southeast Europe. European Journal of Wildlife Research. 2017;63(1):1–6.
110. Dures S, Carbone C, Loveridge AJ, Maude G, Midland N, Gottelli D. Population connectivity across a transboundary conservation network: potential for restoration? ResearchSquare [Preprint]. 2021. Available from: https://www.researchsquare.com/article/rs-452520/v1
111. Schell CJ, Dyson K, Fuentes TL, Des Roches S, Harris N, Miller DS, et al. The ecological and evolutionary consequences of systemic racism in urban environments. Science. 2020;369(6510):eaay4497.
112. Combs M, Byers KA, Ghersi BM, Blum MJ, Caccone A, Costa F, et al. Urban rat races: spatial population genomics of brown rats (*Rattus norvegicus*) compared across multiple cities. Proceedings of the Royal Society B: Biological Sciences. 2018;285(1880):20180245.
113. Combs M, Puckett EE, Richardson J, Mims D, Munshi-South J. Spatial population genomics of the brown rat (*Rattus norvegicus*) in New York City. Molecular Ecology. 2018;27(1):83–98.
114. Vargo EL, Crissman JR, Booth W, Santangelo RG, Mukha DV, Schal C. Hierarchical genetic analysis of German cockroach (*Blattella germanica*) populations from within buildings to across continents. PLoS One. 2014;9(7):e102321.
115. Fan X, Wang C, Bunker D. Population structure of German cockroaches (Blattodea: Ectobiidae) in an urban environment based on single nucleotide polymorphisms. Journal of Medical Entomology. 2022;59(4):1319–27.
116. Meerburg BG, Singleton GR, Kijlstra A. Rodent-borne diseases and their risks for public health. Critical Reviews in Microbiology. 2009;35(3):221–70.
117. Dean KR, Krauer F, Walløe L, Lingjærde OC, Bramanti B, Stenseth NC, et al. Human ectoparasites and the spread of plague in Europe during the Second Pandemic. Proceedings of the National Academy of Sciences of the United States of America. 2018;115(6):1304–9.
118. Goodman H, Egizi A, Fonseca DM, Leisnham PT, LaDeau SL. Primary blood-hosts of mosquitoes are influenced by social and ecological conditions in a complex urban landscape. Parasites & Vectors. 2018;11(1):218.
119. Easterbrook JD, Shields T, Klein SL, Glass GE. Norway rat population in Baltimore, Maryland, 2004. Vector-Borne & Zoonotic Diseases. 2005;5(3): 296–9.
120. Titcomb GC, Jerde CL, Young HS. High-throughput sequencing for understanding the ecology of emerging infectious diseases at the wildlife-human interface. Frontiers in Ecology and Evolution. 2019;7:126.
121. Firth C, Bhat M, Firth MA, Williams SH, Frye MJ, Simmonds P, et al. Detection of zoonotic pathogens and characterization of novel viruses carried by commensal *Rattus norvegicus* in New York City. mBio. 2014;5(5):e01933–14.
122. Weckel M, Wincorn A. Urban conservation: the northeastern coyote as a flagship species. Landscape and Urban Planning. 2016;150(Jun): 10–15.
123. Monzón J, Kays R, Dykhuizen D. Assessment of coyote–wolf–dog admixture using ancestry-informative diagnostic SNPs. Molecular Ecology. 2014;23(1):182–97.
124. Kays R, Curtis A, Kirchman JJ. Rapid adaptive evolution of northeastern coyotes via hybridization with wolves. Biology Letters. 2010;6(1):89–93.
125. Hody JW, Kays R. Mapping the expansion of coyotes (*Canis latrans*) across North and Central America. ZooKeys. 2018;759:81–97.
126. Muntz EM, Patterson BR. Evidence for the use of vocalization to coordinate the killing of a white-tailed deer, *Odocoileus virginianus*, by coyotes, *Canis latrans*. The Canadian Field-Naturalist. 2004;118(2):278–80.
127. Riley SPD, Pollinger JP, Sauvajot RM, York EC, Bromley C, Fuller TK, et al. A southern California freeway is a physical and social barrier to gene flow in carnivores. Molecular Ecology. 2006;15(7): 1733–41.

128. Sugden S, Sanderson D, Ford K, Stein LY, St Clair CC. An altered microbiome in urban coyotes mediates relationships between anthropogenic diet and poor health. Scientific Reports. 2020;10:22207.
129. Acosta-Jamett G, Chalmers WSK, Cunningham AA, Cleaveland S, Handel IG, Bronsvoort BMdC. Urban domestic dog populations as a source of canine distemper virus for wild carnivores in the Coquimbo region of Chile. Veterinary Microbiology. 2011;152(3–4):247–57.
130. Brooks J, Kays R, Hare B. Coyotes living near cities are bolder: implications for dog evolution and human-wildlife conflict. Behaviour. 2020;157(3–4):289–313.
131. Wurth A. Behavior and Genetic Aspects of Boldness and Aggression in Urban Coyotes (Canis latrans) [Ph.D. Thesis]. [Columbus, OH]: The Ohio State University; 2018.
132. Edmands S. Between a rock and a hard place: evaluating the relative risks of inbreeding and outbreeding for conservation and management. Molecular Ecology. 2007;16(3): 463–75.
133. Bradbury D, Smithson A, Krauss SL. Signatures of diversifying selection at EST-SSR loci and association with climate in natural *Eucalyptus* populations. Molecular Ecology. 2013;22(20): 5112–29.

CHAPTER 10

Urban Flagship Umbrella Species and Slender Lorises as an Example for Urban Conservation

Kaberi Kar Gupta[‡,§], Madhusudan Katti[¶], Vidisha Kulkarni[#], Hari Prakash J. Ramesh[‡], Harshitha C. Kumar[‡], Kesang Bhutia[||], Soumya Kori[||], Rajeev Bacchu[‡], and Arun P. Visweswaran[‡]

Introduction

One of the major goals of conservation biology is to conserve species and biodiversity from extinction. However, given the threats to an ecosystem or to a particular species, conservation biologists often use diverse surrogate species to tackle the problems in conservation. Tiger conservation as a proxy for broader landscape conservation in India is one of the earliest examples of using a surrogate species (1). Surrogate species are used to indicate the extent of anthropogenic pressures, monitor the population of the species' health in an ecosystem, locate high-biodiversity areas, design a reserve, or create a conservation plan for the sympatric species in the area. More importantly, some surrogate species are regularly used as flagships in a sociopolitical context to attract public and policymakers' attention and fund large projects for conservation (2). Often conservation biologists use surrogate or proxy species terminology loosely or as a catch-all term; however, there are multiple surrogate species terms that are in consistent use (3). Indicator species are used as proxies for the health of habitats and may be used for assessing pollution and water or air quality. Umbrella species are typically larger in body size, and the viability of their populations is used as a proxy for assessing and managing the viability of other background species. Flagship species are those surrogate species which are used to bolster public awareness on conservation. Proxy or surrogate species are fundamentally social-ecological tools. Conservationists require intricate biological knowledge to ensure the chosen species functions as intended with respect to representing environmental conditions or the needs of other species. But, as or more importantly, the chosen species needs to be charismatic enough to foster social interest and powerful enough to influence decision makers' choices.

In nonurban settings, proxy species have been used with great success. For example, in the Indian context, large mammals such as tigers or elephants—which are endangered megafauna—are used as flagship species for conserving biodiversity in forested landscapes (4). However, in urban biodiversity conservation, such large species cannot be used as surrogate species for urban habitat conservation as they are not typically found in cities or cause too much conflict. Therefore, finding surrogate species of conservation value for

‡ Urban Slender Loris Project, India
§ Biodiversity Lab, North Carolina Museum of Natural Science, USA
¶ North Carolina State University, USA
# Center for Ecological Science, Indian Institute of Science, India
|| Ashoka Trust for Research in Ecology and Environment, India

Kaberi Kar Gupta et al., *Urban Flagship Umbrella Species and Slender Lorises as an Example for Urban Conservation.* In: *Urban Biodiversity and Equity.* Edited by: Max R. Lambert & Christopher J. Schell, Oxford University Press. DOI: 10.1093/oso/9780198877271.003.0010 urban habitats is essential, challenging but especially in urban areas which harbor high biodiversity and which are in close proximity to biodiversity hotspots.

Conservation in and for Bengaluru

Globally, by 2031 urban areas will triple from 2011, and the human urban population will be double what it was in 2000, leading to increased pressures on biodiversity if urban planning continues as it is currently practiced (5). India is one of the world's largest countries and is also experiencing tremendous urban development and population growth. Yet, there are very few studies in India that focus on urban biodiversity (6).

Bengaluru—India's capital and the largest city in the state of Karnataka—is emblematic of Indian and global urban biodiversity challenges because it has high biodiversity which is under pressures from rapid development as it has grown from being a small city to one of the Indian megacities (7). Bengaluru is one of the cities near two biodiversity hotspots—the Western Ghats mountain range and the Nilgiri Biosphere Reserve (8). Because of its geography, the city of Bengaluru also has a high diversity of flora and fauna (e.g., 340 species of birds and 41 mammal species) (9).

Bengaluru is one of the fastest-growing megacities in the Global South. With unplanned growth and development of the city, there is a concern about the loss of green space and wildlife within Bengaluru's urban area (7,10). According to Bruhath Bengaluru Mahanagara Palika (BBMP, i.e., the city's municipal corporation), this once-small city of 69 sq km in 1949 has now expanded to 741 sq km (Figure 10.1) in all directions including to the biodiversity hotspot of Bannerghatta National Park (11). Development activities have included widening roads, constructing densely packed new houses and multistoried buildings, building elevated structures and metro lines, and paving express highways. Concurrently, there has also been a growth in bird and butterfly watching along with a broader interest in nature and wildlife among the residents of Bengaluru. These have included an increased use of the eBird platform and the establishment and high use of the Butterfly Park of Bangalore (11).

Slender loris

Bengaluru is home to a unique primate species endemic to southern India that requires specific habitat conditions to persist. The nocturnal, arboreal slender loris (technically two species: *Loris tardigradus* and *L. lydekkerianus*) is a small nocturnal primate endemic to southern India and Sri Lanka (Figure 10.2) (12). Their diminutive size (less than 200 g in weight and average 20 cm in length) and nocturnal habits has caused the slender loris to be considered elusive and hard to study until dedicated research in the wild in the mid-1990s (13,14). Lorises are hard to see and the only way to find them is by using a light to detect eye shines. They also have distinct shrill-like calls that can be heard from far away. Slender lorises forage solitarily and roost in tree canopies, especially in clumps of thorny bushy structures formed by thorny climbers. In the wild their roosts are typically five meters high but they are typically higher in urban areas, making it even harder to detect them.

Slender lorises pair bond for one or multiple breeding seasons and sleep either in pairs of adults or as groups of a mother and offspring but will occasionally roost alone in the tree canopies and not in nest holes. Three types of adult males have been reported based on morphology, behaviors, and home range sizes. One type of male has smaller home ranges that overlap with that of an adult female, a second has home ranges adjacent to those of adult females, and some adult males roam over large areas covering multiple male and female ranges. Adult females have small ranges compared to sub-adult females. Adult females with infants carry their young for the first several months of life. Then they rest their young in the tree canopy while they forage nearby (14).

Despite this natural history knowledge, there are no studies on slender loris population genetics nor any genetics studies on species or subspecies identities (12). Given the paucity of information about loris populations, the taxon was included on the list of endangered species protected under Schedule 1 of the Indian Wildlife (Protection) Act of 1972, as well as being designated as vulnerable on the International Union for Conservation of Nature's (IUCN's) Red List. Subsequently, several species

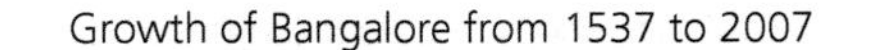
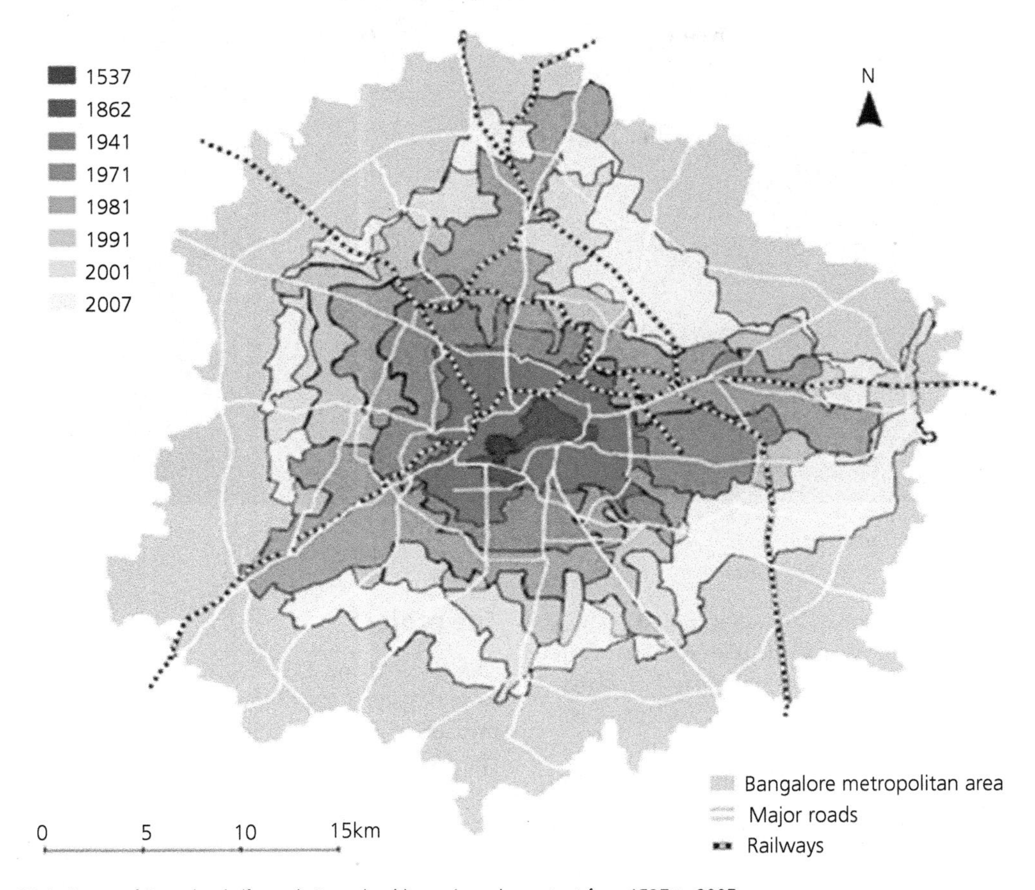

Figure 10.1 A map of Bengaluru's (formerly Bangalore's) growing urban extent from 1537 to 2007.
Map produced courtesy of H.S. Sudhira.

of slender loris have been identified, including the grey slender loris endemic to the southwestern Indian peninsula where Bengaluru is also located.

The urban loris

Lorises need a continuous tree canopy for movement, an abundance of insects, and a dense canopy structure for sleeping sites (14). These are not habitat characteristics usually found in urban areas, although there have been sporadic reports of urban lorises. In 2009–10, lorises were reported from the Indian Institute of Science Campus (IISC) by a group of ecologists at the Center for Ecological Sciences. During a visit in 2011 to Bengaluru we surveyed for lorises on the campus. Interestingly, prior to this time, most of the long-term residents on campus and former students were unaware of the lorises' presence. Since then, regular reports of slender lorises have garnered interest from students and scientists who are interested in this species. By 2013, there were not only regular reports of lorises but also injured lorises found on campus. The lorises were starting to gain attention in the city beyond the campus, which had not occurred before these several years of loris observations. Generally, in rural areas, slender lorises are considered a bad omen, yet their new popularity in the city made lorises a charismatic but enigmatic hard-to-see animal. During this same time, environmentalists were concerned about Bengaluru's urban growth and vanishing wildlife.

Figure 10.2 An urban slender loris.
Photograph courtesy of Kalyan Verma.

To address these concerns, we initiated a study of urban wildlife in Bengaluru that involved community members who were interested in urban conservation. The slender loris was a focus of this work. In 2013, we began surveying for the slender loris at the IISC, which became the starting point for what became the Urban Slender Loris Project (USLP) (15). We started finding lorises not only on that campus but in other parts of the city where a relatively dense and continuous tree canopy was still present. Such a forest canopy was once abundant in the city of Bengaluru but has continued to decline and become fragmented as the city has grown. Local residents of South Bangalore also reported that lorises were rescued from trees that were felled for road extensions. Now the population status of the slender loris in Bengaluru remains uncertain due to urban growth and the disappearance of forest habitats. Besides habitat loss, illegal hunting is also thought to threaten lorises in the wild, as the animals are sought after for their alleged medicinal properties. There was also a report of smugglers caught at Delhi Airport with two slender lorises in September 2012. However, we do not know the source of these animals. Further, there have been several dozen reports over the past few years of slender lorises being rescued and brought into rehabilitation centers in Bengaluru City.

The project

The goal of the USLP is to use the slender loris as a flagship umbrella species to understand the impacts of growth and development on urban green space and the biodiversity that relies on those green spaces. Both public fascination for slender lorises and its specific habitat requirements make the slender loris a good flagship umbrella species for urban conservation by garnering public support for conservation and also protecting habitat for the loris and other co-occurring species.

We initiated a citizen science-based pilot study in January 2014 based on a collaboration between local citizens, scientists, naturalists, environmental educators, nonprofit and government organizations (e.g., forest department, BBMP), and educational institutions. Our partners are the Indian Institute of Science, Gubbi Lab, Dakshin, Eco Edu, Koshy's (a landmark cafe and restaurant at the heart of the city), Save Tiger First, and the National Center for Biological Sciences (NCBS). This project uses an interdisciplinary, participatory, collaborative, community-based conservation effort to study and protect urban slender lorises and their urban habitats. Such an effort has not been done before in India. The aims of the project are (1) to document and map the loris population within Bengaluru's green spaces, (2) to understand loris distribution in the past, and (3) to develop a partnership between organizations, community members, and local governments for the conservation of urban lorises and their habitats in Bengaluru. As a "charismatic mini-fauna" species, the slender loris thus became an umbrella flagship species for urban biodiversity conservation in Bengaluru.

During 2013–16, USLP organized questionnaire-based online surveys, interviews, and social media to advertise and identify participants for the project. We trained a core team of volunteers for nocturnal loris and vegetation surveys. In total, 180 people participated in the first online survey. Of these, 84 people reportedly had seen lorises (mostly on IISC) and the others had never seen any lorises before in Bengaluru or elsewhere.

Ethnographic interviews of selected long-term residents of Bengaluru were conducted to assess their knowledge of slender lorises and the forest patches they remember from 40–50 years ago. We also interviewed community members about perceptions of slender lorises given that lorises are perceived as bad omens and consumed for alleged health reasons (16). We worked with local schools and colleges on outreach programs on the urban lorises (giving popular talks and hosting loris walks). The USLP conducted online surveys initially at the IISC, the NCBS, and the University of Agricultural Sciences (UAS) in Bengaluru. We selected these campuses because they retain a high tree canopy cover with sizable forest patches, include many students and naturalists motivated to participate in citizen science research, and are where loris sightings were frequently reported. This survey helped us identify locations and habitat types of where lorises had putatively been observed in the past 10 years.

During 2015, the team conducted nocturnal surveys in green spaces identified as possible loris habitats within the city, based on the presence of a reasonably continuous tree canopy and the size of the park. The city was divided into a 5 sq km grid. In each grid patch we identified green spaces and then conducted a survey to ground-truth sites and their vegetation structure (Figure 10.3). The USLP trained volunteers for multiple days about loris behavioral ecology and the techniques for surveying for lorises at night and habitat during the daytime. Each volunteer who wanted to participate in the survey was trained for three days. A core team was formed to coordinate and maintain surveys, training, and communication with volunteers/community members, media, and government agencies of the project. USLP surveyors would survey lorises using a visual encounter by seeing loris' eye shine during a survey lasting 1.5 hours between 1830 and 2000 hrs. Subsequently, at each sighting location, the team leaders conducted habitat surveys during the daytime to document tree species identities, canopy cover, height of trees, presence of lianas and climbers, and any visible disturbances or damage to habitat. After the initial training and team building in 2015, the core team continued a variety of activities to build citizen science capacity including training volunteers, maintaining a website and social media, organizing loris walks, filming, presenting, and communicating with local communities, rescue groups, and rehabilitation centers.

Over several years, we surveyed a total of 18 sites within the city, including institutional campuses and city parks. This also included Cubbon Park which is the largest park of Bengaluru and which is currently surrounded by state administrative buildings. Cubbon Park was built in the 1870s by the British Government at the heart of the city on a 100-acre parcel, has since been expanded to over 300 acres in size, and is considered the "lung" of the city. Inside the park there are native and non-native trees, walking paths, motorable roads, playgrounds, and other park amenities (16). Across all of our surveyed sites, we recorded the highest encounter rate in the institutional campuses rather than in the City Parks and reserved forest patches within the city. We documented lorises using trees in the city at heights of one meter in the less disturbed sites and up to 20 meters high in the most disturbed sites. We found that lorises were using a diversity of plant species (including the abundant non-native species) such as bougainvillea, a thorny climber, as well as *Acacia, Parkia, Eucalyptus, Pongamia, Delonix, Tamarindus*, and bamboo clumps. From the online surveys and ethnographic interviews, we identified 10 sites where animals were recorded in the last two to three decades, but which do not have any green cover left (Figure 10.3).

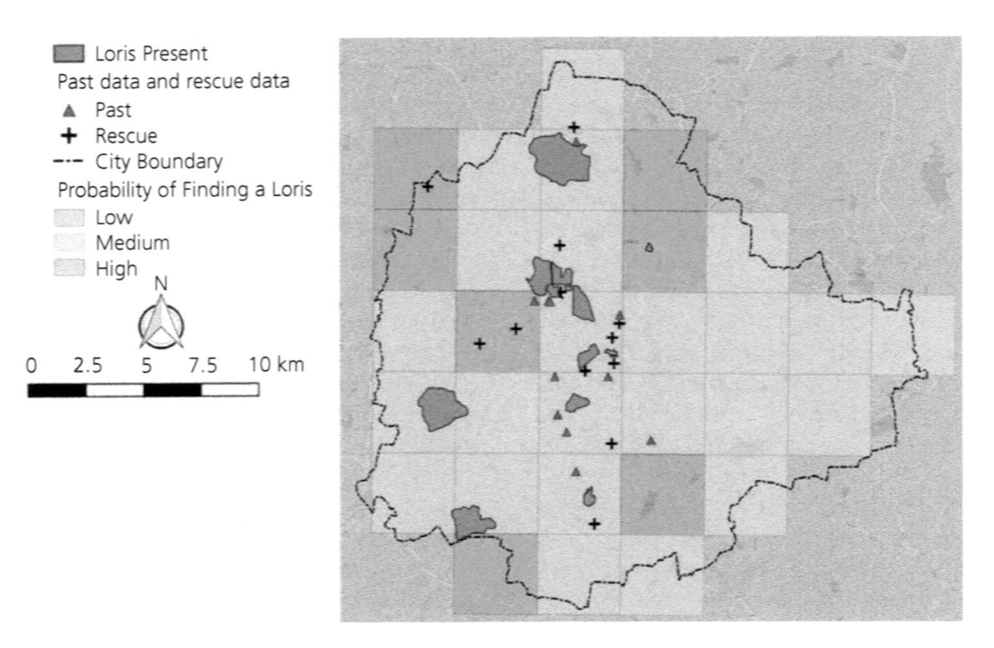

Figure 10.3 Presence and absence of lorises in Bengaluru from the Urban Slender Loris Project.

Social-ecological applications: capacity building and outreach to save loris habitat

It took us many months to build USLP. Substantial time (three months in 2013 and six months in 2014–15) was spent on engaging community members. Meeting community members, building connections with local environmentalists and naturalists, conducting interviews with newspapers, writing articles, giving popular talks in cafes and restaurants and offices, radio interviews, and networking with other organizations took considerable and unanticipated time (Figures 10.6 and 10.7). Our core team trained 176 volunteers from 2013 to 2016 and organized monthly loris walks at night for community members until 2020. The project attracted and continued to receive significant media attention including 42 articles in local and national/international outlets and posts on different blogs (Figure 10.6). Our team has given 10 popular talks led by citizen scientists to schools, colleges, and various public spaces, conducted 10 radio interviews and four podcasts, presented findings at 16 different international scientific meetings and organized loris walks for several locally hosted conferences, and discussed our work in six webinars. And this project has been cited in scientific journal articles, a book, and multiple reports (17–20). Further, a documentary film was made on the USLP by filmmaker Aknksha Sood Singh and Habitat Trust in 2018 that was shown on Animal Planet, Discovery Channel, and in film festivals. Over several years, we not only produced substantial data on the distribution and abundance of lorises throughout Bengaluru, but we built incredible community interest in urban slender lorises. We continue to organize loris walks for the public and students of IISC.

The intensive community interest and awareness we built allowed us to leverage the loris as a flagship umbrella species to protect urban trees. Because of this, the urban slender loris has been instrumental in stopping new development in some prime slender loris habitats within the city of Bengaluru. For instance, in 2016, a developer group was in dispute with the Karnataka State Forest Department over land at the heart of the city. The land was one of our study sites and was part of the one sq km patch of forest adjacent to the Forest Department

Case study Nagawada

The USLP surveyed a Reserve Forest patch at Nagawada, at the city's edge, which was under the jurisdiction of the Karnataka State Forest Department. During our survey from 2015–16, we found a breeding population of six slender lorises in part of Hennur Lake which satellite images show had dried up in the 1980s. The lakebed and adjacent areas were reclaimed by the Karnataka Forest Department as Reserve Forest in the late 1980s and by 1990 this area was protected by the Forest Department. Between 1990 and 2015, this area showed successive stages of increased forest cover (Figure 10.4). However, a highway (Ring Road) was built along this Reserve Forest as the city grew and new development and the information technology industry grew around it in the 2000s. With the increasing pressures from the developers around this forest, the State Forest Department converted this forest patch of 30 acres into a manicured city park, clearing all vegetation, burning certain parts, and concretizing most of the park (Figure 10.5). Our surveys documented a breeding loris population in addition to other wildlife like reptiles, amphibians, birds, and insects. We presented our 2015 findings to the Karnataka Forest Department, suggesting that part of the park be maintained with natural vegetation to preserve the lorises and other biodiversity while the remainder of the park could be managed for public use. Unfortunately, this patch is now mostly cleared of understory vegetation as it has been converted largely to a neighborhood public park for the homeowners in the new developments of the growing city. Because of the near-total loss of trees, lorises are almost certain to have disappeared from this site and USLP core team members reported that, by 2020, the park did not have any loris habitat left. The pressure on government organizations for converting protected areas to manicured urban parks with no native vegetation is a phenomenon that was never seen before in Bengaluru. Despite this, Nagawada is also an important example of how vacant land can grow biodiverse forest cover within 20 years if left relatively unmanaged. The more manicured urban parks are often preferred for recreational use by people in such fast-growing cities, but better-informed planning that also addresses the requirements of a flagship umbrella species like the slender loris can help save some of the local flora and fauna and promote their coexistence with people.

Figure 10.4 A map of forest cover change in the Nagawada Reserve Forest from 2000 to 2014.
Imagery from Google Earth.

Case study *Continued*

Figure 10.5 Nagawada Forest transformation, from urban forest to city park.

headquarters and connected to the Indian Institute of Science, Central Power Research Institute, and Raman Research Institute. Our data showed a population of slender lorises at this site and, with our data, a consortium of nonprofit conservation organizations in the city petitioned the Karnataka High Court. Because the slender loris is a Schedule 1 species under the Indian Wildlife Act, it is guaranteed the highest conservation priorities. The USLP's slender loris data and this petition stopped the development of a dense urban forest patch. Concerned citizens also used USLP data in 2017 to stop construction of a flyover bridge in the city in forest patches where the lorises occurred (20). These examples illustrate the power of community-engaged science to develop a flagship umbrella species that can be used to prioritize and protect urban habitat and biodiversity.

Global perspective: translating urban proxy species for the world's conservation

The USLP approach to reconciliation ecology (21) represents a successful model for similar efforts in other cities in India and around the world. This approach is particularly useful where the public interest in and local knowledge of wildlife and nature can be capitalized on to build community-engaged and community-driven urban wildlife conservation efforts. The USLP also highlights the

Slender Loris thrive in city's green spaces

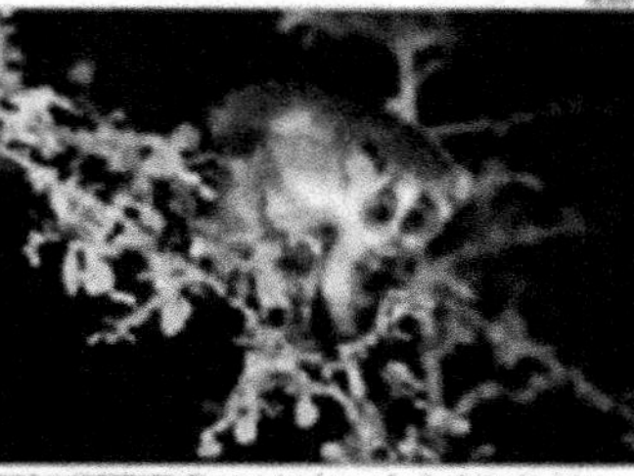

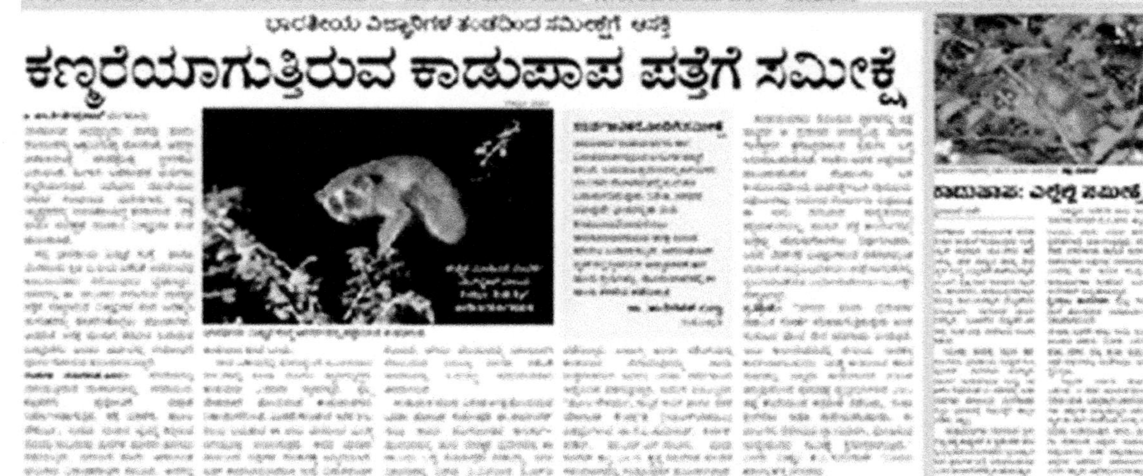

ಕಣ್ಮರೆಯಾಗುತ್ತಿರುವ ಕಾಡುಪಾಪ ಪತ್ತೆಗೆ ಸಮೀಕ್ಷೆ

Figure 10.6 Published newspaper articles on the Urban Slender Loris Project from diverse local, national, and international outlets.

opportunities for using interesting or charismatic urban species as flagship umbrella species to protect urban habitat and biodiversity. Conservation groups have consistently used flagship umbrellas and other proxy species to great success globally. For instance, the World Wildlife Fund (WWF) has used pandas as a charismatic species to generate public interest and funding, conservation groups in the Pacific Northwestern US used the spotted owl to protect old-growth forests from the timber industry, and Project Tiger has used tigers in India to create tiger reserves and protect large areas countrywide. However, there are few examples like the USLP that translate international lessons surrounding flagship umbrella species to urban conservation. There is tremendous opportunity for cities around the world to leverage their own flagship umbrella species or other proxy species to embolden their communities, protect existing urban habitats, and even rewild developed urban areas. Such proxy species have been successful in nonurban settings globally and remain a largely untapped tool for urban conservation.

The past and future of urban slender lorises and conservation in Bengaluru

Below are three stories from local residents of Bengaluru, as reported to USLP:

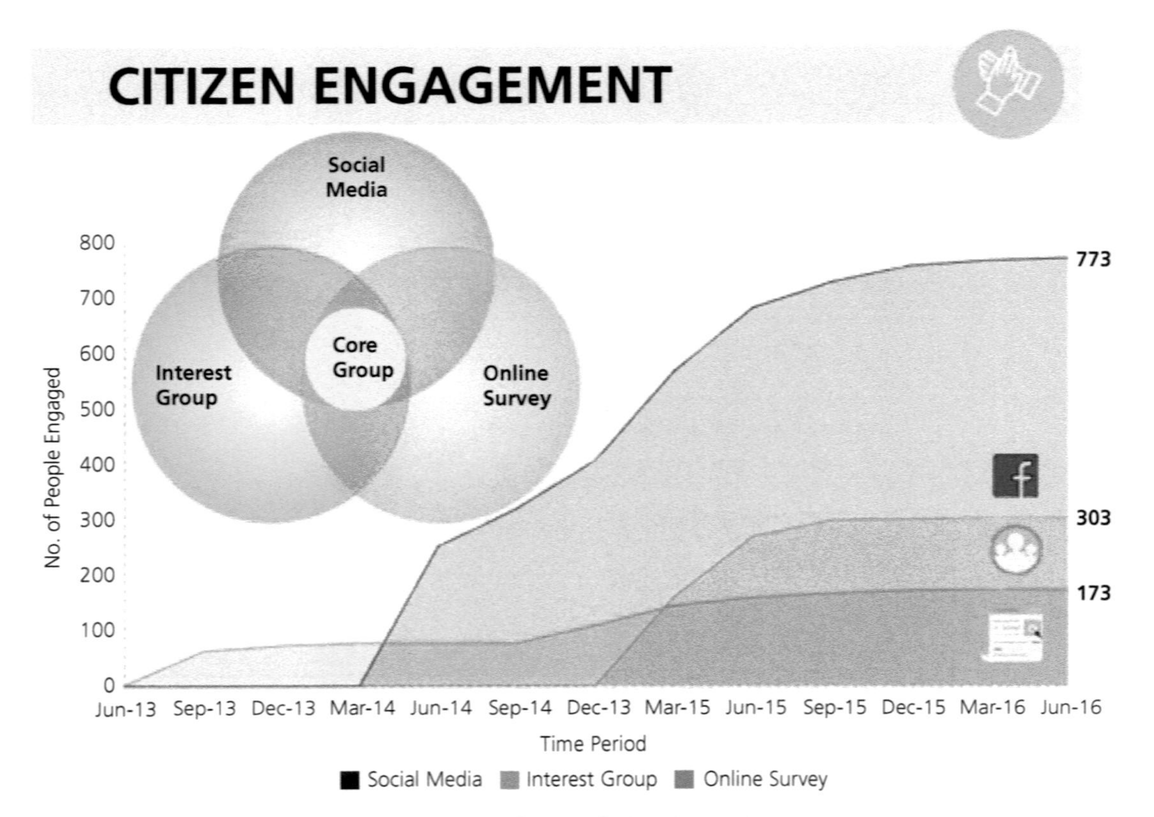

Figure 10.7 Community engagement and capacity building over four years for the Urban Slender Loris Project.

(1) Wildlife illustrator Maya Ramaswamy: "I had a slender loris. Someone saw it was being sold in the Shivaji Nagar Market. It was the 1980s. So, someone saw and rescued it from the seller. He knew I like animals so brought it to me. I had it in our house. Every evening it would go out through the window in our garden. It would forage all night long and then would come back early in the morning. One day it had eaten something, I don't know what. It came back not well. Then it went out in the evening and never came back."

(2) Mr. Prem Koshy of Koshy's Restaurant: "You know we had three lorises in our house when I was young (1960s). That time in our house we had gardens. They would go out at night in the garden and would come back to sleep in the house. Sometimes they would be in the kitchen catching cockroaches and eat them. We liked it because they would eat cockroaches. In those days lorises were so common in Bangalore. They were like squirrels. We never thought much about them. You know I thought they had all gone extinct from Bangalore. We don't have any more jungle or gardens left in the city. You really made my day showing loris and telling me about them. I am so happy to see them in IISC."

(3) Rakesh Kombra, journalist: "When I was in school I had a slender loris. Someone got it from the roadside seller. They thought I would take care of it. I used to have it with me all the time. When I would go to school by bicycle, I would take it in my shirt pocket. In the classroom, it would go under the desk and sleep on the wooden part, underneath the desk. Then when school ended, I would put it in my pocket and come home. I would give it food.

Sometimes it would go out and come back in the morning."

From these and other stories and reports, it is likely that lorises have been in and around Bengaluru City for a long time. In the absence of past records, it is unclear what the historical distribution of slender lorises was before the city's explosive growth in recent decades. Going forward, we may be able to clarify historical occurrences by conducting interviews among the local Indigenous groups who live in small hamlets within the city. Another study on local knowledge of these communities suggests that people do still have ecological knowledge of the area even though the habitats around them have changed (22).

This local knowledge also highlights the importance of equity and inclusion in community science. The survey done by USLP found that urban low-income communities and lower social castes have greater knowledge about lorises than the upper caste. Further, as Bengaluru expanded, many rural areas were subsumed by the city but remained as hamlets within the city (22). In spite of being part of the city, people in these hamlets often do not perceive themselves as part of the city but recognize the loss of access to their sacred sites such as lakes and forest patches (16).

One of the challenges and shortcomings of the USLP so far has been the inability to develop inclusive representation across diverse socioeconomic strata in this urban conservation work. In our early phases, our surveys for slender lorises tended to include primarily upper socioeconomic class members of the community. Many of our volunteers work in the information technology industry or in higher educational institutions. Funding limitations have made it challenging for the USLP to diversify our participants in this citizen science project. In the next phase the USLP is working to include schools and teachers from Bengaluru's marginalized communities. Further, we aim to work with these communities to produce a repository of local biocultural knowledge on flora, fauna, and the local geography before Bangalore became a megacity. The long-term success of urban conservation—both for slender lorises but also more broadly—relies on an equitable and inclusive participation.

Our citizen science project has been a first of its kind for India. Diverse people were able to come together to form an interest group for urban wildlife conservation in a rapidly developing region. Using citizen scientists to study and protect a cryptic, nocturnal animal is challenging, but has also been rewarding. The slender loris is considered an endangered species under the Indian Wildlife Act of 1972 and we have demonstrated that we can protect this species and its habitat in the city by capitalizing on its charisma. Doing so protects not only the lorises but also the many other species which co-occur with it. The USLP represents a model for conservation efforts globally to organize diverse groups and elevate flagship, umbrella, and other proxy species to promote coexistence between people and biodiversity in cities. Ultimately, proxy or surrogate species are social-ecological tools that require powerful biological knowledge as well as tremendous efforts to develop the social capital needed to protect the focal species and use that species to conserve habitat for other species.

References

1. Panwar HS. What to do when you've succeeded: Project Tiger, ten years later. In McNeely JA and Miller KR (eds.) National Parks, Conservation, and Development: the role of protected areas in sustaining society. Washington, DC: Smithsonian Institution Press; 1984. p. 183–9.
2. Caro TM, O'Doherty G. On the use of surrogate species in conservation biology. Conservation Biology. 1999;13(4):805–14.
3. Caro T. Conservation by Proxy: indicator, umbrella, keystone, flagship and other surrogate species. Washington, DC: Island Press; 2010.
4. Johnsingh A, Joshua J. Conserving Rajaji and Corbett National Parks – the elephant as a flagship species. Oryx. 1994;28(2):135–40. Available from: https://doi.org/10.1017/S0030605300028453
5. Secretariat of the Convention on Biological Diversity (CBD). Cities and Biodiversity Outlook: Action and Policy: a global assessment of the links between urbanization, biodiversity, and ecosystem services. Released at the Cities for Life Summit, parallel to the eleventh meeting of the Conference of the Parties to the CBD, October 15, 2012, Hyderabad, India. Montreal, Canada: Secretariat of the Convention on Biological Diversity; 2012.

6. Aronson MF, La Sorte FA, Nilon CH, Katti M, Goddard MA, Lepczyk CA, et al. A global analysis of the impacts of urbanization on bird and plant diversity reveals key anthropogenic drivers. Proceedings of the Royal Society B: Biological Sciences. 2014;281(1780):20133330. Available from: https://doi.org/10.1098/rspb.2013.3330
7. Nagendra H. Nature in the City: Bengaluru in the Past, Present, and Future. New Delhi: Oxford University Press; 2016. 244 p.
8. Mittermeier RA, Turner WR, Larsen FW, Brooks TM, Gascon C. Global biodiversity conservation: the critical role of hotspots. In: Zachos F, Habel J (eds.) Biodiversity Hotspots. Berlin, Heidelberg: Springer; 2011. p. 3–22. Available from: https://doi.org/10.1007/978-3-642-20992-5_1
9. Karthikeyan, S. The Fauna of Bangalore. Bangalore: World Wide Fund for Nature-India; 1999.
10. Nagendra H, Sudhira HS, Katti M, Tengo M, Schweniu M. Urbanization, ecosystems and biodiversity: assessments of India and Bangalore. Released at the Cities for Life Summit, parallel to the eleventh meeting of the Conference of the Parties to the Convention on Biological Diversity (CBD), October 15, 2012, Hyderabad, India. 2012.
11. TV R, Aithal BH, Sanna DD. Insights to urban dynamics through landscape spatial pattern analysis. International Journal of Applied Earth Observation and Geoinformation. 2012;18(Aug):329–43. Available from: https://doi.org/10.1016/j.jag.2012.03.005
12. Kar Gupta K. Slender loris. In: Johnsingh AGT, Manjrekar N (eds.) Mammals of South Asia. Hyderabad, India: Orient BlackSwan Press; 2012. p. 94–108.
13. Kar Gupta K. Slender loris (*Loris tardigradus*) distribution and habitat use in Kalakad-Mundanthurai Tiger Reserve, India. Folia Primatology. 1998;394.
14. Kar Gupta K. Ecology of Slender Loris [Ph.D. Dissertation]. [Tempe, AZ]: Arizona State University; 2007.
15. KarGupta K. Urban slender lorises of Bangalore. JLR Explore; 2014 Nov 15.
16. Bhaskaran V, Nagendra H, Kar Gupta K. The study of people's attitudes towards lorises in Bengaluru. In press. Available from: http://urbanisationjournal.com
17. Pulla P. Indian ecologists turn to crowdsourcing. Nature India. 2013 Nov 20. Available from: https://www.nature.com/articles/nindia.2013.152
18. Soumya E. The night-time hunt for the secretive urban slender loris of Bangalore. Guardian; 2015 Jul 31. Available from: http://www.theguardian.com/cities/2015/jul/31/urban-slender-loris-bangalore-india-animal
19. Beatley T. Citizen science and community engagement. In: Handbook of Biophilic City Planning and Design. Washington, DC: Island Press; 2016. p. 145–7. Available from: https://doi.org/10.5822/978-1-61091-621-9_14
20. Nagendra H, Mundoli S, Nishant V. Report on environmental and ecological impacts of tree felling for proposed steel flyover on Bellary Road and road widening of Jayamahal Main Road, Bengaluru. Bengaluru, India: Azim Premji University; 2017.
21. Rosenzweig ML. Win-Win Ecology: how the Earth's species can survive in the midst of human enterprise. Oxford: Oxford University Press; 2003.
22. Unnikrishnan H, Manjunatha B, Nagendra H. Contested urban commons: mapping the transition of a lake to a sports stadium in Bangalore. International Journal of the Commons. 2015;10(1):265–93. Available from: https://doi.org/10.18352/ijc.616

CHAPTER 11

Animal Behavior, Cognition, and Human–Wildlife Interactions in Urban Areas

Lauren A. Stanton[‡], Christine E. Wilkinson[‡], Lisa Angeloni[§], Sarah Benson-Amram[¶], Christopher J. Schell[‡], and Julie K. Young[#]

Introduction

Interactions between humans and wildlife are accelerating as urban expansion continues (1). Human–wildlife interactions in cities typically arise when humans engage with nature (e.g., visiting green spaces or installing bird feeders) and/or when animals exploit anthropogenic resources and infrastructure for food, travel, and shelter. These interactions subsequently lead to a range of outcomes, including positive, neutral, and negative outcomes that vary in both intensity and frequency (2–4). Though most human–animal interactions are either positive (e.g., derived intrinsic and cultural benefits) or neutral, negative interactions garner considerably more attention than the others (2), with tremendous potential to sway public opinion about individual organisms and biodiversity more broadly. Moreover, given the broad diversity of human perceptions and tolerance of wildlife, a single interaction can yield wildly different outcomes for both the people and wildlife in an interaction. It is thus essential to deconstruct the increasing complexity, subjectivity, and pluralistic perspectives of human–animal interactions (5,6).

These negative interactions, referred to as human–wildlife conflict (HWC), can yield various negative manifestations for people, such as livestock depredation, injury or loss of pets, physical attacks, disease transmission, and traffic accidents (e.g., 7–10). HWC also includes human effects on wildlife, such as habitat destruction, poisoning, disease, physical and chemical pollution, predation by pets (e.g., by domestic cats; *Felis catus*), and poaching (11–14). Although conflict can take many forms, most conflict results from wildlife behaviors that are misaligned with human desires, which can be highly variable. Consuming food or securing shelter on human property, for example, may be tolerated or even encouraged by humans if the species in question is not perceived as a threat or a nuisance (e.g., squirrels; [*Sciurus sp.*] and songbirds [*Passeri*]). However, if feeding or taking shelter is perceived as problematic (e.g., rats [*Rattus sp.*] and coyotes [*Canis latrans*]), this may motivate residents or urban wildlife managers to prevent such behaviors from occurring (Figure 11.1). Thus, the process by which and intensity with which conflicts arise with wildlife in urban environments, and how conflicts are managed, are not only dependent on the needs of a particular species, but also the perception of that species by people (15). Because high human densities compel wildlife to live alongside people much more intimately in urban versus nonurban areas, cities must produce strategies that actively foster coexistence. Doing

[‡] Environmental Science, Policy, and Management, UC Berkeley, USA
[§] Colorado State University, USA
[¶] Simon Fraser University, Canada
[#] Utah State University, USA

Lauren A. Stanton et al., *Animal Behavior, Cognition, and Human–Wildlife Interactions in Urban Areas*. In: *Urban Biodiversity and Equity*. Edited by: Max R. Lambert & Christopher J. Schell, Oxford University Press. © Oxford University Press (2023). DOI: 10.1093/oso/9780198877271.003.0011

Figure 11.1 Illustrations of how foraging in urban environments may lead to negative interactions between people and wildlife. Although seed feeders are typically intended for birds, other species, such as rodents, will capitalize on them as a source of food (top left). Many species, including pigeons, will feed on garbage, especially when it is not contained or easily accessible (bottom left). Gardens and fruit trees provide urban residents with fresh food and beauty, but are also appetizing to generalist species like chacma baboons (top right). Urban carnivores, including coyotes, may prey upon free-ranging domestic species (middle right). Providing bowls of dry kibble for outdoor cats is prohibited in many places due to the attraction of wildlife, including striped skunks and raccoons (bottom right). The degree to which each foraging scenario is considered problematic will depend on the perception of the people involved, and, in most cases, can be mitigated by human actions and behaviors. Images courtesy of: Jonathan Bliss (mouse on bird feeder); Elizabeth Carlen (pigeons on garbage); Gaelle Fehlmann (baboon on fence); Tali Caspi (coyote with cat); Sarah Benson-Amram (skunk and raccoon food bowl).

so provides a mechanistic pathway that will be essential for conserving biodiversity in cities.

Here, we briefly review how human–animal interactions in urban systems shape wildlife behavior and cognition, highlighting how, when, and why HWC arises in urban environments. We illustrate how conservation behavior principles can be applied in wildlife management, and how these same principles can be co-opted to promote human–wildlife coexistence. Finally, we highlight how the integration of wildlife and human behavioral responses into management can help promote equitable coexistence strategies in human-dominated landscapes.

Understanding and managing urban wildlife behavior

Urban wildlife cognition and behavior

Detecting and avoiding danger, finding food and shelter, and securing mates are critical aspects of fitness and survival in any given environment. To overcome these challenges, animals rely on a variety of **sensory** and **cognitive** abilities that successfully guide their **behavior**. This includes **perception**, **attention**, recognition and categorization of cues, as well as several forms of **learning**, memory, and **problem solving** (16). Over time and generations, animals develop cognitive specializations and behavioral adaptations that enable them to overcome challenges experienced in their environment, shaping each individual's **umwelt** (16). Because urban environments are different from the ancestral environments of many species, animals must learn how to adjust to the new, contemporary challenges associated with city life (17,18). For instance, urban environments may not provide the same resources that a species has encountered in its evolutionary past, and species compositions may vary tremendously within and between urban areas (17,19,20). Thus, urban animals may face different predator assemblages and sources of food than those experienced in nonurban or ancestral environments, or increased competition for resources (21). Furthermore, urban environments may have higher temperatures, greater physical and chemical pollution (e.g., lights, noise, lead), and novel anthropogenic stimuli (e.g., various artificial structures and objects) to which animals must respond appropriately (22).

Given the novelty, complexity, and heterogeneity of urban environments, cognition can allow wildlife to better persist in cities by avoiding environmental mismatches and maladaptation (e.g., **evolutionary traps**), and by creating or altering behaviors that facilitate resource acquisition and avoidance of danger (23). Such behaviors will emerge from a cognitive process whereby individual animals recognize opportunities and dangers and weigh the perceived costs and benefits associated with each, which are informed by several factors unique to the species and individual (24). The resulting behavioral adjustments often differentiate urban individuals from their nonurban counterparts.[1]

Common behavioral adjustments typically include shifts in diet, communication, activity patterns, problem-solving, and temperament (43–45). For example, anthropogenic or non-native sources of food have been documented in species across trophic levels including leopards (*Panthera pardus*; 46,28), house sparrows (*Passer domesticus*; 47,29), and American white ibis (*Eudocimus albus*; 48, 30). Urban individuals may also become conditioned to humans, demonstrating less fear (i.e., appearing bolder) or aggression (i.e., appearing more docile) compared to other, nonurban populations (49). In some instances, acquisition of new foods and other resources will require enhanced **behavioral** and **cognitive flexibility** or problem-solving skills (20,22). Indeed, some research suggests that urban species and individuals have larger relative brain sizes (50) and cognitive experiments with birds (e.g., *Parus major*; 33,34,51,52) and rodents (e.g., *Apodemus agrarius*; 53,35) have found superior learning and problem-solving abilities in urban individuals compared to their nonurban counterparts (Figure 11.2). Although not as widely studied, within-city heterogeneity can also shape wildlife behaviors and should be considered when managing urban wildlife. Urban cockatoos (*Cacatua galerita*) in Sydney, for instance, demonstrate neighborhood-specific bin-raiding techniques and may be in a cognitive arms race with humans despite tireless work to prevent the development of these undesirable behaviors (54,55) (Figure 11.3).

In these ways, cognition may serve as a mechanism that buffers individuals against environmental challenges within urban environments (i.e., **cognitive buffer hypothesis** (30,31,38,39)) and, paradoxically, places them at odds with humans (56).

[1] Avilés-Rodríguez et al.'s chapter discusses how modern molecular approaches demonstrate situations where urban wildlife populations are genetically diverging from nearby nonurban populations, including by spatial isolation and adaptation.

Glossary table

Term	Definition
Attention	The focus of an animal's interest or concern. Because an animal cannot attend equally to all of the stimuli that it perceives at a given time, attention is a limited commodity. An animal must, therefore, selectively attend to whatever is most salient. (77,16,25)
Behavior	The coordinated responses (i.e., actions or inactions) of living organisms to internal and/or external stimuli. Typically excludes responses more easily characterized as developmental changes. Unlike cognition, the behavior of an animal is directly observable. (16,26)
Behavioral flexibility	A broad term that refers to an individual organism's ability to modify its behavior in response to change and variation in its environment. Often used interchangeably with the term cognitive flexibility, which additionally considers the neural and cognitive mechanisms that underpin flexibility in behavior (e.g., inhibition control). (27,28)
Classical conditioning	A form of associative learning in which a relationship is formed between a novel stimulus and an existing stimulus, such that an animal learns one external cue predicts another (may also be referred to as Pavlovian conditioning). (16,29)
Cognition	The mechanisms by which individual animals acquire, process, store, and act on information from their environment. Unlike behavior, the cognition of an animal is difficult to observe and is often, therefore, inferred. (16)
Cognitive buffer hypothesis	Suggests that the primary adaptive function of a large brain is to buffer individuals against environmental challenges by facilitating the construction of behavioral responses. (30,31)
Ecological trap	A specific type of evolutionary trap where an organism makes a maladaptive habitat choice (despite the availability of higher-quality habitat). (32)
Evolutionary trap	A case where, often due to human activity, formerly reliable environmental cues no longer indicate high-quality resources and, consequently, lead an organism to make maladaptive behavioral or life-history choices that yield reduced fitness (despite the availability of higher-quality options). (18,32)
Habitat selection	A hierarchical decision-making process that results in an individual's disproportionate use of certain habitats over others, ultimately influencing their survival and fitness. (33)
Habituation	A form of nonassociative learning that leads to decreased responsiveness to a stimulus that is repeatedly encountered and not followed by any kind of reinforcement. Generally considered to be the opposing counterpart to sensitization. (16,34,35)
Human shield hypothesis	The idea that prey use areas with humans and human infrastructure as a buffer against predation risk. (36)
Landscape of fear	The spatially explicit distribution of predation risk perceived by individuals in a given population. (37)
Learning	A change in an animal's state that is gained through experience. There are many forms of learning, including associative (e.g., operant and classical conditioning) and nonassociative (e.g., habituation and sensitization), as well as asocial (i.e., gained through an individual's own experiences) and social (i.e., gained through observation of, or interaction with, another individual or their by-products). (16,38)
Operant conditioning	A form of associative learning in which a relationship is formed between a stimulus and a response, such that an individual learns to associate its behavior with a particular event or outcome (may also be referred to as instrumental conditioning). For example, aversive conditioning, a common wildlife management strategy, is a form of operant conditioning that creates an association between an undesirable behavior and a negative cue (e.g., fear, pain, illness). (16,29)
Perception	An individual's interpretation of sensory information within its environment. (16,39)
Problem solving	The ability to overcome challenges and obstacles in order to achieve a goal. This is often assessed by presenting animals with novel, but ecologically relevant, operant foraging problems (e.g., puzzles). (40,41)
Sensitization	A form of nonassociative learning that leads to increased responsiveness to a stimulus that is repeatedly encountered and followed by some kind of reinforcement. Generally considered to be the opposing counterpart to habituation. (16,34,35)
Sensory	Refers to an organ or system that conveys, or procures, sense impulses that allow an animal to collect information about its environment. (16,39)
Sensory modalities	Receptors of various stimuli (e.g., light, sound, smell, taste, touch, magnetism) that give rise to particular sensations (e.g., vision, audition, olfaction, gustation). (25)
Tolerance	Capacity to endure continued exposure to a stimulus or environmental condition (e.g., intensity of disturbance) before responding in a defined way. (35)
Umwelt	The integration of an individual's perceptual world and effector world. In other words, it is an animal's own self-world formed by the kinds of information its sensory modalities can process and it acts as the subject. (16,42)

Figure 11.2 Research on how humans and urban environments affect the behavior and cognition of wildlife with implications for human–wildlife coexistence. Left side (top to bottom): Marmosets approaching a researcher making behavioral observations at an urban park in Belo Horizonte (Minas Gerais, Brazil). Assessing the effects of human gaze on herring gull behavior in a coastal town (Cornwall, UK). Setting up a camera trap and novel object to assess animal boldness at an urban park in Oakland (California, US). Placing a Global Positioning System (GPS) collar on an urban leopard outside of Mumbai (Maharashtra, India). Right side: Image of a GPS-collared bobcat taken by a trail camera. Novel object testing to assess boldness in urban grackles. Giving-up density testing to assess risk perception of humans in California ground squirrels. A raccoon standing next to an automated testing device used to assess learning and cognitive flexibility. Puzzle tasks used to compare the problem-solving abilities of urban and nonurban Barbados bullfinches (white plastic cylinder) and striped field mice (colorful LEGO house).
Images courtesy of: Marina Duarte (marmosets), Madeline Goumas (gulls); Cesar Estien (camera trap); Nikit Surve (leopard); Kevin Crooks/CSU/USGS (bobcat); Alison Greggor (grackle); Jennifer E. Smith (ground squirrel); Lauren Stanton (raccoon); Louis Lefebvre (bullfinch); Valeria Mazza (mouse).

Figure 11.3 Examples of urban wildlife management that are based on an understanding of animal behavior and cognition. From left to right: images of humans performing aversive conditioning with a herd of elk by running with a hockey stick covered with flagging material (top) and with a coyote by throwing a tennis ball covered with flagging tape (bottom). Habitat modifications that illustrate the addition of artificial structures to encourage basking of Western pond turtles (top) and the installation of underpasses to provide corridors for mountain lions (bottom). Next, a Cooper's hawk sits next to a plastic owl effigy used to deter avian occupancy (top) and multiple objects placed on trash bins to prevent opening by sulfur-crested cockatoos (bottom). Note: stationary effigies, like plastic owls, are largely ineffective when used alone as depicted in the photo (i.e., as opposed to being paired with other frightening stimuli like movement, sound, lights, etc.). Signs that warn people to avoid interacting with an aggressive mother goose (top) and to refrain from feeding wildlife (bottom).
Images courtesy of: Elsabé Kloppers/Banff Elk Aversive Conditioning Project (elk); Sean Clarkson (coyote); Max Lambert (turtle); Winston Vickers/Karen C. Drayer Wildlife Health Center, UC Davis (mountain lions); Marie Cerda (hawk and owl effigy); Barbara Klump/Max Planck Institute of Animal Behavior (cockatoo); Gabby Barnas (aggressive mother goose sign); Lauren Stanton (procyonid and primate feeding sign).

Managing urban wildlife behavior

While there are various strategies for alleviating HWC, most have historically centered around lethally removing or altering the behavior of urban wildlife. Lethal management, such as poisoning and trapping, may be an effective means of reducing population numbers locally or removing individuals deemed problematic (e.g., 55). However, there are often undesirable consequences of widespread culling and targeted lethal removals (15). New individuals can migrate back into these areas and potentially exacerbate management issues further via animal social instability or disease, as is seen in European badgers (*Meles meles*; 42,43,57,58). Individual animals may also show evolutionary and plastic responses that reduce the efficacy of such lethal strategies, thereby worsening conflict issues. For instance, rats repeatedly develop resistance to anticoagulant rodenticides (59), and some species, like coyotes, demonstrate compensatory reproduction (i.e., increased breeding and litter sizes) when breeding individuals are lethally removed from a population (60). Moreover, animals will avoid toxic baits and traps over time via individual and social learning processes (e.g., 46,47,61,62). Because lethal management is only partially effective and can be tumultuous with the public (63), nonlethal, behavior-based solutions are essential for mitigating urban HWC (Figure 11.3).

To prevent HWC in cities, it is imperative to know the natural history and ecology of a focal species, and how this intersects with various local, urban environmental features and patterns (Figure 11.2). For example, vehicular collisions with ungulates are typically highest during breeding seasons and daily foraging times (64) and animals may behave more aggressively when accompanied by vulnerable offspring (65). Within cities, low-income neighborhoods encompassing higher building vacancy and reduced municipal services may experience increased populations of "pest" species and the potential for disease spread (66,67). Furthermore, certain individual animals may be more dependent on anthropogenic food resources than their conspecifics due to conditioning (e.g., via direct or indirect feeding by humans (68,69) or competition (70,71)). Because urban wildlife behavior is influenced by the spatiotemporal distribution of resources within a city, as well as other neighborhood-specific attributes and local human behaviors, it is critical to avoid static, one-dimensional management strategies and, instead, recognize each individual and species within a given city as a distinct entity.

Behavioral-based solutions to HWC are growing and have been extensively discussed in both academic and management-focused literature (e.g., (48,57,60,63,72–74) (Table 11.1). For situations where animals are seasonally or consistently making use of a particular location or resource that is deemed problematic, oftentimes the most straightforward solution is to identify and either remove or exclude animals from the attractant. For example, removing bird feeders or capping chimneys can prevent unwanted visitation and residency from striped skunks (*Mephitis mephitis*), raccoons (*Procyon lotor*), and opossums (*Didelphis virginiana*) (72,75). When removal or exclusion is not possible or insufficient, use of nonlethal deterrents or aversive conditioning strategies may be necessary. These strategies entail harassment, lights, flavors, sounds, smells, and

Table 11.1 Examples of how knowledge of animal behavior and cognition has been leveraged in urban wildlife management.

Issue	Strategy
Western pond turtles (*Emys marmorata*) are an imperiled species endemic to the North American western coast and are poised to be listed under the Endangered Species Act soon. They are skittish around humans and will immediately abandon their basking sites if they see people. This is a problem in urban areas because turtles are ectotherms that rely on basking for warmth in order to metabolize their food, clear parasites and pathogens, and become sexually active. Furthermore, urban pond management typically involves the removal of large pieces of wood because it is not considered to be aesthetically pleasing and makes pond maintenance more challenging. However, this practice reduces potential basking sites for turtles.	Observations of Western pond turtle behavior in an urban waterway of Davis, CA allowed researchers to estimate the distance that turtles can see humans and subsequently flee by abandoning their basking sites. Practitioners in urban areas of California have since installed logs or artificial basking platforms that were far enough away from walking paths where people can still see the turtles but the turtles are not disturbed by people and will continue basking (Figure 11.3). This simple habitat manipulation addresses a key limiting habitat feature for this imperiled species in urban areas. (76) (MR Lambert, personal experience)
Road and railway collisions with wildlife pose economic losses and safety concerns for humans and are a major source of wildlife mortality around the globe (especially for large carnivores and ungulates). Previous strategies, including fences and visual warning systems (e.g., flashing lights), are reported to be only marginally effective at preventing train collisions. Although acoustic warning systems using sirens have shown more promise, animals will habituate to the sirens over time, meaning that these systems may only be temporarily effective at preventing train collisions.	Researchers in Japan and Poland have recently implemented train warning systems that emit recordings of alarm calls of local animal species. Preliminary findings suggest that animals are highly responsive to the alarm calls, and these "natural" warning systems are effective at reducing train collisions. Importantly, because these signals are ecologically relevant, it is less likely that habituation to the alarm calls will occur, providing a more long-term solution. (77,78,79)
In urban areas, animals may become habituated to humans and learn to capitalize on anthropogenic resources. In cases where habituation and associative learning become problematic, such as through overpopulation, loss of natural behaviors, disruption of trophic relationships, or threats to human safety, managers may elect to use aversive conditioning to sensitize animals to humans and deter the use of urban spaces and resources. Although aversive conditioning can successfully alter behavior, there is substantial individual variation in animal response to aversive conditioning treatments, and its desired effects may be lost (i.e., extinguished) over time.	Researchers have applied principles from animal personality and learning theory to better understand the efficacy of aversive conditioning treatments in elk (*Cervus canadensis*) that congregate in towns adjacent to Canadian protected areas (Banff and Jasper National Parks). Bold elk demonstrate greater responsiveness to aversive conditioning treatments (e.g., humans chasing elk with hockey sticks covered in flagging tape; Figure 11.3), but also faster extinction of learned wariness compared to shy elk, which may help explain the individual variation in responses to aversive conditioning. In accordance with learning theory, researchers also found that the frequency of aversive conditioning treatment matters: when aversive conditioning is too mild or infrequent, wariness of humans will not be learned by elk, but when implemented too frequently, it becomes predictable and increases the likelihood of habituation to the treatment. It is, therefore, suggested that aversive conditioning be conducted at intermittent frequencies (e.g., once every two weeks), as this will allow wariness of humans to be learned and maintained over longer periods of time. (80–82)

continued

Table 11.1 *Continued*

Issue	Strategy
Many urban habitat features, such as lawns, ledges, and utility poles, provide unintentional harborage for various avian species. Group foraging, nesting, and roosting can lead to property damage and accumulation of feces around homes, buildings, and public spaces, prompting the removal of the birds to be desired by urban residents. In agricultural landscapes, managers and farmers often employ avian frightening devices that are species- and context-specific, typically incorporating a combination of lights, sounds, movement, and even taxidermies applied at unpredictable intervals. However, the use of such frightening devices may be too time-consuming and disruptive for conflict mitigation in urban areas. Although stationary effigies, like plastic owls (Figure 11.3), serve as popular alternatives in urban areas, they are largely ineffective due to rapid habituation by birds.	Dynamic frightening devices that can recognize and respond to specific species and behaviors may be a promising new management tool in urban areas. For instance, a Denmark-based research group is developing adaptive scaring technology that uses automatic recognition of barnacle goose (*Branta leucopsis*) vocalizations to identify undesirable behaviors (e.g., foraging) and subsequently activate auditory frightening stimuli (e.g., distress calls). Such selective application of aversive conditioning can help reduce the likelihood of habituation by wildlife and disturbance to the public. Other conflict scenarios with urban birds can be resolved with simpler habitat modifications using physical barriers, such as nets and spikes. For example, researchers in Pretoria, South Africa found that bird spikes placed on buildings were more effective at reducing pigeon numbers on a university campus compared with several commercially available visual deterrents. Integration of alternative roosting and nesting options alongside exclusion measures can bolster coexistence in urban areas. Examples of such innovative, bird-friendly designs date back to 16th-century Persia (e.g., "pigeon towers") and can be seen in other works of modern architecture, including Oscar Niemeyer's "Pombal" in Brasília (Brazil) and Antoni Gaudí's "Parc Güell" in Barcelona (Spain). (86–85)
In South Africa, troops of chacma baboons (*Papio ursinus*) will enter urban spaces in search for anthropogenic food. During these brief, high-activity "raids," baboons will forage in garbage bins, cars, and homes, and try to take food directly from people. Such interactions threaten the health, safety, and food security of residents, and often result in the killing of baboons. The adaptive responses of baboons to various mitigation attempts over time have made the management of this negative interaction between humans and baboons very difficult.	In Simon Town, researchers attempted to alter a troop's raiding behavior by providing a supplemental feeding patch located away from people and urban areas. Although food provisioning alone did not significantly reduce urban space use by the troop, it became more effective when paired with exclusion from the main food waste sites (i.e., when enclosed in wire mesh). Another study in Cape Town found that field rangers engaged in the monitoring and deterrence of baboon troops reduced the amount of time baboons spent in urban areas by 70%, and that the intensity of deterrence had a significant effect on baboon activity as well. New research focused on individual and social behavior of baboons is providing additional insights on the adaptive responses of baboons to field rangers that can be used to improve aversive conditioning strategies (e.g., identify which individual baboon(s) to target; frequency and consistency of strategy). The implementation of multiple management strategies informed by behavioral research is effectively reducing the frequency of conflict between people and baboons with additional solutions on the horizon. (86–88)

startling movements (or preferably a combination thereof) to repel animals from a given attractant or to **sensitize** animals to humans and other anthropogenic dangers (e.g., roads and railways). To ensure that animals will respond successfully, these strategies must be tailored to the species' sensory modalities such that it can be perceived and attended to by the focal individual(s). For instance, scent-based deterrents are more salient for species with greater olfactory capabilities, such as rodents and carnivores, whereas visual deterrents may be most effective for species that are more reliant on visual information, such as birds and primates (16).

A critical point of note is that eventual **habituation** or **tolerance** to behavioral-based strategies may undermine mitigation efforts. Animals typically habituate to the continued use of nonlethal tools that lack a negative stimulus, and thus additional frightening stimuli may need to be periodically administered. Alternating practices and employing multiple deterrents that target different **sensory modalities** at once may also be more effective than when left permanently or used singularly (72,29,34). This may be particularly applicable for urban individuals that are likely to be behaviorally flexible and/or highly conditioned to the use of anthropogenic resources (56). Importantly, such strategies will only be effective at deterring or repelling wildlife if alternative resources and attractants are available (59,73; Table 11.1). As such, it is equally important to design and manage urban habitat in ways that will reduce conflict and promote the well-being of both people and wildlife living there.

Managing urban habitats

Heterogeneity in urban habitats

Legacies of residential segregation and ongoing social inequity have created unequal environmental services and benefits within cities, and this affects the well-being of, and relationship between, people and animals (15,89).[2] For example, socioeconomically advantaged groups living in greener, more biodiverse neighborhoods (i.e., the luxury effect; 64,90) generally experience a higher frequency of positive interactions with a diversity of wildlife, whereas disadvantaged groups living in impoverished neighborhoods may not only experience fewer positive interactions, but also have more negative interactions with pest species (6,91). Nevertheless, residents of marginalized communities may be less likely to report conflict with wildlife, possibly due to distrust of local governments (92). They may also harbor general apprehension toward greening initiatives due to the perceived dangers associated with vegetation (93) and fear of displacement or erasure via neighborhood gentrification (94). Such differences in urban resident experiences and perspectives toward wildlife should be taken into account when managing urban habitats (95).

Although the effects of environmental disparities on human health and well-being have long been articulated in the environmental justice literature, we currently do not know the extent of racial oppression and social inequality on animal behavior, cognition, and human–wildlife interactions (89). For instance, environmental disamenities like heavy metal pollution (e.g., affecting European honey bees [*Apis mellifera*]; 96), roadway noise (e.g., affecting rats; 71,97), and use of pesticides (e.g., affecting bobcats [*Lynx rufus*]; 72,98) are typically higher in neighborhoods of low-socioeconomic status (e.g., 99–101) and can disrupt physiological, behavioral, and cognitive development. Furthermore, diets high in anthropogenic foods may increase animal boldness through complex pathways in the gut–brain axis (e.g., 76,102), indicating that increased access to garbage via poor municipal services could have reverberating consequences on wildlife behavior. Vacant, unmaintained structures are hospitable for many "pest" species, which can thereby increase population numbers and the potential for disease transmission in neighborhoods with greater vacancy and general disrepair (66,77,78,79,92,103,184). Similarly, increased consumption of anthropogenic subsidies by wildlife due to a lack of municipal services and improper waste management can be distressing to people and can negatively impact urban wildlife health (e.g., periodontal disease and hyperglycemia in raccoons (104,105); compromised immune function in coyotes (106).

Thus, the same environmental disamenities that disproportionately harm minoritized communities might spillover to jeopardize wildlife and human–wildlife relationships.[3] Providing environmental education, equitable green space access and infrastructure, efficient waste management, sound housing integrity, and maintaining/modernizing transit routes are a few of the many justice-centered habitat management strategies that can promote environmental health and more positive interactions with urban wildlife across cityscapes (89).

Animal behavior and cognition research is central to urban design and planning

There is growing recognition that urban areas can play a role in biodiversity conservation by attracting and supporting animal populations (107,108).[4] Some of the features that attract animals to cities include protection from predators that may

[2] Hoover's and Scarlett's chapter discusses contemporary impacts and historical legacies of systemic racism and classism on urban nature.

[3] Byers et al.'s chapter outlines the intricate relationships between equity and human and wildlife health in cities.

[4] Lambert's and Schell's chapter details the emergence of urban biodiversity conservation in science and society.

be excluded by urbanization (i.e., **human shield hypothesis**, (36,84,85,109)), built structures that serve as shelter and nest sites, and food sources that are provided both intentionally and unintentionally (107,110,111). It is important to recognize that these attractants can produce unintentional negative conservation outcomes by enhancing non-native populations, increasing HWC, or reducing the fitness of native species, yet they also have the potential to contribute to conservation. For instance, nest boxes can be designed with optimal dimensions to attract native birds, with guards to protect against predators, and without the perches that tend to attract more aggressive species (112). Thus, a behavioral approach to urban biodiversity conservation involves understanding the specific cues animals use in **habitat selection**, providing those cues, and ensuring that they lead native species to places where they can be successful while avoiding any unintentional negative outcomes.

Green spaces are an important tool for promoting urban biodiversity, particularly when they are restored or designed with cues that encourage settling, shelter, food, and breeding resources used by native wildlife (107,108,113). Designing green spaces that are attractive to native wildlife often involves planting specific trees and plants, mimicking the habitat structure of nearby wildlands, incorporating riparian areas into city parks, and clustering them with other green areas and gardens to allow connectivity within the urban matrix[5] (114–117). Indeed, connectivity of green spaces presents a major challenge for wildlife and can be difficult to achieve, especially when it conflicts with other management goals. For instance, fences and other linear barriers can be useful for mitigating HWC in and near cities (15), yet they can have varying effectiveness (118) and can shift problematic human–wildlife interactions elsewhere (119). Fences also produce ecological "winners" and "losers" depending on context and scale, impacting factors as varied as habitat structure, community structure, animal behavior, and gene flow (120). Thus, careful planning and integration of green spaces and fencing should be undertaken when designing urban habitats.

For species that use social information to select habitat, conspecific cues—including acoustic calls, odors, or visual models—may help bring animals to high-quality habitat (113,121). The mainstreaming of such approaches may be useful in future, cutting-edge biodiversity management strategies, such as urban translocations.[6] In addition to incorporating cues that attract animals to urban green spaces, it is also critical to limit cues that repel them, which may include chemical, noise, and light pollution, or the presence of humans (73). If minimizing human disturbance is not possible, the impacts of those stimuli may be gradually reduced over time with repeated exposure through deliberate strategies to promote habituation (113,122). Of particular concern is the potential for cities and restored habitats to become **ecological traps** (123) if animals are attracted to urban settings but populations fail to persist because of excess predation, low-quality food, low reproductive success, or mortality associated with HWC (124). Hence, wildlife populations in cities must be studied and managed carefully, typically with a species-specific approach, to detect and disarm ecological traps (125,126).

There have been repeated calls for mechanistic research on how urban systems affect HWC and biodiversity in order to inform urban planning (111,123,124). For example, we need a better understanding of the factors that influence individual colonization, dispersal, breeding success, mortality, and responses to human disturbance (124). Understanding how various wildlife species perceive cues and make decisions that guide their behavior in urban environments will enable us to predict how wildlife should respond to urban habitat designs and manipulations (Figures 11.2 and 11.3). This will help inform the optimal size, type, and connectedness of green spaces, determine when they become

[5] Stanford et al.'s chapter outlines key urban ecological design principles that enhance habitat quality for city planning.

[6] Spotswood et al.'s chapter begins creating an urban conservation toolbox and emphasizes the need to report standard conservation tools, innovate tools, and create new approaches for urban biodiversity given the unique nature of urban environments.

ecological traps, and suggest ideal buffer areas and limits to human visitation, with the ultimate goal of increasing biodiversity, decreasing HWC, and minimizing the homogenizing effects of urban environments (123,124).

One issue that has been studied extensively is how road crossing structures can be designed to target and enhance use by different taxa, depending on features like location, size, vegetation, nearby ponds, and elements that block noise and light (127–129) (Figure 11.3). Wildlife overpasses are not typically implemented in urban settings because of their high cost and space requirements compared to underpasses (127). However, the world's largest wildlife overpass, the Wallis Annenberg Liberty Canyon Wildlife Crossing (LCOC), is currently being constructed over US-101 in Los Angeles County, largely due to a successful public campaign to protect mountain lions (*Puma concolor*) in this urban biodiversity hotspot (127,128). Previous research on the behavior of mountain lions and other wildlife species has demonstrated that excessive noise and light can inhibit the use of wildlife crossings (129). Therefore, to increase its use by wildlife, several structural features are being integrated into the LCOC, including strategically placed noise barriers and berms, that will provide functional reduction of anthropogenic noise and glare and contribute to a more attractive approach zone for mountain lions and other urban species (128).

Many of the urban improvements that benefit wildlife can also benefit human health and well-being (95).[7] However, not all urban residents will support environmental initiatives that increase green space or wildlife populations. Furthermore, without education and participation in wildlife management strategies, residents may continue engaging in behavior that unknowingly contributes to HWC. Thus, it is essential that urban planning, ecological restoration, and neighborhood improvements be equitable and accompanied by educational outreach and partnership with local communities.

[7] Byers et al.'s chapter's approach to One Health in cities outlines how to improve wildlife and human well-being together.

Managing human expectations and behavior

Urban wildlife behavior as a function of human behavior

Managing animal behavior and habitats are important components for fostering urban human–wildlife coexistence, yet urban wildlife behavior is fundamentally a function of human behavior. Individual animals and species demonstrate differentiated behaviors depending on their perception of humans, which can be measured using assessments like flight initiation distances (e.g., 106,130), giving-up densities (e.g., (107,131), and playbacks (e.g., 108,132) (Figure 11.2). For example, one study found that residents of Seattle were more discouraging toward birds compared to residents of Berlin and, correspondingly, found higher flight initiation distances in Seattle vs. Berlin with the highest scores exhibited by species typically considered to be a nuisance (i.e., crows [*Corvus brachyrhynchos*] and starlings [*Sturnus vulgari*]; 133). Species like pigeons (Columba livia) or squirrels that are generally ignored by humans are likely to habituate to the presence of humans over time, whereas species that fear humans or are repelled by anthropogenic pollution (e.g., noise) may become sensitized to human presence. Urban species ranging from small primates (marmosets; *Callithrix penicillata*) to large carnivores (mountain lions) will avoid humans by adjusting their activity (134–137) and movement (i.e., **landscape of fear hypothesis**; 138) around predictable human cues (e.g., diurnal vs. nocturnal patterns, weekday vs. weekend activities). Indeed, pulses and pauses in human activity will affect animal activity, which was most clearly illustrated during the recent stay-at-home orders during the Covid-19 pandemic, when many animals altered their movement and habitat use patterns in response to reduced human activity (138). Interestingly, species that receive frequent, mixed feedback from humans (e.g., sometimes fed, ignored, and/or harassed) may attend to certain cues that allow them to recognize and differentiate among individual humans, and may categorize humans as "safe" or "dangerous" (22,139). Thus, urban wildlife behavior is increasingly understood as a reciprocal function with human behavior (140,141), highlighting the importance of understanding what drives human attitudes and behaviors toward wildlife.

Perceptions and attitudes toward urban wildlife

Human–wildlife interactions in cities and elsewhere are heavily influenced by people's perceptions, experiences, values, and attitudes (e.g., 142,143).[8] Importantly, human attitudes and tolerance are both commonly used proxies for predicting human behaviors toward wildlife, such as whether a person will support or actively participate in (legal or illegal) lethal removal of certain species or "problem individuals" (e.g., 10,144,145). As such, human behavior, situated within sociocultural and political contexts, is a critical component of all human interactions with wild animals (146). These behaviors can be driven by many factors, including experience and emotion, relationships with the sociopolitical surroundings, and resulting risk perceptions and attitudes (147). For example, residents of Singapore were more likely to exhibit tolerance of, and positive attitudes toward, nuisance wildlife if they had had more childhood nature experiences (148). Sociocultural and religious feeding practices in cities around the world may provide personal enjoyment and connection with nature (69,149) despite it sometimes contributing to poor health (e.g., 126,150) or increased aggression in urban wildlife (e.g., 151,152). People's perceived (even if not real) risks from certain species can also be strong predictors of whether they have negative attitudes toward the conservation of other species (143).

Just as wildlife perceptions of their environment are important drivers of human–wildlife interactions, so too are people's perceptions of their environments. Across the global spectrum of human–wildlife interactions, human perceptions of wildlife and of each other have a strong influence on how humans interact with and manage certain wildlife species. "Human–human conflicts," such as inequities, cultural differences, and top-down policies, underlie nearly all human–wildlife conflicts (153,154).[9] People's perceptions can vary across scales and locations, differ from what is scientifically or ecologically recorded, and influence wildlife conservation and coexistence efforts (155–157).

For example, in many cities there exists a debate over whether coyotes "belong" within cities (see Case study: The ubiquitous urban coyote), especially as they pose a threat to people's outdoor cats, pets, and poultry (9). People's propensities to support lethal control of urban coyotes can be influenced by their gender identity, level of fear toward coyotes, where they live, and willingness to interact with their local government agencies (e.g., 134,158). Yet, people who express concern about conflicts with urban wildlife, stemming from both actualized conflict instances (e.g., crop raiding and loss of domestic animals) and intangible factors (e.g., personal anxiety), can still support nonlethal interventions for the wildlife in question (159). This dichotomy of opinion between advocates for domestic animals and advocates of wildlife conservation is common in cities and can impact urban wildlife conservation policies. Human perceptions of particular species are also key to understanding how to promote coexistence (155). For instance, people are more supportive of conservating species for which they have an aesthetic appreciation (e.g., red-tailed hawks [*Buteo jamaicensis*]; 136,160) than those that they fear or deem aesthetically unpleasing (e.g., bats [*Chiroptera* spp.]; 137,161).

Improving attitudes and behaviors toward urban wildlife

Urban human populations are not uniformly welcoming to wildlife in their backyards and many people do not have equitable access to nature, positive interactions with wildlife, and the benefits of conservation policies (89). Fortunately, there are many strategies currently in use to foster pro-environmental perspectives, such as tolerance and local environmental stewardship, in urban areas.

[8] Larson's and Brown's chapter details the perceptions and motivations of people toward biodiversity in cities and suburbs.

[9] Kar Gupta et al.'s chapter illustrates the value of developing flagship or umbrella species for urban conservation as ways to bridge human–human conflict.

For instance, providing equitable access to environmental resources and services is key to addressing within-city heterogeneity in experiences with and risk perceptions about urban wildlife. People from low socioeconomic and ethnic minority backgrounds have been found to have access to fewer acres of urban parks, and access to parks with lower quality and safety than more privileged groups (162). Because childhood nature experiences and other forms of place-based nature experiences for all age groups can have positive effects on attitudes toward wildlife and the environment (148,163,164), cities should prioritize bolstering and perpetuating nature access programs and organizations, such as City Parks Alliance (https://cityparksalliance.org/) and Groundwork USA (https://groundworkusa.org/). Additionally, to be most impactful, cities should foster educational programs on urban wildlife that provide experiential training about wildlife behaviors, reducing wildlife attractants, familiarization with green spaces and safety, and employing nonlethal wildlife management techniques.

Case study The ubiquitous urban coyote

Coyotes have rapidly expanded their range and now live in most urban areas across North America (165). They can be of concern to humans because of the risks they may pose, including attacking people or their pets. While incidences of attacks are low relative to the abundance of people and coyotes living in urban areas and the frequency of human–coyote encounters (166), coyotes require management strategies that ensure human safety (and perceived safety) while allowing for coexistence. Research has long focused on how coyotes navigate urban environments, and we have learned that urban coyotes may be bolder than rural coyotes (41,167,168), descend from a few individuals (169), temporally avoid humans (170–172), and take advantage of anthropogenic food resources (106).

Research into coyote cognition is limited to only a few studies, but coyotes have demonstrated behavioral and cognitive flexibility across multiple tasks and experimental paradigms (173–175). This flexibility can facilitate the behavioral adjustment and spread of coyotes in urban environments. For example, urban coyotes primarily consume prey similar to that of rural coyotes (176) but have more diverse diets than rural coyotes, which is largely caused by the addition of anthropogenic food into the diets of urban coyotes (177). Urban coyotes are also bolder and more exploratory than rural coyotes (178), which allows them to discover and access more resources.

This same flexibility, however, may also be bringing coyotes into greater conflict with humans (56). For example, some urban residents intentionally feed coyotes or may provide food unintentionally, such as how urban coyotes are attracted to compost piles (179). In both scenarios, coyotes are more likely to make contact with humans which increases the spread of zoonotic diseases and rates of human–wildlife conflict. Similarly, emboldened coyotes may be more likely to attack humans and their pets (56,178). Conflicts with pets can be especially difficult to navigate because there are perceptions and beliefs associated with both coyotes and pets. For example, many cat owners believe their cats should be allowed to roam freely outdoors (180,181), and consider outdoor cats as family members (e.g., 181,192), and therefore may become vocal advocates for more intensive management of coyotes and other mesopredators even though domestic cats can cause considerable losses to native wildlife (182).

Despite the widespread presence of coyotes in urban areas across North America, observational and experimental research has only recently been used to inform management decisions that mitigate risks. For example, a study in the Denver Metropolitan area found that teaching residents to haze coyotes (183,194) as a nonlethal tool to reduce conflict may not be effective reactively (i.e., once a coyote has become too emboldened or involved in conflicts with humans) but can be used proactively (165,176). This work suggests that selective lethal removal may also be needed. In the same region, surveys of human views about lethal and nonlethal tools found that an individual's perceptions (e.g., fear) and beliefs about coexistence predicted support around lethal control (134,158). Researchers have also experimentally looked at how common practices can be used as nonlethal tools (e.g., 184,195) and how experiential learning can create more positive perceptions about coyote encounters (185,196). Cities are beginning to use these findings to create management and outreach (186,197), but additional work is still needed to improve the language used to describe coyotes (187,188,222,223) that addresses the complex beliefs and perceptions of people sharing spaces with coyotes (134,158,185,196).

Future directions in human–wildlife coexistence in cities

There are several exciting new frontiers in the study of urban wildlife behavior and cognition that will further inform our ability to promote coexistence between humans and wildlife. Recent research, for example, has begun to uncover the effects of urban diets and the gut microbiome on behavior (e.g., 189,141) and identified potential feedback loops that may be influencing human–wildlife interactions (e.g., 106,184). The distribution and abundance of food resources contribute to higher densities and more frequent social interactions for some species in urban environments (190), which has implications for not only zoonotic disease transmission, but also the evolution of animal sociality and behavior (e.g., raccoons) (191,192). Furthermore, studies that link behavior and cognitive ability to specific features of urban environments, such as human activity vs. human footprint, green space, and competition (e.g., 41,111,135,209), as well as fitness (51), are providing new insights and needed clarity on the role of cognition in urban living. However, many knowledge gaps remain. For instance, most studies discussed here focus on dichotomous comparisons between urban and nonurban cognitive and behavioral responses. Few within- and between-city comparisons of behavior exist (e.g., 109,133), which are essential for deconstructing the fine-scale contributions of societal inequity to the emergence of behavioral traits often associated with conflict in cities.

Future research on urban wildlife behavior can also be leveraged to gauge the well-being of urban populations, assess the effectiveness of various management strategies, increase positive perceptions of and experiences with wildlife, and predict the occurrence of conflict with urban species. For example, to understand whether greening efforts are increasing biodiversity and coexistence, investigating if and how animals are using resources within green spaces (e.g., food, shelter) and how they perceive and interact with humans (e.g., attraction, repulsion, indifference) will be more informative for long-term population establishment compared to single measures of abundance and richness. Furthermore, the ubiquity of common urban species often leads to the assumption that these individuals are "thriving" in cities, yet several studies demonstrate that urban populations may be in poor health, highly inbred, physiologically stressed, and combating ecological traps (13,193–195). Thus, quantifying wildlife health (e.g., body condition, diet, disease) and species-specific behaviors (e.g., movement, habitat use, communication, competition, reproduction) in urban environments can provide a useful baseline by which we gauge population trends and the success of our management strategies over time. In addition, because the temporal and spatial occurrence of human–wildlife interactions are moderately predictable, and conflict is hugely influenced by human perceptions and behaviors, quantification of social and ecological factors unique to a neighborhood can allow us to predict the occurrence of conflict and create more positive experiences with wildlife. For example, HWC in Chicago was most likely to occur in areas where humans and wildlife overlap; however, complaints of raccoons and opossums were more likely to be reported in high- versus low-income neighborhoods, despite higher occupancy of these species in low-income neighborhoods (92). Thus, pairing data on the behavior of urban wildlife across species, seasons, cities, and neighborhoods with participatory surveys on local human attitudes and interactions with wildlife will allow us to build predictive models that prevent impending conflicts based on: (1) how various wildlife utilize urban space and (2) how human residents vary in their perceptions of those wildlife.

Engaging urban residents in surveys and other participatory science efforts may provide extraordinary insight to help mitigate future conflicts. Participatory science[10] can be a productive method for gathering data about biodiversity and wildlife behavior (196), understanding and addressing HWC (197), and engaging community members with wildlife conservation issues (198). Researchers can actively develop participatory research projects

[10] Perkins et al.'s chapter details how to enhance the value of participatory science methods for better biodiversity data, improving environmental equity, and enhancing urban wildlife opportunities.

in collaboration with urban community members to manage motion-activated camera traps, assess water quality and other indicators, look for wildlife tracks, and undertake various other activities that can support management and coexistence goals coproduced with local communities. Researchers can also utilize freely available data collected through participatory science social networks in which people log organism sightings, such as iNaturalist (https://www.inaturalist.org/) and eBird (https://ebird.org/), both of which house substantial numbers of observations in urban areas. Additionally, urban residents are keen to express their experiences with wildlife, and often do so via neighborhood communication platforms such as NextDoor (https://nextdoor.com/). Despite the inherent reporting biases of all participatory social science networks, they are increasingly being viewed by scientists as a valuable resource for both public engagement and data collection about human–wildlife interactions (199,200). Finally, many local government agencies and institutions have options for reporting wildlife sightings and interactions (e.g., Carnivore Spotter: https://carnivorespotter.org/, and San Francisco Animal Care and Control coyote observation report: https://www.sfanimalcare.org/), which provide yet more avenues of both data collection and community engagement.

Perhaps the greater challenge is developing a tangible consensus around what coexistence entails. One recent definition of coexistence suggests it is "a dynamic but sustainable state in which humans and wildlife co-adapt to living in shared landscapes, where human interactions with wildlife are governed by effective institutions that ensure long-term wildlife population persistence, social legitimacy, and tolerable levels of risk" (147,201, p. 787). With so much variation in human attitudes and societal disparity present in any given city, let alone a region or nation, how can we determine what social legitimacy and tolerable levels of risk are? Although complex, the answer lies in partnering with communities to bolster environmental education and justice so that residents better understand local urban wildlife and have the tools that will allow people to actively participate in biodiversity conservation and coexist with wildlife.

Global perspectives

Both large-scale and within-city patterns and consequences of urban human–wildlife interactions have been studied worldwide but such studies have mostly been conducted in "Global North" nations, particularly the US (202). However, human values, culture, politics, and urban design show tremendous variation around the world. As such, our ability to apply wildlife behavior research and translate inferences about human–wildlife interactions from one part of the world to another requires a globally representative body of work. For example, human social factors, such as culture, politics, and religion, and ecological factors, such as season, vegetation structure, water availability, and green space sizes, all vary within and across urban areas worldwide (203). This results in variations in human–wildlife interactions across spatial and temporal scales, as well as variations in the types and levels of biodiversity that can be maintained (204). In a broader-scale example, different distributions of people and birds across multiple urban areas in England were correlated to ecosystem services and disservices provided by birds, where people in the lowest socioeconomic groups experienced the same level of disservices but fewer services (205).

Global research also demonstrates that these contextual factors can vary at a finer scale, within cities. Cities are usually heterogeneous in multiple ways, often as a function of socioeconomic histories and present experience. Socioeconomics and other cultural histories can lead to spatial differences in availability of green space and ecological health, which then determine where and how people and wildlife interact, as well as within-city differences in biodiversity outcomes (e.g., 89,63). For example, the size of long-tailed macaque (*Macaca fascicularis*) troops, and the relative number of infants within them, varied with active provisioning of food by tourists among three sites in Padang, Sumatra (206). The researchers noted that managing human provisioning could reduce macaque population growth and the growing rates of human–macaque conflict, such as crop-raiding and aggressive behavior toward visitors (207,208). Similar variation in human–wildlife interactions was seen in Belo Horizonte, Brazil, where the people who called the environmental police to report human-caused injuries to birds were more likely to have high salaries (209).

Variations in green space allocation and other types of land management within cities can also affect how vegetation is impacted by typical wildlife behavior. For example, there was little variation across 15 sites within the city of Wellington, New Zealand, in tree damage caused by sap-sucking behavior of North Island kākā birds (*Nestor meridionalis septentrionalis*, (166,210)). Instead, the most important predictors of damage were tree characteristics (e.g., tree species and diameter), which varied depending on the city's history of protecting and creating green space. To alleviate conflicts and dangers such as treefall, while maintaining high human tolerance for kākā, researchers recommended that managers prioritize planting and managing tree species that are more resilient to sap-sucking behavior. In short, diverse sociopolitical contexts and histories of biodiversity and wildlife management within cities can have consequences for human–wildlife interactions, human opinions on wildlife management, and resulting policies.

Social-ecological applications

While urban landscapes are rapidly changing, so are the perceptions, beliefs, and equity of people living in cities, as well as the way information is disseminated and absorbed. These changes will likely influence urban wildlife behavior, interactions with humans, and the positive and negative effects of cognition on human–wildlife interactions. It is critical that studies of urban wildlife behavior and cognition are intersectional (211), accounting for the sociopolitical landscape. This has the potential to improve human–wildlife coexistence via informed wildlife management, create opportunities for community involvement, and enhance environmental justice (89).

While a deep understanding of animal behavior and cognition can improve our ability to develop policies and management practices that improve coexistence, management actions and human behavior are often unrelated to policies and are instead based on human experience and histories (212,213). In some scenarios, outreach and education may be the best strategies for reducing human–wildlife conflicts, but in other cases better enforcement of policies may be needed (214). Further, there may be scenarios where efforts to reduce conflict are difficult to achieve because humans are unwilling to change their behavior, such that resources would be better invested elsewhere (215,224). However, these scenarios could improve by considering within-city differences that allow for better community engagement.

For example, for several decades leopards (*Panthera pardus fusca*) in Mumbai were involved in conflict after translocations (216) likely because of challenges acquiring food and shelter, especially when a hard release was used (217,218). However, researchers also recognized the methods communities used to capture leopards for translocations were varied, with many causing injury and stress to the leopards (219). At the same time, journalistic coverage of these capture events and human–leopard conflicts used negative and incendiary language (220). Subsequently, researchers worked with local officials from various parts of the greater Mumbai area on humane capture and handling (*personal communication to author Young*), while also hosting information clinics with journalists that challenged perceptions and beliefs (220). Today, these efforts have resulted in fewer deaths and injuries to leopards and people during and after translocations. This example combining an understanding of leopard behavior and management needs (i.e., post-translocation conflict propensity due to lack of resources) with the beliefs and perceptions of people illustrates the power in considering all aspects of urban communities to inform policies and actions.

Conclusion

Wildlife use cognitive abilities to guide their behavioral responses to environmental conditions (16). In urban environments, humans play an especially outstretched role in this process across multiple scales. From city design and resource distribution, to individual encounters with nature, to wildlife management and environmental policy, our actions have a prominent influence on the way animals perceive and behave in urban environments (91,140,141). To manage wildlife in a way that promotes coexistence and biodiversity, we must consider not only which ecological attributes make cities attractive, safe, and hospitable for a variety of species with unique needs and umwelten, but also how human attitudes and actions impact urban wildlife behavior and eventually feed back into human society. Like any wildlife conservation or management program, our success hinges on the support and participation of interested parties (5). Thus, we must work to become culturally competent and cognizant of the preferences and expectations of a diverse public, and thereby strive to develop and employ multiple urban management strategies that can serve all communities equally.

In many cases, urban habitat and wildlife management actions can improve the health and quality of life for both people and animals. Making cities more equitable, such as by increasing presence of and access to natural resources like wild habitats and alleviating exposure to disamenities like pollution, has the potential to increase biodiversity and shape more desirable behaviors in wildlife. Providing equitable green space access and infrastructure, efficient waste management, sound housing integrity, and supporting ecocultural relationships with nature are a few of many justice-centered management strategies that can promote more positive perceptions of and interactions with urban wildlife (89). Justice-centered management efforts will, therefore, create a more resilient system by which humans and wildlife can coexist.

References

1. Lambert MR, Brans KI, Des Roches S, Donihue CM, Diamond SE. Adaptive evolution in cities: progress and misconceptions. Trends in Ecology and Evolution. 2021;36(3):239–57.
2. Soulsbury CD, White PCL. Human–wildlife interactions in urban areas: a review of conflicts, benefits and opportunities. Wildlife Research. 2015;42(7):541–53.
3. Cram DL, van der Wal JEM, Uomini N, Cantor M, Afan AI, Attwood MC, et al. The ecology and evolution of human-wildlife cooperation. People and Nature. 2022 Aug;4(4):841–55.
4. Harris NC, Wilkinson CE, Fleury G, Nhleko ZN. Responsibility, equity, justice, and inclusion in dynamic human–wildlife interactions. Frontiers in Ecology and the Environment. 2023. Available from: https://esajournals.onlinelibrary.wiley.com/doi/full/10.1002/fee.2603
5. Riley SJ, Decker DJ, Carpenter LH, Organ JF, Siemer WF, Mattfeld GF, et al. The essence of wildlife management. Wildlife Society Bulletin. 2002;30(2): 585–93.
6. Soga M, Gaston KJ. The ecology of human–nature interactions. Proceedings of the Royal Society B: Biological Sciences. 2020 Jan 15;287(1918):20191882.
7. DeStefano S, DeGraaf RM. Exploring the ecology of suburban wildlife. Frontiers in Ecology and the Environment. 2003 Mar;1(2):95–101.
8. Lukasik VM, Alexander SM. Human–coyote interactions in Calgary, Alberta. Human Dimensions of Wildlife. 2011 Mar 29;16(2):114–27.
9. Hunold C, Lloro T. There goes the neighborhood: urban coyotes and the politics of wildlife. Journal of Urban Affairs. 2022 Feb 7;44(2): 156–73.
10. Mormile JE, Hill CM. Living with urban baboons: exploring attitudes and their implications for local baboon conservation and management in Knysna, South Africa. Human Dimensions of Wildlife. 2017 Mar 4;22(2):99–109.
11. Treves A, Naughton-Treves L. Evaluating lethal control in the management of human–wildlife conflict. In: Woodroffe R, Thirgood S, Rabinowitz A (eds.) People and Wildlife, Conflict or Co-existence? Cambridge: Cambridge University Press; 2005. p. 86–106.
12. Loss SR, Boughton B, Cady SM, Londe DW, McKinney C, O'Connell TJ, et al. Review and synthesis of the global literature on domestic cat impacts on wildlife. Journal of Animal Ecology. 2022;91(7): 1361–72.
13. Murray MH, Sánchez CA, Becker DJ, Byers KA, Worsley-Tonks KEL, Craft ME. City sicker? A meta-analysis of wildlife health and urbanization. Frontiers in Ecology and the Environment. 2019;17(10):575–83.
14. Mateo-Tomás P, Olea PP, Sánchez-Barbudo IS, Mateo R. Alleviating human–wildlife conflicts: identifying the causes and mapping the risk of illegal poisoning of wild fauna. Journal of Applied Ecology. 2012;49(2):376–85.
15. Schell CJ, Stanton LA, Young JK, Angeloni LM, Lambert JE, Breck SW, et al. The evolutionary consequences of human–wildlife conflict in cities. Evolutionary Applications. 2021;14(1):178–97.
16. Shettleworth SJ. Cognition, Evolution, and Behaviour. New York: Oxford University Press; 2010.

17. Sih A, Ferrari MCO, Harris DJ. Evolution and behavioural responses to human-induced rapid environmental change. Evolutionary Applications. 2011;4(2):367–87.
18. Robertson BA, Rehage JS, Sih A. Ecological novelty and the emergence of evolutionary traps. Trends in Ecology and Evolution. 2013;28(9):552–60.
19. Shochat E, Warren PS, Faeth SH, McIntyre NE, Hope D. From patterns to emerging processes in mechanistic urban ecology. Trends in Ecology and Evolution. 2006;21(4):186–91.
20. Griffin AS, Netto K, Peneaux C. Neophilia, innovation and learning in an urbanized world: a critical evaluation of mixed findings. Current Opinion in Behavioral Sciences. 2017 Aug 1;16:15–22.
21. Grimm NB, Faeth SH, Golubiewski NE, Redman CL, Wu J, Bai X, et al. Global change and the ecology of cities. Science. 2008;319(5864):756–60.
22. Lee VE, Thornton A. Animal cognition in an urbanised world. Frontiers in Ecology and Evolution. 2021;9:633947.
23. Sol D, Lapiedra O, Ducatez S. Cognition and adaptation to urban environments. In: Szulkin M (ed.) Urban Evolutionary Biology. Oxford: Oxford University Press; 2020. p. 253–67.
24. Owen MA, Swaisgood RR, Blumstein DT. Contextual influences on animal decision-making: significance for behavior-based wildlife conservation and management. Integrative Zoology. 2017 Jan;12(1):32–48.
25. Immelmann K, Colin B. A Dictionary of Ethology. Cambridge, MA: Harvard University Press; 1989.
26. Levitis DA, Lidicker WZ, Freund G. Behavioural biologists do not agree on what constitutes behaviour. Animal Behaviour. 2009 Jul;78(1):103–10.
27. Audet JN, Lefebvre L. What's flexible in behavioral flexibility? Behavioral Ecology. 2017;28(4): 943–7.
28. Lea SEG, Chow PKY, Leaver LA, McLaren IPL. Behavioral flexibility: a review, a model, and some exploratory tests. Learning and Behavior. 2020 Mar 1;48(1):173–87.
29. Greggor AL, Clayton NS, Phalan B, Thornton A. Comparative cognition for conservationists. Trends in Ecology and Evolution. 2014;29(9):489–95.
30. Sol D. Revisiting the cognitive buffer hypothesis for the evolution of large brains. Biology Letters. 2009;5(1):130–3.
31. Allman J, McLaughlin T, Hakeem A. Brain weight and life-span in primate species. Proceedings of the National Academy of Sciences of the United States of America. 1993;90(1):118–22.
32. Schlaepfer MA, Runge MC, Sherman PW. Ecological and evolutionary traps. Trends in Ecology and Evolution. 2002 Oct;17(10):474–80.
33. Jones J. Habitat selection studies in avian ecology: a critical review. The Auk. 2001;118(2):557–62.
34. Blumstein DT. Habituation and sensitization: new thoughts about old ideas. Animal Behaviour. 2016;120:255–62.
35. Bejder L, Samuels A, Whitehead H, Finn H, Allen S. Impact assessment research: use and misuse of habituation, sensitisation and tolerance in describing wildlife responses to anthropogenic stimuli. Marine Ecology Progress Series. 2009 Dec 3;395: 177–85.
36. Berger J. Fear, human shields and the redistribution of prey and predators in protected areas. Biology Letters. 2007 Dec 22;3(6):620–3.
37. Bleicher SS. The landscape of fear conceptual framework: definition and review of current applications and misuses. PeerJ. 2017 Sep 12;5:e3772.
38. Hoppitt W, Laland K. Social Learning: an introduction to mechanisms, methods, and models. Princeton, NJ: Princeton University Press; 2013.
39. Barrow E. Animal Behavior Desk Reference. Boca Raton, FL: CRC Press; 1995.
40. van Horik JO, Madden JR. A problem with problem solving: motivational traits, but not cognition, predict success on novel operant foraging tasks. Animal Behaviour. 2016;114:189–98.
41. Chow PKY, Uchida K, Von Bayern AMP, Koizumi I. Characteristics of urban environments and novel problem-solving performance in Eurasian red squirrels. Proceedings of the Royal Society B: Biological Sciences. 2021;288(1947):1–9.
42. Uexküll J. Umwelt und Innenwelt der Tiere. Berlin: Springer; 1909.
43. Lowry H, Lill A, Wong BBM. Behavioural responses of wildlife to urban environments. Biological Reviews. 2013;88(3):537–49.
44. Sol D, Lapiedra O, González-Lagos C. Behavioural adjustments for a life in the city. Animal Behaviour. 2013;85(5):1101–12.
45. Ritzel K, Gallo T. Behavior change in urban mammals: a systematic review. Frontiers in Ecology and Evolution. 2020 Nov 16;8:576665.
46. Athreya V, Odden M, Linnell JDC, Karanth KU. Translocation as a tool for mitigating conflict with leopards in human-dominated landscapes of India. Conservation Biology. 2011 Feb;25(1):133–41.
47. Teyssier A, Rouffaer LO, Saleh Hudin N, Strubbe D, Matthysen E, Lens L, et al. Inside the guts of the city: urban-induced alterations of the gut microbiota in

a wild passerine. Science of the Total Environment. 2018;612:1276–86.

48. Murray MH, Kidd AD, Curry SE, Hepinstall-Cymerman J, Yabsley MJ, Adams HC, et al. From wetland specialist to hand-fed generalist: shifts in diet and condition with provisioning for a recently urbanized wading bird. Philosophical Transactions of the Royal Society B: Biological Sciences. 2018;373(1745):20170100.
49. Geffroy B, Samia DSM, Bessa E, Blumstein DT. How nature-based tourism might increase prey vulnerability to predators. Trends in Ecology and Evolution. 2015;30(12):755–65.
50. Sayol F, Sol D, Pigot AL. Brain size and life history interact to predict urban tolerance in birds. Frontiers in Ecology and Evolution. 2020;8(Mar):1–9.
51. Preiszner B, Papp S, Pipoly I, Seress G, Vincze E, Liker A, et al. Problem-solving performance and reproductive success of great tits in urban and forest habitats. Animal Cognition. 2017 Jan;20(1): 53–63.
52. Audet JN, Ducatez S, Lefebvre L. The town bird and the country bird: problem solving and immunocompetence vary with urbanization. Behavioral Ecology. 2016;27(2):637–44.
53. Mazza V, Guenther A. City mice and country mice: innovative problem solving in rural and urban noncommensal rodents. Animal Behaviour. 2021 Feb;172:197–210.
54. Klump BC, Major RE, Farine DR, Martin JM, Aplin LM. Is bin-opening in cockatoos leading to an innovation arms race with humans? Current Biology. 2022 Sep;32(17):R910–11.
55. Klump BC, Martin JM, Wild S, Hörsch JK, Major RE, Aplin LM. Innovation and geographic spread of a complex foraging culture in an urban parrot. Science. 2021 Jul 23;373(6553): 456–60.
56. Barrett LP, Stanton L, Benson-Amram S. The cognition of "nuisance" species. Animal Behaviour. 2019;147:167–77.
57. Ham C, Donnelly CA, Astley KL, Jackson SYB, Woodroffe R. Effect of culling on individual badger *Meles meles* behaviour: potential implications for bovine tuberculosis transmission. Journal of Applied Ecology. 2019;56(11):2390–9.
58. Woodroffe R, Donnelly CA, Jenkins HE, Johnston WT, Cox DR, Bourne FJ, et al. Culling and cattle controls influence tuberculosis risk for badgers. Proceedings of the National Academy of Sciences of the United States of America. 2006 Oct 3;103(40): 14713–17.
59. Haniza MZH, Adams S, Jones EP, MacNicoll A, Mallon EB, Smith RH, et al. Large-scale structure of brown rat (*Rattus norvegicus*) populations in England: effects on rodenticide resistance. PeerJ. 2015 Dec 7;3:e1458.
60. Gese E. Demographic and spatial responses of coyotes to changes in food and exploitation. In: Nolte DL, Fagerstone KA (eds.) Proceedings of the 11th Wildlife Damage Management Conference. Bethesda, MD: Wildlife Damage Management Working Group; 2005. p. 271–85. Available from: https://digitalcommons.unl.edu/icwdm_wdmconfproc/131/
61. Stanton LA, Bridge ES, Huizinga J, Johnson SR, Young JK, Benson-Amram S. Variation in reversal learning by three generalist mesocarnivores. Animal Cognition. 2021;24:555–68.
62. Young J, Schultz J, Jolley B, Basili N, Draper J. Social learning of avoidance behaviors: trap aversion in captive coyotes. Animal Behavior and Cognition. 2022 Aug 1;9(3):336–48.
63. Conover M. Resolving Human–Wildlife Conflicts: the science of wildlife damage management. Boca Raton, FL: CRC Press; 2001.
64. Cunningham CX, Nuñez TA, Hentati Y, Sullender B, Breen C, Ganz TR, et al. Permanent daylight saving time would reduce deer-vehicle collisions. Current Biology. 2022 Nov;32(22):4982–8.e4.
65. Quigley H, Herrero S. Characterization and prevention of attacks on humans. In: Woodroffe R, Thirgood S, Rabinowitz A (eds.) People and Wildlife, Conflict or Co-existence? Cambridge: Cambridge University Press; 2005. p. 27–48.
66. Peterson AC, Ghersi BM, Campanella R, Riegel C, Lewis JA, Blum MJ. Rodent assemblage structure reflects socioecological mosaics of counter-urbanization across post-Hurricane Katrina New Orleans. Landscape and Urban Planning. 2020;195(Mar):103710.
67. Camacho-Rivera M, Kawachi I, Bennett GG, Subramanian SV. Associations of neighborhood concentrated poverty, neighborhood racial/ethnic composition, and indoor allergen exposures: a cross-sectional analysis of Los Angeles households, 2006–2008. Journal of Urban Health. 2014 Aug;91(4): 661–76.
68. Lamb CT, Mowat G, McLellan BN, Nielsen SE, Boutin S. Forbidden fruit: human settlement and abundant fruit create an ecological trap for an apex omnivore. Journal of Animal Ecology. 2017;86(1): 55–65.

69. Fehlmann G, O'riain MJ, Fürtbauer I, King AJ. Behavioral causes, ecological consequences, and management challenges associated with wildlife foraging in human-modified landscapes. BioScience. 2021 Jan 11;71(1):40–54.
70. Scholz C, Firozpoor J, Kramer-Schadt S, Gras P, Schulze C, Kimmig SE, et al. Individual dietary specialization in a generalist predator: a stable isotope analysis of urban and rural red foxes. Ecology and Evolution. 2020 Aug;10(16): 8855–70.
71. Newsome SD, Garbe HM, Wilson EC, Gehrt SD. Individual variation in anthropogenic resource use in an urban carnivore. Oecologia. 2015 May 12;178(1): 115–28.
72. VerCauteren KC, Dolbeer RA, Gese EM. Identification and management of wildlife damage. In: Silvy NJ (ed.) The Wildlife Techniques Manual. Baltimore, MD: Johns Hopkins University Press; 2010. p. 232–69.
73. Greggor AL, Berger-Tal O, Blumstein DT. The rules of attraction: the necessary role of animal cognition in explaining conservation failures and successes. Annual Review of Ecology, Evolution, and Systematics. 2020;51:483–503.
74. Snijders L, Greggor AL, Hilderink F, Doran C. Effectiveness of animal conditioning interventions in reducing human–wildlife conflict: a systematic map protocol. Environmental Evidence. 2019 Jun;8(S1):10.
75. Theimer TC, Clayton AC, Martinez A, Peterson DL, Bergman DL. Visitation rate and behavior of urban mesocarnivores differs in the presence of two common anthropogenic food sources. Urban Ecosystems. 2015;18:895–906.
76. Lambert MR, Nielsen SN, Wright AN, Thomson RC, Shaffer HB. Habitat features determine the basking distribution of introduced red-eared sliders and native Western pond turtles. Chelonian Conservation and Biology. 2013 Jul;12(1):192–9.
77. St. Clair CC, Backs J, Friesen A, Gangadharan A, Gilhooly P, Murray M, et al. Animal learning may contribute to both problems and solutions for wildlife–train collisions. Philosophical Transactions of the Royal Society B: Biological Sciences. 2019 Sep 16;374(1781):20180050.
78. Babińska-Werka J, Krauze-Gryz D, Wasilewski M, Jasińska K. Effectiveness of an acoustic wildlife warning device using natural calls to reduce the risk of train collisions with animals. Transportation Research Part D: Transport and Environment. 2015 Jul;38:6–14.
79. Shimura M, Ushiogi T, Ikehata M. Development of an acoustic deterrent to prevent deer-train collisions. Quarterly Report RTRI. 2018;59(3):207–11.
80. Found R, St. Clair CC. Personality influences wildlife responses to aversive conditioning: elk personality and response to management. Journal of Wildlife Management. 2018 May;82(4):747–55.
81. Found R, Kloppers EL, Hurd TE, St. Clair CC. Intermediate frequency of aversive conditioning best restores wariness in habituated elk (*Cervus canadensis*). PLoS ONE. 2018 Jun 25;13(6):e0199216.
82. Found R, St. Clair CC. Influences of personality on ungulate migration and management. Frontiers in Ecology and Evolution. 2019;7(Nov):438.
83. Steen KA, Therkildsen OR, Karstoft H, Green O. An adaptive scaring device. International Journal of Sustainable Agricultural Management and Informatics. 2015;1(2):130–41.
84. Vantassel S, Groepper S. A survey of wildlife damage management techniques used by wildlife control operators in urbanized environments in the USA. In: Angelici FM (ed.) Problematic Wildlife: a cross-disciplinary approach. Cham, Switzerland: Springer International; 2016. p. 175–204.
85. bin Shabib R, bin Shabib A. Buildings for the birds [Internet]. The Common Table; 2022 Jan 27. Available from: https://thecommontable.eu/buildings-for-the-birds/
86. Doorn AC, O'Riain MJ. Nonlethal management of baboons on the urban edge of a large metropole. American Journal of Primatology [Internet]. 2020 Aug [cited 2023 Jan 13];82(8):e23164. Available from: https://onlinelibrary.wiley.com/doi/10.1002/ajp.23164
87. Kaplan BS, O'Riain MJ, van Eeden R, King AJ. A low-cost manipulation of food resources reduces spatial overlap between baboons (*Papio ursinus*) and humans in conflict. International Journal of Primatology. 2011 Dec;32(6):1397–412.
88. Fehlmann G, O'Riain MJ, Kerr-Smith C, King AJ. Adaptive space use by baboons (*Papio ursinus*) in response to management interventions in a human-changed landscape. Animal Conservation. 2017;20(1):101–9.
89. Schell CJ, Dyson K, Fuentes TL, Des Roches S, Harris NC, Miller DS, et al. The ecological and evolutionary consequences of systemic racism in urban environments. Science. 2020;369(6510): eaay4497.
90. Hope D, Gries C, Zhu W, Fagan WF, Redman CL, Grimm NB, et al. Socioeconomics drive urban plant diversity. Proceedings of the National

Academy of Sciences of the United States of America. 2003;100(15):8788–92.

91. Des Roches S, Brans KI, Lambert MR, Rivkin LR, Savage AM, Schell CJ, et al. Socio-eco-evolutionary dynamics in cities. Evolutionary Applications. 2021 Jan 1;14(1):248–67.
92. Fidino M, Lehrer EW, Kay CAM, Yarmey NT, Murray MH, Fake K, et al. Integrated species distribution models reveal spatiotemporal patterns of human–wildlife conflict. Ecological Applications [Internet]. 2022 Oct [cited 2023 Jan 13];32(7):e2647. Available from: https://onlinelibrary.wiley.com/doi/10.1002/eap.2647
93. Brownlow A. An archaeology of fear and environmental change in Philadelphia. Geoforum. 2006 Mar;37(2):227–45.
94. Anguelovski I. From toxic sites to parks as (green) LULUs? New challenges of inequity, privilege, gentrification, and exclusion for urban environmental justice. Journal of Planning Literature. 2016 Feb;31(1):23–36.
95. Murray MH, Buckley J, Byers KA, Fake K, Lehrer EW, Magle SB, et al. One Health for all: advancing human and ecosystem health in cities by integrating an environmental justice lens. Annual Review of Ecology, Evolution, and Systematics. 2022 Nov 2;53:403–26.
96. Monchanin C, Blanc-Brude A, Drujont E, Negahi MM, Pasquaretta C, Silvestre J, et al. Chronic exposure to trace lead impairs honey bee learning. Ecotoxicology and Environmental Safety. 2021 Apr;212:112008.
97. Berg EL, Pedersen LR, Pride MC, Petkova SP, Patten KT, Valenzuela AE, et al. Developmental exposure to near roadway pollution produces behavioral phenotypes relevant to neurodevelopmental disorders in juvenile rats. Translational Psychiatry. 2020;10(1):289.
98. Serieys LEK, Lea AJ, Epeldegui M, Armenta TC, Moriarty J, Vandewoude S, et al. Urbanization and anticoagulant poisons promote immune dysfunction in bobcats. Proceedings of the Royal Society B: Biological Sciences. 2018;285(1871):20172533.
99. Adamkiewicz G, Zota AR, Fabian MP, Chahine T, Julien R, Spengler JD, et al. Moving environmental justice indoors: understanding structural influences on residential exposure patterns in low-income communities. American Journal of Public Health. 2011 Dec;101(S1):S238–45.
100. Casey JA, Morello-Frosch R, Mennitt DJ, Fristrup K, Ogburn EL, James P. Race/ethnicity, socioeconomic status, residential segregation, and spatial variation in noise exposure in the contiguous United States. Environmental Health Perspectives. 2017 Jul 24;125(7):077017.
101. Donley N, Bullard RD, Economos J, Figueroa I, Lee J, Liebman AK, et al. Pesticides and environmental injustice in the USA: root causes, current regulatory reinforcement and a path forward. BMC Public Health. 2022 Dec;22(1):708.
102. Sugden S, Murray M, Edwards MA, St. Clair CC. Inter-population differences in coyote diet and niche width along an urban–suburban–rural gradient. Journal of Urban Ecology. 2021 Jan 22;7(1): juab034.
103. Feng AYT, Himsworth CG. The secret life of the city rat: a review of the ecology of urban Norway and black rats (*Rattus norvegicus* and *Rattus rattus*). Urban Ecosystems. 2014;17(1):149–62.
104. Schulte-Hostedde AI, Mazal Z, Jardine CM, Gagnon J. Enhanced access to anthropogenic food waste is related to hyperglycemia in raccoons (*Procyon lotor*). Conservation Physiology. 2018;6(1):1–6.
105. Hungerford LL, Mitchell MA, Nixon CM, Esker TE, Sullivan JB, Koerkenmeier R, et al. Periodontal and dental lesions in raccoons from a farming and a recreational area in Illinois. Journal of Wildlife Diseases. 1999;35(4):728–34.
106. Sugden S, Sanderson D, Ford K, Stein LY, Cassady C, Clair S. An altered microbiome in urban coyotes mediates relationships between anthropogenic diet and poor health. Scientific Reports. 2020 [cited 2022 Feb 23];10:22207. Available from: https://doi.org/10.1038/s41598-020-78891-1
107. Savard JPL, Clergeau P, Mennechez G. Biodiversity concepts and urban ecosystems. Landscape and Urban Planning. 2000 May;48(3–4): 131–42.
108. Nielsen AB, van den Bosch M, Maruthaveeran S, van den Bosch CK. Species richness in urban parks and its drivers: a review of empirical evidence. Urban Ecosystems. 2014 Mar;17(1):305–27.
109. Møller AP. Urban areas as refuges from predators and flight distance of prey. Behavioral Ecology. 2012 Sep 1;23(5):1030–5.
110. Ryan AM, Partan SR. Urban wildlife behavior. In: McCleery RA, Moorman CE, Peterson MN (eds.) Urban Wildlife Conservation: theory and practice. Boston, MA: Springer US; 2014. p. 149–73.
111. Knapp S, Aronson MFJ, Carpenter E, Herrera-Montes A, Jung K, Kotze DJ, et al. A research agenda for urban biodiversity in the global extinction crisis. BioScience. 2021 Mar 1;71(3):268–79.

112. Moorman C. Managing urban wildlife habitat at the local scale. In: McCleery RA, Moorman CE, Peterson MN (eds.) Urban Wildlife Conservation: theory and practice [Internet]. Boston, MA: Springer US; 2014 [cited 2023 Jan 14]. p. 303–21. Available from: https://link.springer.com/10.1007/978-1-4899-7500-3_14
113. Hale R, Swearer SE. When good animals love bad restored habitats: how maladaptive habitat selection can constrain restoration. Journal of Applied Ecology. 2017 Oct;54(5):1478–86.
114. McKinney ML. Urbanization, biodiversity, and conservation. BioScience. 2002;52(10):883–90.
115. Goddard MA, Ikin K, Lerman SB. Ecological and social factors determining the diversity of birds in residential yards and gardens. In: Murgui E, Hedblom M (eds.) Ecology and Conservation of Birds in Urban Environments [Internet]. Cham, Switzerland: Springer International Publishing; 2017 [cited 2023 Jan 14]. p. 371–97. Available from: https://link.springer.com/10.1007/978-3-319-43314-1_18
116. Litteral J, Shochat E. The role of landscape-scale factors in shaping urban bird communities. In: Murgui E, Hedblom M (eds.) Ecology and Conservation of Birds in Urban Environments [Internet]. Cham, Switzerland: Springer International Publishing; 2017 [cited 2023 Jan 14]. p. 135–59. Available from: https://link.springer.com/10.1007/978-3-319-43314-1_8
117. Panlasigui S, Spotswood E, Beller E, Grossinger R. Biophilia beyond the building: applying the tools of urban biodiversity planning to create biophilic cities. Sustainability. 2021 Feb 24;13(5):2450.
118. Wilkinson CE, McInturff A, Kelly M, Brashares JS. Quantifying wildlife responses to conservation fencing in East Africa. Biological Conservation. 2021 Apr;256:109071.
119. Osipova L, Okello MM, Njumbi SJ, Ngene S, Western D, Hayward MW, et al. Fencing solves human-wildlife conflict locally but shifts problems elsewhere: a case study using functional connectivity modelling of the African elephant. Journal of Applied Ecology. 2018 Nov;55(6):2673–84.
120. McInturff A, Xu W, Wilkinson CE, Dejid N, Brashares JS. Fence ecology: frameworks for understanding the ecological effects of fences. BioScience. 2020 Nov;70(11):971–85.
121. Ward MP, Schlossberg S. Conspecific attraction and the conservation of territorial songbirds. Conservation Biology. 2004 Apr;18(2):519–25.
122. Nisbet I. Disturbance, habituation, and management of waterbird colonies. Waterbirds. 2000;23(2):312–32.
123. Lepczyk CA, Aronson MFJ, Evans KL, Goddard MA, Lerman SB, MacIvor JS. Biodiversity in the city: fundamental questions for understanding the ecology of urban green spaces for biodiversity conservation. BioScience. 2017 Sep;67(9):799–807.
124. Jokimäki J, Kaisanlahti-Jokimäki ML, Suhonen J, Clergeau P, Pautasso M, Fernández-Juricic E. Merging wildlife community ecology with animal behavioral ecology for a better urban landscape planning. Landscape and Urban Planning. 2011 Apr;100(4):383–5.
125. Robertson BA, Blumstein DT. How to disarm an evolutionary trap. Conservation Science and Practice. 2019;1(11):e116.
126. Greggor AL, Trimmer PC, Barrett BJ, Sih A. Challenges of learning to escape evolutionary traps. Frontiers in Ecology and Evolution. 2019 Oct 25;7:408.
127. Riley SPD, Brown JL, Sikich J, Schoonmaker C, Boydson E. Wildlife friendly roads: the impacts of roads on wildlife in urban areas and potential remedies. In: McCleery RA, Moorman CE, Peterson MN (eds.) Urban Wildlife Conservation: theory and practice. Boston, MA: Springer US; 2014. p. 323–60.
128. Shilling F, Waetjen D, Longcore T, Vickers W, McDowell S, Oke A, et al. Improving light and soundscapes for wildlife use of highway crossing structures. Davis, CA: UC Davis Institute of Transportation Studies; 2022 [cited 2023 Jan 14]. Available from: https://escholarship.org/uc/item/4vk0m9cs
129. Glista DJ, DeVault TL, DeWoody JA. A review of mitigation measures for reducing wildlife mortality on roadways. Landscape and Urban Planning. 2009 May;91(1):1–7.
130. Carlen EJ, Li R, Winchell KM. Urbanization predicts flight initiation distance in feral pigeons (*Columba livia*) across New York City. Animal Behaviour. 2021 Aug;178:229–45.
131. Bowers MA, Breland B. Foraging of gray squirrels on an urban–rural gradient: use of the GUD to assess anthropogenic impact. Ecological Applications. 1996 Nov;6(4):1135–42.
132. Reilly CM, Suraci JP, Smith JA, Wang Y, Wilmers CC. Mesopredators retain their fear of humans across a development gradient. Behavioral Ecology. 2022 Apr 18;33(2):428–35.
133. Clucas B, Marzluff JM. Attitudes and actions toward birds in urban areas: human cultural differences influence bird behavior. The Auk. 2012;129(1):8–16.

134. Gaynor KM, Hojnowski CE, Carter NH, Brashares JS. The influence of human disturbance on wildlife nocturnality. Science. 2018;360(6394):1232–5.
135. Nickel BA, Suraci JP, Allen ML, Wilmers CC. Human presence and human footprint have non-equivalent effects on wildlife spatiotemporal habitat use. Biological Conservation. 2020;241(Jan):108383.
136. Wilmers CC, Nisi AC, Ranc N. COVID-19 suppression of human mobility releases mountain lions from a landscape of fear. Current Biology. 2021 Sep;31(17):3952–5.e3.
137. Duarte MHL, Vecci MA, Hirsch A, Young RJ. Noisy human neighbours affect where urban monkeys live. Biology Letters. 2011 Dec 23;7(6):840–2.
138. Bates AE, Primack RB, Biggar BS, Bird TJ, Clinton ME, Command RJ, et al. Global COVID-19 lockdown highlights humans as both threats and custodians of the environment. Biological Conservation. 2021 Nov;263:109175.
139. Goumas M, Boogert NJ, Kelley LA. Urban herring gulls use human behavioural cues to locate food. Royal Society Open Science. 2020;7(2):191959.
140. Liu J, Dietz T, Carpenter SR, Alberti M, Folke C, Moran E, et al. Complexity of coupled human and natural systems. Science. 2007;317(5844):1513–16.
141. Clucas B, Marzluff JM. Coupled relationships between humans and other organisms in urban areas. In: Niemelä J (ed.) Urban Ecology: patterns, processes, and applications. Oxford: Oxford University Press; 2011. p. 135–47.
142. Kellert SR. American attitudes toward and knowledge of animals: an update. In: Fox MW, Mickley LD (eds.) Advances in Animal Welfare Science 1984 [Internet]. Dordrecht: Springer Netherlands; 1985 [cited 2023 Jan 14]. p. 177–213. Available from: https://link.springer.com/10.1007/978-94-009-4998-0_11
143. Dickman AJ, Hazzah L, Carbone C, Durant SM. Carnivores, culture and "contagious conflict": multiple factors influence perceived problems with carnivores in Tanzania's Ruaha landscape. Biological Conservation. 2014 Oct;178:19–27.
144. Reiter D, Brunson M, Schmidt R. Public attitudes toward wildlife damage management and policy. Wildlife Society Bulletin. 2023;27(3):746–58.
145. Koval MH, Mertig AG. Attitudes of the Michigan public and wildlife agency personnel toward lethal wildlife management. Wildlife Society Bulletin. 2004 Mar;32(1):232–43.
146. Lischka SA, Teel TL, Johnson HE, Reed SE, Breck S, Don Carlos A, et al. A conceptual model for the integration of social and ecological information to understand human-wildlife interactions. Biological Conservation. 2018 Sep;225:80–7.
147. Carter NH, Riley SJ, Liu J. Utility of a psychological framework for carnivore conservation. Oryx. 2012 Oct;46(4):525–35.
148. Ngo KM, Hosaka T, Numata S. The influence of childhood nature experience on attitudes and tolerance towards problem-causing animals in Singapore. Urban Forestry & Urban Greening. 2019 May;41: 150–7.
149. Young JK, Coppock DL, Baggio JA, Rood KA, Yirga G. Linking human perceptions and spotted hyena behavior in urban areas of Ethiopia. Animals. 2020 Dec 15;10(12):2400.
150. Wilcoxen TE, Horn DJ, Hogan BM, Hubble CN, Huber SJ, Flamm J, et al. Effects of bird-feeding activities on the health of wild birds. Conservation Physiology. 2015;3(1):cov058.
151. Kumar N, Jhala YV, Qureshi Q, Gosler AG, Sergio F. Human-attacks by an urban raptor are tied to human subsidies and religious practices. Scientific Reports. 2019;9:2545.
152. Saraswat R, Sinha A, Radhakrishna S. A god becomes a pest? Human-rhesus macaque interactions in Himachal Pradesh, northern India. European Journal of Wildlife Research. 2015 Jun;61(3): 435–43.
153. Peterson MN, Birckhead JL, Leong K, Peterson MJ, Peterson TR. Rearticulating the myth of human–wildlife conflict. Conservation Letters. 2010 Apr;3(2):74–82.
154. Redman CL, Grove JM, Kuby LH. Integrating social science into the Long-Term Ecological Research (LTER) Network: social dimensions of ecological change and ecological dimensions of social change. Ecosystems. 2004;7(2):161–71.
155. Dickman AJ. Complexities of conflict: the importance of considering social factors for effectively resolving human–wildlife conflict. Animal Conservation. 2010;13(5):458–66.
156. Siex KS, Struhsaker TT. Colobus monkeys and coconuts: a study of perceived human–wildlife conflicts. Journal of Applied Ecology. 1999 Dec;36(6):1009–20.
157. Wilkinson CE, Brashares JS, Bett AC, Kelly M. Examining drivers of divergence in recorded and perceived human-carnivore conflict hotspots by integrating participatory and ecological data. Frontiers in Conservation Science. 2021 Jul 20;2: 681769.

158. Draheim MM, Parsons ECM, Crate SA, Rockwood LL. Public perspectives on the management of urban coyotes. Journal of Urban Ecology [Internet]. 2019 [cited 2022 Jul 10];5(1):juz003. Available from: https://academic.oup.com/jue/article/doi/10.1093/jue/juz003/5424021
159. Basak SM, Hossain MS, O'Mahony DT, Okarma H, Widera E, Wierzbowska IA. Public perceptions and attitudes toward urban wildlife encounters – a decade of change. Science of the Total Environment. 2022 Aug;834:155603.
160. White J, Kemmelmeier M, Bassett S, Smith J. Human perceptions of an avian predator in an urban ecosystem: close proximity to nests increases fondness among local residents. Urban Ecosystems. 2018;21(2):271–80. Available from: https://link.springer.com/10.1007/s11252-017-0713-y
161. Boso À, Álvarez B, Pérez B, Imio JC, Altamirano A, Lisón F. Understanding human attitudes towards bats and the role of information and aesthetics to boost a positive response as a conservation tool. Animal Conservation. 2021 Dec;24(6):937–45.
162. Rigolon A. A complex landscape of inequity in access to urban parks: a literature review. Landscape and Urban Planning. 2016 Sep;153:160–9.
163. Warkentin T. Cultivating urban naturalists: teaching experiential, place-based learning through nature journaling in Central Park. Journal of Geography. 2011 Nov;110(6):227–38.
164. Larson LR, Cooper CB, Stedman RC, Decker DJ, Gagnon RJ. Place-based pathways to proenvironmental behavior: empirical evidence for a conservation–recreation model. Society & Natural Resources. 2018 Aug 3;31(8):871–91.
165. Poessel SA, Gese EM, Young JK. Environmental factors influencing the occurrence of coyotes and conflicts in urban areas. Landscape and Urban Planning. 2017 Jan;157:259–69.
166. Drake D, Dubay S, Allen ML. Evaluating human–coyote encounters in an urban landscape using citizen science. Journal of Urban Ecology. 2021 Jan 22;7(1):juaa032.
167. Breck SW, Poessel SA, Mahoney P, Young JK. The intrepid urban coyote: a comparison of bold and exploratory behavior in coyotes from urban and rural environments. Scientific Reports. 2019;9(Feb):2104.
168. Brooks J, Kays R, Hare B. Coyotes living near cities are bolder: implications for dog evolution and human-wildlife conflict. Behaviour. 2020 Mar 20;157(3–4):289–313.
169. Henger CS, Herrera GA, Nagy CM, Weckel ME, Gormezano LJ, Wultsch C, et al. Genetic diversity and relatedness of a recently established population of eastern coyotes (*Canis latrans*) in New York City. Urban Ecosystems. 2020 Apr;23(2):319–30.
170. Wurth AM, Ellington EH, Gehrt SD. Golf courses as potential habitat for urban coyotes. The Wildlife Society Bulletin. 2020 Jun;44(2):333–41.
171. Gámez S, Harris NC. Living in the concrete jungle: carnivore spatial ecology in urban parks. Ecological Applications [Internet]. 2021 Sep [cited 2022 Jul 14];31(6):e02393. Available from: https://onlinelibrary.wiley.com/doi/10.1002/eap.2393
172. Thompson CA, Malcolm JR, Patterson BR. Individual and temporal variation in use of residential areas by urban coyotes. Frontiers in Ecology and Evolution. 2021 Jun 1;9:687504.
173. Gilbert-Norton LB, Shahan TA, Shivik JA. Coyotes (*Canis latrans*) and the matching law. Behavioural Processes. 2009 Oct 1;82(2):178–83.
174. Stanton LA, Bridge ES, Huizinga J, Benson-Amram S. Environmental, individual and social traits of free-ranging raccoons influence performance in cognitive testing. Journal of Experimental Biology. 2022 Sep 15;225(18):jeb243726.
175. Van Bourg J, Young JK, Alkhalifah R, Brummer S, Johansson E, Morton J, et al. Cognitive flexibility and aging in coyotes (*Canis latrans*). Journal of Comparative Psychology. 2022 Feb;136(1):54–67.
176. Poessel SA, Mock EC, Breck SW. Coyote (*Canis latrans*) diet in an urban environment: variation relative to pet conflicts, housing density, and season. Canadian Journal of Zoology. 2017 Apr;95(4): 287–97.
177. Murray M, Cembrowski A, Latham ADM, Lukasik VM, Pruss S, St Clair CC. Greater consumption of protein-poor anthropogenic food by urban relative to rural coyotes increases diet breadth and potential for human–wildlife conflict. Ecography. 2015 Dec;38(12):1235–42.
178. Breck S, Poessel S, Bonnell MA. Evaluating lethal and non-lethal management options for urban coyotes. In: Timm RM, Baldwin RA (eds.) Proceedings of the 27th Vertebrate Pest Conference [Internet]. Davis, CA: Universty of California Davis; 2016 [cited 2022 Jul 14]. p. 103–11. Available from: https://escholarship.org/uc/item/0g83m9km
179. Murray MH, Hill J, Whyte P, St. Clair CC. Urban compost attracts coyotes, contains toxins, and may promote disease in urban-adapted wildlife. EcoHealth. 2016 Jun;13(2):285–92.
180. Wald DM, Jacobson SK, Levy JK. Outdoor cats: identifying differences between stakeholder beliefs, perceived impacts, risk and management. Biological Conservation. 2013 Nov;167:414–24.

181. Nattrass N, O'Riain MJ. Contested natures: conflict over caracals and cats in Cape Town, South Africa. Journal of Urban Ecology. 2020 Jan 1;6(1):juaa019.
182. Loss SR, Will T, Marra PP. The impact of free-ranging domestic cats on wildlife of the United States. Nature Communications. 2013 Jun 26;4(1):1396.
183. Bonnell MA, Breck SW. Using resident-based hazing programs to reduce human-coyote conflicts in urban environments. Human–Wildlife Interactions. 2017;11(2):5.
184. McLellan BA, Walker KA. Efficacy of motion-activated sprinklers as a humane deterrent for urban coyotes. Human Dimensions of Wildlife. 2021 Jan 2;26(1):76–83.
185. Sponarski CC, Miller CA, Vaske JJ, Spacapan MR. Modeling perceived risk from coyotes among Chicago residents. Human Dimensions of Wildlife. 2016 Nov;21(6):491–505.
186. Sampson L, Van Patter L. Advancing best practices for aversion conditioning (humane hazing) to mitigate human–coyote conflicts in urban areas. Human–Wildlife Interactions. 2020;14(2): 166–83.
187. Alexander SM, Quinn MS. Portrayal of interactions between humans and coyotes (Canis latrans): content analysis of Canadian print media (1998-2010). Cities and the Environment. 2011;4(1):1–24.
188. Draheim MM, Crate SA, Parsons ECM, Rockwood LL. The impact of language in conflicts over urban coyotes. Journal of Urban Ecology. 2021;7(1):juab036. Available from: https://doi.org/10.1093/jue/juab036
189. Davidson GL, Raulo A, Knowles SCL. Identifying microbiome-mediated behaviour in wild vertebrates. Trends in Ecology and Evolution. 2020 Nov 1;35(11):972–80.
190. Jones TB, Evans JC, Morand-Ferron J. Urbanization and the temporal patterns of social networks and group foraging behaviors. Ecology and Evolution. 2019;9(8):4589–602.
191. Prange S, Gehrt SD, Hauver S. Frequency and duration of contacts between free-ranging raccoons: uncovering a hidden social system. Journal of Mammalogy. 2011;92(6):1331–42.
192. Hirsch BT, Prange S, Hauver SA, Gehrt SD. Raccoon social networks and the potential for disease transmission. PLoS ONE. 2013;8(10): 4–10.
193. Westall TL, Cypher BL, Ralls K, Wilbert T. Observations of social polygyny, allonursing, extrapair copulation, and inbreeding in urban San Joaquin kit foxes (*Vulpes macrotis mutica*). The Southwestern Naturalist. 2019 Oct 7;63(4):271.
194. Vlaschenko A, Kovalov V, Hukov V, Kravchenko K, Rodenko O. An example of ecological traps for bats in the urban environment. European Journal of Wildlife Research. 2019 Apr;65(2):20.
195. Strasser EH, Heath JA. Reproductive failure of a human-tolerant species, the American kestrel, is associated with stress and human disturbance. Journal of Applied Ecology. 2013 Aug;50(4):912–19.
196. Frigerio D, Pipek P, Kimmig S, Winter S, Melzheimer J, Diblíková L, et al. Citizen science and wildlife biology: synergies and challenges. Ethology. 2018 Jun;124(6):365–77.
197. Ostermann-Miyashita E, Pernat N, König HJ. Citizen science as a bottom-up approach to address human–wildlife conflicts: from theories and methods to practical implications. Conservation Science and Practice [Internet]. 2021 Mar [cited 2023 Jan 14];3(3):e385. Available from: https://onlinelibrary.wiley.com/doi/10.1111/csp2.385
198. Lasky M, Parsons A, Schuttler S, Mash A, Larson L, Norton B, et al. Candid critters: challenges and solutions in a large-scale citizen science camera trap project. Citizen Science: Theory and Practice. 2021 Feb 26;6(1):4.
199. Swanson AC, Conn A, Swanson JJ, Brooks DM. Record of an urban ringtail (*Bassariscu astutus*) outside of its typical geographic range. Urban Naturalist Notes. 2022;9(4):1–6.
200. Carter NH, Linnell JDC. Co-adaptation is key to coexisting with large carnivores. Trends in Ecology and Evolution. 2016 Aug;31(8):575–8.
201. König HJ, Kiffner C, Kramer-Schadt S, Fürst C, Keuling O, Ford AT. Human–wildlife coexistence in a changing world. Conservation Biology. 2020 Aug;34(4):786–94.
202. Collins MK, Magle SB, Gallo T. Global trends in urban wildlife ecology and conservation. Biological Conservation. 2021 Sep;261:109236.
203. Parker SS. Incorporating critical elements of city distinctiveness into urban biodiversity conservation. Biodiversity and Conservation. 2015 Mar;24(3):683–700.
204. Xie S, Wang X, Zhou W, Wu T, Qian Y, Lu F, et al. The effects of residential greenspace on avian biodiversity in Beijing. Global Ecology and Conservation. 2020 Dec;24:e01223.
205. Cox DTC, Hudson HL, Plummer KE, Siriwardena GM, Anderson K, Hancock S, et al. Covariation in urban birds providing cultural services or disservices and people. Journal of Applied Ecology. 2018 Sep;55(5):2308–19.
206. Ilham K, Rizaldi, Nurdin J, Tsuji Y. Status of urban populations of the long-tailed macaque (*Macaca fas-*

cicularis) in West Sumatra, Indonesia. Primates. 2017 Apr;58(2):295–305.

207. Sha JCM, Gumert MD, Lee BPY-H, Jones-Engel L, Chan S, Fuentes A. Macaque–human interactions and the societal perceptions of macaques in Singapore. American Journal of Primatology. 2009 Oct;71(10):825–39.
208. Chen M, Tan ADJ, Quek WL, Chahed H. A proposal for a technology-assisted approach to wildlife management in Singapore. Pacific Conservation Biology [Internet]. 2022 [cited 2022 Jul 15];29(1):1–16. Available from: http://www.publish.csiro.au/?paper=PC21055
209. Hinchcliffe DL, Young RJ, Teixeira CP. Callout analysis in relation to wild birds in a tropical city: implications for urban species management. Urban Ecosystems. 2022 Jun 21;25(6):1643–52.
210. Charles KE, Linklater WL. Selection of trees for sap-foraging by a native New Zealand parrot, the Kaka (*Nestor meridionalis*), in an urban landscape. Emu - Austral Ornithology. 2014 Dec;114(4): 317–25.
211. Crenshaw K. On Intersectionality: essential writings. New York: The New Press; 2017.
212. Baruch-Mordo S, Breck SW, Wilson KR, Broderick J. A tool box half full: how social science can help solve human–wildlife conflict. Human Dimensions of Wildlife. 2009 Jun 3;14(3):219–23.
213. St John FAV, Edwards-Jones G, Jones JPG. Conservation and human behaviour: lessons from social psychology. Wildlife Research. 2010;37(8): 658–67.
214. Baruch-Mordo S, Breck SW, Wilson KR, Broderick J. The carrot or the stick? Evaluation of education and enforcement as management tools for human-wildlife conflicts. PLoS ONE. 2011 Jan 12;6(1):e15681.
215. Johnson HE, Lewis DL, Lischka SA, Breck SW. Assessing ecological and social outcomes of a bear-proofing experiment. Journal of Wildlife Management. 2018;82(6):1102–1114.
216. Athreya V, Thakur S, Chaudhuri S, Belsare A. Leopards in human-dominated areas: a spillover from sustained translocations into nearby forests? Journal of the Bombay Natural History Society. 2007;104(1):45–50.
217. Massei G, Quy RJ, Gurney J, Cowan DP. Can translocations be used to mitigate human–wildlife conflicts? Wildlife Research. 2010;37(5):428–39.
218. Resende PS, Viana-Junior AB, Young RJ, Azevedo CS. What is better for animal conservation translocation programmes: soft- or hard-release? A phylogenetic meta-analytical approach. Journal of Applied Ecology. 2021 Jun;58(6): 1122–32.
219. Athreya V, Odden M, Linnell JDC, Krishnaswamy J, Karanth U. Big cats in our backyards: persistence of large carnivores in a human dominated landscape in India. PLoS ONE. 2013 Mar 6;8(3):e57872.
220. Bhatia S, Athreya V, Grenyer R, MacDonald DW. Understanding the role of representations of human–leopard conflict in Mumbai through media-content analysis. Conservation Biology. 2013 Jun;27(3): 588–94.

SECTION 3

Emergent Urban Planning and Management Frameworks for Addressing Societal and Conservation Goals

How we integrate the diverse needs of people and biodiversity in existing and future cities requires adapting existing planning paradigms in ways that create synergistic benefits and minimize conflict. Because cities are unique social-ecological systems, we will also need to create new approaches tailored to situations that are not seen in nonurban systems. Building on existing tools and approaches and designing new ones will be central to creating urban spaces where people and other species can thrive together. This section begins providing the new toolboxes and frameworks that can push forward equity- and justice-centered conservation in cities.

CHAPTER 12

Urban Places Create Unique Health Spaces for Wildlife, People, and the Environment

Kaylee A. Byers[§*], Maureen H. Murray[‡*], and Joanne E. Nelson[§]

Introduction

The health of animals, people, and the environment are intimately interconnected. The Covid-19 pandemic has been a stark reminder of these connections. The likely animal origins of the SARS-CoV-2 virus (1,2) and its subsequent spread to animals such as mink and deer demonstrated how viruses and bacteria are spread among people and animals (3,4). These animal species-crossing pathogens are called "zoonotic," with 75% of emerging infectious diseases in people originating in other animals (5–8). The Covid-19 pandemic also illustrated how changes in human behaviors affect environmental health. Efforts to reduce the spread of SARS-CoV-2 led to less air and vehicle traffic, lower carbon emissions (9), and decreased disturbance to wildlife (10). These unintended outcomes reveal the power of collective human behaviors to bolster health across species and geographies.

"One Health" is a holistic framework that envisions our shared health (Figure 12.1). It encompasses infectious and noninfectious issues, recognizing that health is just one part of a complex, constantly changing system of environmental and sociological factors that intersect (11). And although the term "One Health" is relatively recent in the dominant science discourse, this concept of interconnected human, animal, and environmental health is far from new. The One Health framework aligns with many Indigenous perspectives of environmental health (12). The need to integrate knowledges and perspectives across cultures is increasingly recognized as vital to applying One Health approaches across contexts.

Here, we explore the unique One Health challenges posed by cities, and share examples from our collective experience. For that reason, we want to introduce ourselves and our connection to this work. Two of the authors (Byers and Murray) have worked in the field of wildlife health for over a decade. Byers (settler, European ancestry) has studied urban bird health and urban rat ecology, disease transmission, and its links to mental health in an underserved neighborhood of Canada. Murray (settler, European ancestry) has studied disease dynamics and human–wildlife interactions in diverse urban wildlife species, including coyotes, wading birds, and rats. Nelson (Ts'msyen) is an emerging scholar in the field of Indigenous water governance whose recent fieldwork and research focus on arts-informed methods with urban Indigenous communities. She also has a background in qualitative public health research methodologies. Together, they share an interest in applying One Health to improve shared human, animal, and environmental health in cities, but are also interested in re-envisioning this framework to integrate more holistic and interdisciplinary perspectives.

[§] Simon Fraser University, Canada
[‡] Lincoln Park Zoo, USA
[*] Co-first author.

Kaylee A. Byers, Maureen H. Murray, and Joanne E. Nelson, *Urban Places Create Unique Health Spaces for Wildlife, People, and the Environment.* In: *Urban Biodiversity and Equity.* Edited by: Max R. Lambert & Christopher J. Schell, Oxford University Press. © Oxford University Press (2023).
DOI: 10.1093/oso/9780198877271.003.0012

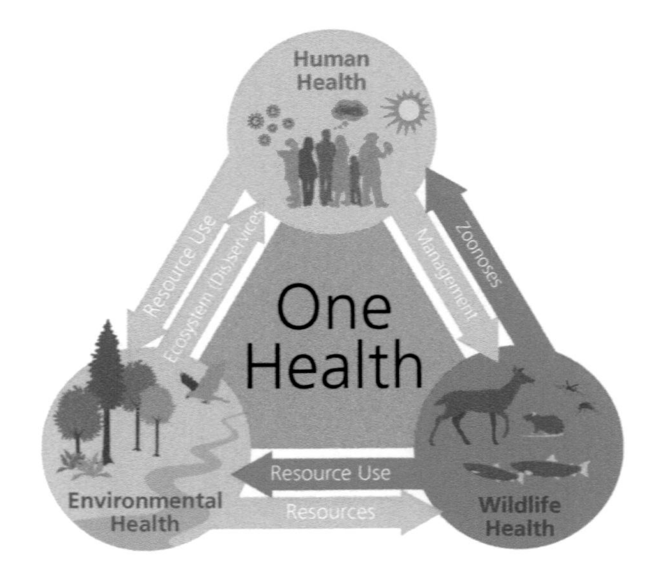

Figure 12.1 The One Health framework emphasizes the interconnectedness of human, animal, and environmental health. Humans (orange) control nearly all aspects of urban environments, which affects the quantity and quality of resources available to wildlife (red) such as clean air, water, and food from the environment (blue). Text labels on arrows are some examples of reciprocal relationships.
Figure adapted with permissions from (11). Image courtesy of Simone Des Roches.

To describe One Health in cities, we focus on how changes in land use and species composition affect health (Figure 12.2). Importantly, while these health challenges vary among cities, they are not uniformly distributed within a single city. Further, the patterns that exist today will not be the same 5 or 10 years from now due to rapid urbanization, densification, and climate change (13), and so deriving a single, static understanding of wildlife health in cities is impossible. This requires continuously reflecting on and re-evaluating the One Health landscape in cities to adapt to changing needs. We discuss how social-ecological One Health approaches can be used to form solutions that adapt to changing local and global dynamics.

What we talk about when we talk about health

A quick Internet search of One Health issues might lead you to believe that One Health is only referring to **infectious diseases**. And while they are the most studied—particularly those that can be spread between animals and people ("**zoonotic**")—One Health also includes noninfectious health outcomes including impacts on physiology, immunity, nutrition, and stress.

Infectious diseases are spread among organisms and caused by **pathogens**. They include living organisms such as microscopic **bacteria** (e.g., *Salmonella*, *Escherichia coli*), **single-celled parasites** (e.g., *Plasmodium* spp., the causative agent of malaria), as well as **multicellular parasites** (e.g., tapeworms) and fungi (e.g., white nose fungus caused by *Pseudogymnoascus destructans*). Pathogens can also be nonliving, such as **viruses** (e.g., Dengue virus) and **prions**—perhaps the most bizarre—proteins that become pathogenic when bent out of their normal shape (e.g., the causative agent of chronic wasting disease). Each infectious agent has its own mode of **transmission** (i.e., moving from one individual to another); they can spread among animals directly through close contact or indirectly through feces or urine. Transmission dynamics are further complicated by arthropods such as mosquitoes and ticks that vector some pathogens among animals in their saliva (e.g., *Borrelia burgdorferi*, the causative agent of Lyme disease) or feces (e.g., *Trypanosoma cruzi*, the causative agent of Chagas disease).

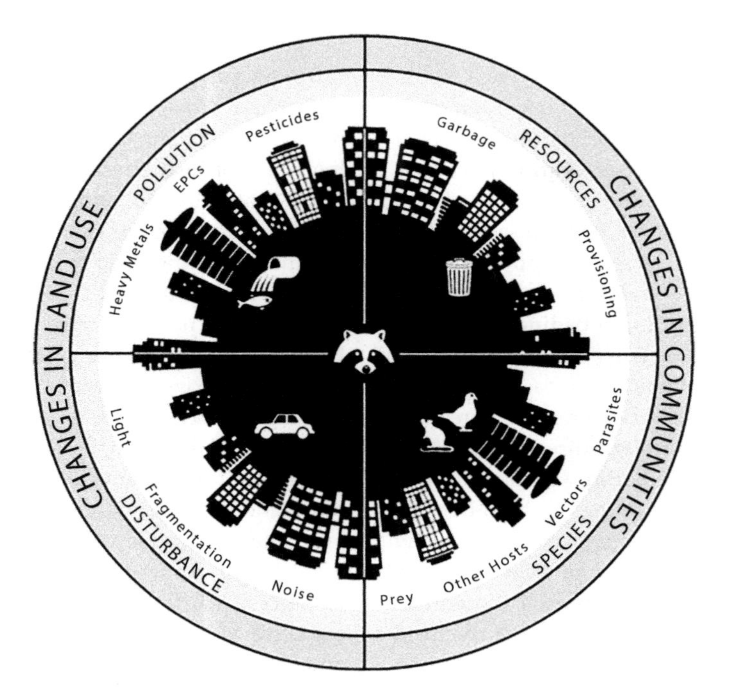

Figure 12.2 Urban environments have unique characteristics that impact wildlife health. Changes in land use can lead to habitat fragmentation, light and noise pollution that disturb animal behaviors, pathogen transmission, and increased stress. Pollution (e.g., heavy metals, pesticides) can result in toxicant levels that exceed Estimated Permissible Concentrations (EPCs). Acute and chronic exposure to pollution can impact physiological and immune function and increase susceptibility to infection. Anthropogenic foods (e.g., food waste, bird feeders) can promote aggregation of wildlife and the spread of pathogens. Urban-tolerant species such as rats and pigeons tend to be in higher abundance than other urban wildlife, altering how pathogens are transmitted among animals. Habitat changes also influence which disease vectors (e.g., mosquitos and ticks) are present. These numerous processes result in a health landscape that varies across space and time.
Rat and pigeon images by J. Dunham and raccoon image by C. Please, through the Noun Project. Image courtesy of Brandon Doty.

These types of infections are called **vector-borne diseases**.

Noninfectious health outcomes are not spread among organisms but can occur through exposure to physical and social environments. Noninfectious health outcomes include: **stress** (acute or chronic) which is measured through stress hormone concentrations (i.e., glucocorticoids) and heterophil-lymphocyte blood counts; measures of **nutrition** (e.g., cholesterol and blood sugar); dysfunction of normal **immunity** or physiology; and **body condition** (e.g., determined by body mass, length, or fat score) which can be negatively impacted through poor nutrition or a lack of resources.

Infectious and noninfectious health outcomes are interrelated, and they are sometimes correlated. For example, animals in more polluted areas may experience higher levels of stress, a suppressed immune response, and increased susceptibility to infection. This reiterates the complexity of untangling ecosystem, human, and animal health.

Changes in land use

Fragmented landscapes: boisterous and burning brighter

The most obvious difference between urban and nonurban settings is in the physical space. Trees are transformed to towers. Green grasses are paved in gray concrete. Rivers are bypassed by bridges. And while green spaces are integrated within cities, access to nature varies dramatically, with patterns of industrialization and urbanization determined by practices dictated by histories of environmental injustice and systemic racism (14).[1] As a result, these transformations result in habitat fragmentation and reductions in green space that are unevenly distributed across communities, with implications for wildlife and human health.

As roadways are built to connect and transport people, they often have the opposite effect on wildlife. Animals traversing roads may be killed or injured by vehicles, impeding movement across roads and disconnecting populations ((15,16) but see (17)). Fragmentation increases aggregation of wildlife, which can heighten the spread of pathogens transmitted through direct contact and aerosols. For example, in Berlin, Germany, urban red foxes (*Vulpes vulpes*) living in high-density, urban natural areas near water were more likely to have canine distemper than in heavily developed areas of the city (18). This was thought to be due to there being fewer foxes in urbanized areas and barriers to movement that limited virus spread. Barriers to pathogen spread can significantly impact local prevalence at fine spatial scales. In Vancouver, Canada, the prevalence of the bacterial pathogen *Leptospira interrogans* among urban rats varied from 0% to 60% across city blocks (19). This bacterium can cause a fever-like illness in people and can be life-threatening (20). Genetic analysis of related rats suggested that this clustering was likely due to minimal movement of rats among city blocks due to roadways (21). These patterns demonstrate how pathogen-associated disease risks for animals and people are location-specific and might even differ from one side of the street to the other.

Urban infrastructure also influences the abiotic features of urban environments. Reduced canopy cover and a greater abundance of heat-absorbing surfaces, such as asphalt and concrete, make cities hotter than their rural counterparts (urban heat island effect) (22). Heat can have several detrimental effects, including heat stress and death (23,24). And although some species adapt to heat stress, these adaptations may impose their own health challenges. For example, brief spikes in temperatures can raise mortality of embryos among urban lizards (25). The Mediterranean lizard (*Psammodromus algirus*) can manage higher temperatures by increasing their thermoregulatory activity during times of heat stress. However, they were also more susceptible to ectoparasites, suggesting that this adaptation had its own health costs (26).

Land-use change also influences the presence and abundance of arthropods. Because many pathogens are vector-borne (e.g., Lyme disease, malaria, West Nile Virus (27)), changing arthropod distributions can alter pathogen transmission. Standing water in storm drains, car tires, and garbage cans as well as accumulated rainwater in plants such as bromeliads provide important mosquito habitats (28). Managing standing water is essential to reduce habitat available for disease vector persistence and proliferation (29). As with waste (discussed in Reliable resources, worrisome waste), effectively managing standing water requires community engagement and behavior change to identify and reduce vector access to this key resource.

Urban centers don't abide by natural daily systems. Human activity at all times of the day and night creates substantial noise and light pollution which causes stress and altered behaviors to human and nonhuman animals alike (30). For people, light and noise pollution are linked to negative mental and physical health outcomes (31,32). Light at night, particularly for people employed in shift work, has been linked to greater risks of some types of cancer, including breast cancer (33). For wildlife, light pollution can impact natural behaviors with health consequences. For example, great tits (*Parus major*) roosting in white light had increased nighttime activity, decreased sleep, and a higher likelihood of infection with malaria (34). For song sparrows (*Melospiza melodia*), birds exposed to broad-spectrum artificial light at night had higher rates of mortality due to West Nile Virus than did birds exposed to amber-hue light, suggesting that even differences in light wavelengths may have differential health effects (35). Similarly, noise pollution can disrupt normal sleep-wake cycles (36), which can be stressful. For example, traffic noise increased stress hormones in tree frogs (*Hyla arborea*), and was linked to changes in coloration of their vocal sac (37). These changes can lower an individual's

[1] Hoover's and Scarlett's chapter reviews how systemic racism and classism have shaped access to urban nature.

attractiveness to mates and their overall fitness. Thus, the pressures placed on populations by noise and light pollution go beyond immediate increases in stress, extending to chronic impacts such as immune suppression, infection risks, and fitness.

Pervasive pollutants

Urban settings often have more pollutants and toxins due to human activities. These pollutants are in the air, land, water, food, and infrastructure. Concentrations of these pollutants accumulate near roadways and high-traffic and -density areas (38,39), which often leads to higher levels in low-income and underserved communities (40). Indeed, pollutants are often disproportionately placed in marginalized communities, furthering health inequities (41).

Ambient air pollution is among the most pervasive of pollutants. Air pollution exacerbates asthma, respiratory, and cardiovascular diseases, and increases mortality rates in people (42,43). While less is known about the effects of air pollution on wildlife health, our meta-analysis found significantly higher toxicant loads in the tissues of urban wildlife as compared to nonurban wildlife (44). Toxicants include heavy metals, pesticides, and other toxins which wildlife can be exposed to in the environment (e.g., through pesticide application to managed lawns (45) and through consumption of contaminated foods). These pesticides leach into urban waterways, resulting in downstream exposures (Figure 12.3).

Environmental contamination with heavy metals such as lead poses significant, but understudied, health risks. Lead is a toxic metal that commonly enters the environment in lead-based paint applied to older homes (46), drinking water from lead-containing pipes (47), and through auto and air transport where leaded gasoline is used (48). Lead exposure is particularly concerning for children, increasing in more urbanized settings (49); in some areas, soil lead contamination levels exceed state and federal thresholds for healthy soils (50). Wildlife are exposed when they consume leaded paint (51) and through contaminated soil (52). Urban adult birds and fruit bats have been found to have significantly higher blood lead concentrations than their counterparts in nonurban settings (52,53). Although the impacts of heavy metals on wildlife health are still unclear, lead exposure has been reported to negatively impact breeding success and survival as well as to disrupt endocrine function (54,55). Interestingly, a genomics-based study found that house sparrows (*Passer domesticus*) in areas with higher lead changed their expression levels of genes associated with transporting heavy metals across cell membranes, suggesting some potential for evolutionary adaptation to contaminated environments (56). The advancement of genomics technologies will be important for furthering our understanding of wildlife responses to environmental contaminants.

Wildlife can also be exposed to toxicants through consumption of poisons intended for pest species. For example, anticoagulant rodenticides are frequently applied to reduce populations of urban rats. However, their impacts extend beyond rats, as predators such as coyotes, mountain lions, and birds of prey consume these animals, transmitting rodenticides through the food chain (57–59). Exposure to rodenticides among rats and nontarget species can impact normal immune function which may increase risks to infection with pathogens (60,61). In recognition of the negative health effects of rodenticides on nontarget wildlife, groups such as "Raptors Are The Solution (RATS)" provide education to communities about the harms of rodenticides and the role that birds of prey can play in reducing pests in the absence of rodenticide application (62).

Urban foods can also be contaminated. Provisioned food such as bird seed or compost piles can be contaminated with toxins produced by molds—known as mycotoxins—which induce vomiting and impair immune functioning (63). In fact, in one survey nearly one-fifth of commercially available bird seed was contaminated with mycotoxins (64). Industrial activities are a more problematic source of toxicants. Wildlife can consume plastic products when foraging for food in landfills (65). At a smaller scale, microplastics are increasingly detected in aquatic and terrestrial systems (66). Wildlife can be exposed to microplastics simply from consuming organisms exposed to contaminated soil such as fungi and invertebrates and can experience a range of physiological harms

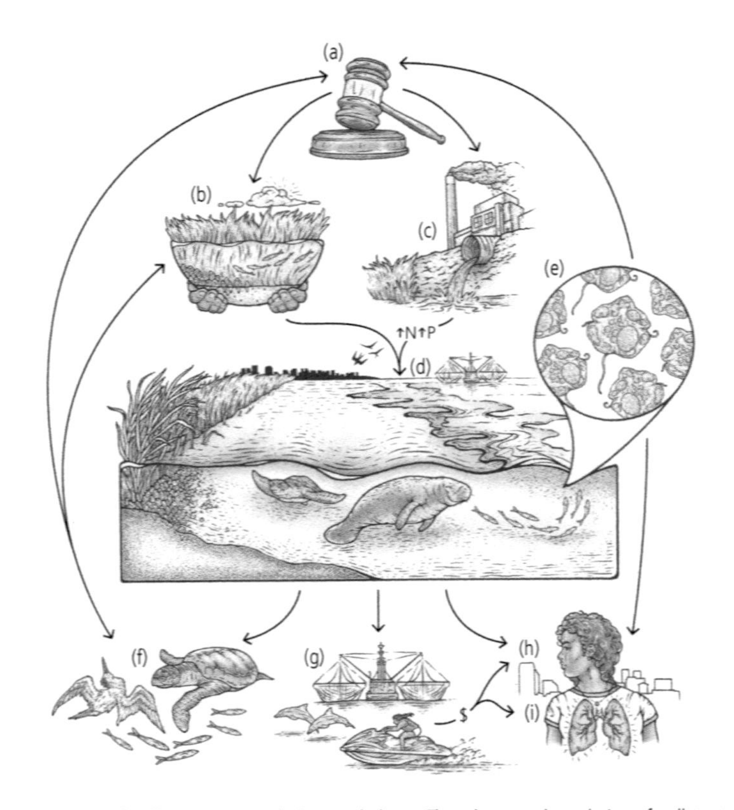

Figure 12.3 Downstream detriments of pollutants on populations and places. The release and regulation of pollutants into the environment reveal the interconnected socio-ecological processes at the heart of One Health. Policies and regulations govern the application, testing, and treatment of toxicants (a) and their enforcement impacts environmental health and protection (b) because they determine how and how much pollution is released into the environment (c). Increasing concentrations of wastewater containing high levels of phosphorus (p) and nitrogen (n) in concert with warming temperatures can create ideal environments for harmful algal blooms (d) caused by the microorganism *Karenia brevis*. The neurotoxin created by *K. brevis* (brevetoxin; E) can cause paralysis, seizures, and death in wildlife (f). Negative impacts on marine ecosystems disrupt local fishing, food services, and tourism (g). Reduced income further affects government funding for multisectoral programs and protections including education, infrastructure, health care, and conservation (h). Harmful algal blooms can also affect human health (i). These events can disproportionately impact lower-income communities who live and work near coastlines, widening financial and health inequities.
Adapted with permission from (11). Image courtesy of Henry Adams.

such as inflammation and oxidative stress (66). The full extent of health harms from urban toxicants is unknown but this knowledge is crucial for maintaining urban biodiversity.

Changes in communities and resources

Reliable resources, worrisome waste

Cities are rich with human-associated foods. From discarded snacks to backyard fruit trees, high densities of people produce large amounts of waste, unintentionally providing novel foods for wildlife. Food waste is the most abundant item in household garbage, amounting for one-fifth of municipal solid waste in the US (67). To curb food waste, many cities in North America collect compost; in Canada up to one-third of households maintain a garden compost pile (68). Many urbanites also intentionally feed wildlife, either for recreation or due to a sense of helping nature (69). For example, up to two-thirds of households feed birds in Europe, the UK, the US, Australia, and New Zealand (70). These novel foods can affect wildlife and environmental health through changes in wildlife nutrition and behavior (71).

Many species change their diets in urban landscapes to incorporate human-associated foods (72). These foods can benefit wildlife by buffering against nutritional stress and improving body condition (73,74). However, provisioned foods, especially those originally intended for human consumption, are typically higher in sugar, salt, and fat, and lower in protein and micronutrients than more natural diets ((73); Figure 12.4). Even well-intentioned provisioning of waterfowl with bread in parks can cause deficiencies in key micronutrients, leading to wing deformities known as "angel wing" (75). Nutrient deficiencies are especially damaging for young animals that require protein-rich foods for proper growth (76). However, too much low-quality food appears to affect adult wildlife similarly to humans. Urban raccoons (*Procyon lotor*), who are notorious for accessing even secured garbage cans, can exhibit hyperglycemia (i.e., high blood sugar) from consuming food waste (77). Urban birds who consume human foods can also have higher cholesterol (78). These conditions in urban wildlife mirror the health harms experienced by under-resourced communities where fresh produce is inaccessible or too expensive (79). Urban humans and nonhuman animals alike exhibit altered gut microbiome composition and diversity compared to nonurban populations. Urban guts tend to have higher abundances of bacteria reflecting a high-carbohydrate diet (80–83). Although our knowledge of the health impacts of the microbiome is still growing, changes to gut flora can impact the immune system, metabolism, and even behavior (84). These parallels underscore the importance of diet quality, not simply quantity, for all urban communities.

Anthropogenic resources can benefit or harm wildlife health based on their distribution. Because people replenish them regularly, food sources from garbage cans, compost piles, and bird feeders are more reliable than natural foods. This stability can increase wildlife survival, especially during periods when more natural foods are scarce (85). However,

Figure 12.4 Urban foods may be abundant, but can have inappropriately high sugar, fat, and salt content. In this image, an Eastern gray squirrel (*Sciurus carolinensis*) is consuming a chocolate bar near a garbage bin in Vancouver, Canada.
Photo courtesy of author Kaylee Byers.

increased survival of animals that would otherwise die (e.g., because they are too sick to forage effectively) could lead to greater proportions of sick animals and crowded conditions (86). Reliable resources can also lead to locally high densities of foraging animals, promoting the transmission of pathogens and parasites spread through close contact (44). For example, coyotes (*Canis latrans*) can aggregate around urban compost piles, particularly in winter, and compost has been found to attract coyotes with sarcoptic mange (87). Mangy coyotes could have been attracted to compost because it is an easy and reliable food source; however, coyotes that make frequent use of compost could also be more likely to be diseased because of increased transmission of mange mites (*Sarcoptes scabiei*) through close contact. Wildlife might also exhibit behaviors that promote disease transmission at anthropogenic food sources. For example, house finches (*Haemorhous mexicanus*) with conjunctivitis spent more time on feeders and in closer proximity to other birds (88). Similarly, white ibises (*Eudocimus albus*) fed by people in parks were less nomadic than their nonurban counterparts, potentially increasing their exposure to *Salmonella* in local soil and water (89).

Species that thrive in cities may be especially likely to exploit novel urban foods compared to other species because they are more likely to reach high densities and use a wide variety of habitats and foods (90,91). Because cities tend to benefit non-native generalists, species that typically would not interact while foraging might cluster around garbage cans, compost piles, and bird feeders (92), which could promote the spread of generalist pathogens. Bird feeders can also disproportionately benefit invasive species such as the house sparrow (93), influencing local host–pathogen dynamics (see Shifting species).

The consequences of provisioned anthropogenic resources extend beyond wildlife health. People who spend time in alleys and near waste disposal are more likely to be exposed to foraging animals and their waste. For example, unhoused people are disproportionately vulnerable to zoonotic diseases because of exposure to wildlife and vectors such as fleas and mosquitoes (94). Restaurants are a prominent source of food waste and workers who take out the garbage or sanitation workers emptying dumpsters may contend with raccoons and rats. At home, residents who cannot afford the servicing or equipment to exclude wildlife from food attractants on their property may be more vulnerable to zoonotic diseases from animal waste. These examples underscore how socioeconomic status and occupational hazards can increase health risks.

Shifting species

In cities, humans—whether intentionally or unintentionally—have created intensely altered, dynamic environments comprised of unique communities and novel species interactions (Figure 12.5). These changes in community composition have important consequences for disease ecology.

Because cities tend to filter out species that cannot tolerate human disturbance, urban wildlife communities can be less diverse than in undisturbed landscapes (95). If these remaining species are competent hosts for a particular pathogen (i.e., can effectively transmit a pathogen to another host or vector), then the risk of pathogen transmission will be higher in less diverse wildlife communities (i.e., more diverse communities dilute disease risk, known as the dilution effect hypothesis, (96)). For example, urban wetlands are inhabited by a more diverse community of birds and have a lower prevalence of West Nile Virus in local mosquitoes than nearby urban residential areas (97). Although there has been mixed support for the dilution effect (98), land-use change associated with biodiversity loss, such as urbanization, is consistently associated with increased disease risk (99).

Within the city, more affluent neighborhoods often have access to more green space—and, as a result—more biodiversity (i.e., the luxury effect (100)). There is growing evidence that engaging with biodiverse environments can promote positive mental health for people, such as increased pleasure and reduced stress (reviewed in (101)). One continent-wide study in Europe found that bird diversity was positively correlated with life satisfaction across Europe at a level comparable with income (102). Conversely, encounters with

Figure 12.5 Generalist wildlife species such as coyotes (*Canis latrans*) consume a variety of anthropogenic foods allowing them to survive in diverse environments. This coyote is foraging at a landfill.
Photo courtesy of author Maureen Murray.

perceived pest species such as rats (103–105) and bedbugs (106), which are often more abundant in low-income neighborhoods, can negatively impact mental health. This relationship between biodiversity, zoonotic infection risk, and mental health may be a powerful motivator to conserve biodiversity to promote human health, but we must also consider how to distribute these benefits equally among people.

Novel interactions between humans, wildlife, and introduced species in cities can create unique disease dynamics. Wildlife hosts for zoonotic pathogens are both more abundant and make up a larger proportion of urban communities relative to undisturbed habitats (107). These shifts in community structure could increase the risk of zoonotic diseases as well as diseases transmitted from people to wildlife such as influenza (108). Introduced and invasive species that thrive in urban environments bring their associated pathogens, increasing disease risk for native species (i.e., pathogen pollution (109)). For example, cats are important hosts for the protozoan parasite *Toxoplasma gondii*, which can cause disease in the heart and central nervous system of young or immunocompromised wildlife, including marine mammals near urban coasts (110,111). Humans have also intentionally introduced raccoons to Germany and Japan, inadvertently introducing zoonotic raccoon roundworm (*Baylisascaris procyonis*) to local mammal communities. Although human cases of raccoon roundworm are still rare, over 70% of Germany's raccoons are infected (112). Common urban exploiters can also expose native predators to pathogens. For example, urban raptors become infected with the protozoan parasite *Trichomonas galinae*, the causative agent of trichomoniasis, from predating on urban pigeons (*Columba livia*) (113), which can be fatal (114,115). These disease risks may be mitigated with a better understanding of how humans and other urban processes shape urban communities.

Social-ecological applications

Cities are shared, multispecies spaces. Yet, any efforts to create healthier environments rest on the actions of people. While doing so is daunting, there are many options for actions at the individual, neighborhood, municipal, and federal levels,[2] and people can enact measures across scales to support healthy environments, wildlife, and human communities (Table 12.1). We highlight two important considerations for furthering social-ecological approaches within the scope of One Health initiatives.

Integrating Indigenous knowledges within One Health

As mentioned earlier, One Health is certainly not a "new" concept among most Indigenous cultures. One of the three coauthors is an Indigenous Ts'msyen woman that had lived for 30 years in the urban area known as Vancouver on the territory shared by the xʷməθkʷəy̓əm (Musqueam), Sḵwx̱wú7mesh (Squamish), and səlilwətaɬ (Tsleil-Waututh) Nations since time immemorial. It is important here to introduce some concepts—namely, Indigenous knowledge systems, Two-Eyed Seeing, and resurgence.

Holistic health and the interconnectedness of human, animal, and environmental health can be seen in most, if not all, Indigenous knowledge systems. For example, the term *Mino-pimatisiwin*, which roughly translates to "living the good life," is used by Indigenous peoples of the Algonquian language family in North America to express a state of harmony, well-being, and comprehensive health based on relationships, cultural identity, and relationship with the earth (116). Many Indigenous cultures view themselves as part of the land, not separate from it, with living and nonliving entities being "relatives" with their own sense of agency (117).

The concept of *Two-Eyed Seeing* was first introduced in 2004 by Albert Marshall (Mi'kmaw). *Two-Eyed Seeing* refers to the ability to see with one eye using the strengths of *Indigenous* knowledges and ways of knowing and the other using the strengths of *western* knowledges and ways of knowing, and to combine the two eyes for the benefit of all (118).

To make way for these concepts in mainstream discourse necessitates that we include the concept of Indigenous resurgence. Resurgence can be broadly understood as "movements and embodied practices focused on rebuilding nation-specific Indigenous ways of being and actualizing self-determination" (119, p. 7). Here the concept of Indigenous resurgence goes hand in hand with *Two-Eyed Seeing* which seeks to hold Indigenous knowledges alongside western knowledges as distinct and whole. For if mainstream systems were to answer the call to use Two-Eyed Seeing, it would require resurgence of Indigenous knowledge systems.

So, what can Indigenous knowledge systems and their strengths in a Two-Eyed Seeing approach offer One Health? As previously mentioned, One Health is seen as "holistic" in nature and "integrative," therefore including Indigenous knowledge systems would strengthen the concept. Urban settings are no different, despite some colonial notions that Indigenous people only exist in rural, "Reserve," or "Reservation" settings. As Michi Saagiig Nishnaabeg scholar Leanne Betasamosake Simpson states in her book (120):

> The beauty of culturally inherent resurgence is that it challenges settler colonial dissections of our territories and our bodies into reserve/city or rural/urban dichotomies. All Canadian cities are on Indigenous lands. Indigenous presence is attacked in all geographies. In reality, the majority of Indigenous peoples move regularly through reserves, cities, towns, and rural areas. We have found ways to connect to the land and our stories and to live our intelligences no matter how urban or how destroyed our homelands have become.
>
> (p. 173)

Environmental health research, including One Health research, could be strengthened by grounding in a theoretical framework that includes Indigenous resurgence and Two-Eyed Seeing (Case study 1).

[2] Larson's and Brown's chapter discusses motivations and governance for urban biodiversity across scales.

Table 12.1 Social-ecological applications across scales and knowledge gaps at the One Health interface.

Urban process	Individual actions	Neighborhood actions	Municipal actions	Knowledge gaps for One Health solutions
Land-use planning	Reduce standing water on properties by improving drainage and removing or reducing infrastructure that collects water (i.e., containers and car tires)	Support community initiatives to reduce standing water	Manage urban green spaces throughout cities to create refuge for mosquito predators in ways that do not exacerbate the luxury effect	How to "green" neighborhoods without creating habitat for disease vectors, inducing gentrification, and excluding marginalized communities?
Pollution	Reduction or elimination of pesticides on managed lawns and rodenticides to manage pests	Reduction or elimination of pesticides on community green spaces (e.g., school yards, parks) and rodenticides to manage pests	Policy changes to ban or reduce industrial application of pesticides and pollutants	How can demonstrated health harms in humans, wildlife, and the environment inform regulatory policies and practices in industry?
Anthropogenic resources	Minimize food waste Ensure waste bins are sealed and secured until pickup	Management and securing of waste bins near local businesses and restaurants	Improve waste pickup infrastructure (e.g., wildlife-proof garbage cans) and timing (e.g., more frequent pickups)	How can we promote positive human behavior change to reduce accessible food waste among individual households and municipalities?
Species composition	Plant endemic and food-producing plants that support healthy wildlife populations and people Keep pets indoors	Plant endemic and food-producing plants that support healthy wildlife populations	Support green space development and preservation	In what contexts does species diversity predict disease risk in cities?

Case study 1 Indigenous resurgence, One Health, and urban water

Recently coauthor Nelson conducted a virtual research sharing with urban Indigenous people in so-called Vancouver that sought to explore Indigenous ways of knowing and water governance in an urban context. The coproduction of knowledge in this research project took place by means of arts-informed research methods that sought to explore Indigenous knowledges and ways of knowing in a creative manner that allowed for their stories to emerge (Figure 12.6).

Results indicated that the research group was quite eager to participate in environmental research since they observed that "land-based" Indigenous research is often confined to rural and remote areas. They expressed a wealth of knowledge about the environment and water that could help to advance water governance debates, and environmental debates more broadly.

The results were accepted eagerly by stakeholders such as the City of Vancouver with the intent to incorporate the results into their Rain City Strategy (121). This strategy aims to reimagine and transform rainwater management to improve water quality, resilience, and sustainability in cities. For example, a key outcome of the project indicated that Indigenous peoples would love to see the City of Vancouver build infrastructure that filtered contaminants from runoff that feeds into local streams and rivers. They were also eager to see the inclusion of more streams and rivers in the city that are clean so that they can conduct ceremonies, pass on Traditional Knowledge, connect to the land, and see thriving wildlife without the need for long treks to rural locales.

Such visions common in Indigenous world views include a deeply embedded sense of reciprocity with the land and our nonhuman kin which can benefit the concept of One Health tremendously.

Figure 12.6 This knowledge cocreator participated in the arts-informed urban Indigenous water study. She is an urban Metis woman who is reconnecting and reinforcing her cultural knowledge. The image portrays the cocreator as a caretaker of the land and water, including a topographic map of a lake and lily pad with flower.
Photo courtesy of author Joanne Nelson.

Promoting health, reducing harms

To expand strategies for addressing One Health issues, we could leverage other well-established frameworks employed in public health such as Health Promotion and Harm Reduction. Health Promotion is rooted in resilience. It recognizes that health is impacted by factors external to the health system (e.g., social interactions, income inequality, living conditions; (122)). Harm Reduction goes one step further to reduce the effects of specific harms incrementally (123–125). Applying both strategies within One Health is a way to integrate holistic (Health Promotion) and targeted (Harm Reduction) actions to reveal a multitude of solutions that are not solely focused on harm eradication (125).

For example, a Harm Reduction approach to mosquito-borne zoonotic diseases might involve education campaigns which encourage individual measures to reduce risk, like applying mosquito repellents, removing standing water on private properties, and installing screens on windows. A Health Promotion approach would recognize mosquito-borne illness as part of a complex system including social determinants of health such as housing, waste management, and access to community-level preventative measures. It would also acknowledge that broad pesticide application to manage mosquitoes results in negative downstream effects that reduce food sources for insectivorous wildlife including birds and bats. Health Promotion activities might include improving housing infrastructure and subsidies to assist residents in purchasing and installing preventative window screens, as well as enhancing neighborhood waste services to reduce habitat for mosquitoes.

Another Health Promotion activity is to add, expand, or restore green space such as urban forests, prairies, wetlands, and even native plantings in yards. Such green spaces help to counteract health harms from pollution and heat (11) and promote physical and mental health in people (126). However, new parks or street trees can raise property values, displacing the original residents through "green gentrification" (127,128). For example, a massive restoration project in Prospect Park in Brooklyn, NY, USA in the 1980s led to higher incomes and Whiter neighborhoods surrounding the park (129). To avoid displacing residents, ecological stewardship in partnership with local communities can help improve parks and nontraditional green spaces such as vacant lots and community gardens (130).

Health Promotion and Harm Reduction strategies are inextricably linked with environmental justice and community engagement because they recognize that harms and perceptions of these harms vary across communities. For example, in Australia, interventions to control rabies outbreaks in domestic dogs include vaccination campaigns, efforts to restrict the movement of free-living dogs among communities, as well as the culling of dogs. However, dogs and dingos have cultural and societal significance in some Indigenous cultures. Because differences in attitudes toward measures such as culling exist, working with local communities has been recommended to ensure culturally relevant and effective approaches (131). Therefore, Health Promotion and Harm Reduction approaches must be community informed to ensure that activities are relevant and feasible.

Global perspectives

The processes we have described throughout occur in cities globally. Cities vary in building construction and population density, industrial processes and pollution, food waste and management, and species composition. Despite these differences, urbanization invariably creates ecosystems that differ significantly from nonurban settings and provides habitat for global urban adapters such as rats (Case study 2). Cities around the world also struggle with air pollution (132) and vector-borne diseases, with climate change shifting and shaping these dynamics (133). Understanding the nuances of local disease hosts, vectors, pathogens, and stressors will help One Health initiatives to think globally but act locally.

Even so, local contexts and processes underpin the One Health issues that are unique to a city, resulting in different health risks, human–wildlife interactions, and community experiences. For example, while human food wastes are ubiquitous in cities, garbage receptacles may attract raccoons in North America, red foxes in Europe, hyenas in Ethiopia, and langurs in Sri Lanka (Figure 12.7), all of which carry unique pathogens. As highlighted throughout, the infectious and non-infectious health issues impacting people, animals, and environments vary across geographic scales due to differences in habitat (e.g., tropical, temperate, arid), species biodiversity (e.g., both pathogen and animal host diversity), and human behaviors (e.g., political, cultural, economic). These differences also determine how people perceive wildlife and the types of management actions to address these issues that are feasible and acceptable. For example, residents who perceived a threat from baboons in Cape Town, South Africa, were less concerned about their conservation and more supportive of lethal measures to remove them (134).

Figure 12.7 Anthropogenic food wastes are accessed by wildlife all over the world, despite different types of food and waste management practices. Gray langurs (*Semnopithecus priam*), Anuradhapura, Sri Lanka.
Photo courtesy of author Kaylee Byers.

And while efforts to promote healthy urban wildlife must be rooted in the needs, resources, and interests of individual communities, global movements toward climate change adaptation and reductions in pollution are important steps to support One Health globally. Together, this points to the need for actions that support both individual and collective health security.

Case study 2 Urban rats: global travelers, backyard neighbors

Over the past decade, two of the authors (Byers and Murray) have spent a lot of time studying urban rats and their associated health harms (Figure 12.8). Urban rats are an excellent model species for investigating the nexus of human, animal, and environmental health because they live in close association with people on all continents except Antarctica (135). Two species are especially widespread and well documented in cities, including brown/Norway/sewer rats (*Rattus norvegicus*) and black/roof rats (*Rattus rattus*). Rats are prevalent in cities in part because they are generalists, with the ability to survive on all types of foods (136). They also require very little space to burrow or nest, with an ability to survive in both green spaces and urban infrastructure (137,138). While found throughout cities, they are frequently in higher abundance in under-resourced areas (139,140). People who are unhoused may be disproportionately more likely to be exposed to zoonotic diseases from rat bites and fleas relative to other people (94,141).

Rats can impact the mental and physical health of people. People who frequently encounter rats may experience unease and anxiety (103–105) which can lead to feelings of helplessness and hopelessness. Rats also carry a number of zoonotic diseases (e.g., *Yersinia pestis* (the causative agent of the plague), *L. interrogans*, *Rickettsia typhi*, *Streptobacillus moniliformis*) and viruses (Seoul hantavirus) that can cause significant morbidity and mortality in people globally (142–145). Despite the global distribution of these species, the pathogens carried by rats differ not only by city, but the prevalence of these pathogens can even vary by city block, with some city blocks having many infected rats and adjacent blocks having none (19).

Figure 12.8 Urban rats among trash bags (*Rattus norvegicus*). Photo courtesy of Cecilia Sánchez.

Because rats pose a variety of health risks, they are also the target for intensive municipal management campaigns. These approaches vary by city (146,147). Most rat management has focused on eradication through lethal methods such as trapping or poison baiting. In addition to posing toxic health risks for nontarget wildlife, these lethal approaches can also have unintended and unpredictable impacts. For example, in Vancouver, Canada, lethal pest control measures were found to increase the prevalence of *L. interrogans* among rats (148), while they appeared to have the opposite impact on rats' carriage of *Bartonella* spp. (149). Further, rats exposed to poison were more likely to carry *L. interrogans* bacteria (61). These studies underscore how human behaviors can impact risks in a way that may exacerbate health issues.

Although health risks and approaches to rats differ across cities, systems-based management is a unifying approach which can be applied across contexts to address rat-associated health risks. By integrating rat management into urban planning and municipal initiatives such as housing and waste management, cities can tailor their response to the needs, resources, and interests of communities while improving social services across sectors and scales.

Conclusion

Conservation in urban ecosystems requires a careful understanding of the health of the organisms being managed. The key takeaway here is that human, animal, and environmental health are interconnected by social and ecological dimensions. There are numerous dimensions (physical, social, emotional) and metrics (disease, stress, body condition, immunity) of health. But these health outcomes can be managed at various scales at the socio-ecological interface by using a One Health lens.

And while there is much more to uncover about One Health in cities, what is certain is that promoting environmental health creates healthier spaces for all of us to coexist. With climate change, biodiversity loss, and continued globalization, the health impacts and outcomes that exist today will be different tomorrow. To address these challenges, we must acknowledge the interdependence of human, animal, and environmental health and integrate knowledges to create solutions that are sustainable now and into the future.

Acknowledgements

This work is based upon work supported by the National Science Foundation under grant no. 1,923,882; the Natural Sciences and Engineering Research Council of Canada, under grant no. RGPIN-2015-05058; and the Pacific Institute on Pathogens, Pandemics, and Society at Simon Fraser University.

References

1. Pekar JE, Magee A, Parker E, Moshiri N, Izhikevich K, Havens J, et al. The molecular epidemiology of multiple zoonotic origins of SARS-CoV-2. Science. 2022;377(6609):960–6.
2. Keusch GT, Amuasi JH, Anderson DE, Saif L. Pandemic origins and a One Health approach to preparedness and prevention: solutions based on SARS-CoV-2 and other RNA viruses. Proceedings of the National Academy of Sciences of the United States of America. 2022;119(42):e2202871119.
3. Oude Munnink BB, Sikkema RS, Nieuwenhuijse DF, Jan Molenaar R, Munger E, Molenkamp R, et al. Transmission of SARS-CoV-2 on mink farms between humans and mink and back to humans. Science. 2020;371(6525):172–7.
4. Kuchipudi SV, Surendran-Nair M. Ruden RM, Kapur V. Multiple spillovers from humans and onward transmission of SARS-CoV-2 in white-tailed deer. Proceedings of the National Academy of Sciences of the United States of America. 2022;119(6):e2121644119.
5. Daszak P, Cunningham AA, Hyatt AD. Emerging infectious diseases of wildlife—threats to biodiversity and human health. Science. 2000;287(5452): 443–9.
6. Taylor LH, Latham SM, Woolhouse MEJ. Risk factors for human disease emergence. Philosophical Transactions of the Royal Society B: Biological Sciences. 2001;356(1411):983–9.
7. Gibbs EPJ. The evolution of One Health: a decade of progress and challenges for the future. The Veterinary Record. 2014;174(4):85–91.
8. Cunningham AA, Daszak P, Wood JLN. One Health, emerging infectious diseases and wildlife: two decades of progress? Philosophical Transactions of the Royal Society B: Biological Sciences. 2017;372(1725):20160167.
9. Prasad Vadrevu K, Eaturu A, Biswas S, Lasko K, Sahu S, Garg JK, et al. Spatial and temporal variations in air pollution over 41 cities of India during the COVID-19 lockdown period. Scientific Reports. 2020;10:16574.
10. Łopucki R, Kitowski I, Perińska-Teresiak M, Klich D. How is wildlife affected by the COVID-19 pandemic? Lockdown effect on the road mortality of hedgehogs. Animals. 2021;11(3):868.
11. Murray MH, Buckley J, Byers KA, Fake K, Lehrer EW, Magle SB, et al. One Health for all: advancing human and ecosystem health in cities by integrating an environmental justice lens. Annual Reviews in Ecology, Evolution, and Systematics. 2022;53:18.1–18.24.
12. Copper Jack J, Gonet J, Mease A. Traditional knowledge underlies One Health. Science. 2020;369(6511):1576.
13. Hobbie SE, Grimm NB. Nature-based approaches to managing climate change impacts in cities. Philosophical Transactions of the Royal Society B: Biological Sciences. 2020;375(1794):20190124.
14. Schell CJ, Dyson K, Fluentes TL, des Roches S, Harris NC, Miller DS, et al. The ecological and evolutionary consequences of systemic racism in urban environments. Science. 2020;369(6510):eaay4497.
15. LaPoint S, Balkenhol N, Hale J, Sadler J, van der Ree R. Ecological connectivity research in urban areas. Functional Ecology. 2015;29(7):868–78.
16. de Rivera CE, Bliss-Ketchum LL, Lafrenz MD, Hanson AV, McKinney-Wise LE, Rodriguez AH, et al. Visualizing connectivity for wildlife in a world without roads. Frontiers in Environmental Science. 2022;10:757954.
17. Hill JE, DeVault TL, Belant JL. A review of ecological factors promoting road use by mammals. Mammal Review. 2021;51(2):214–22.
18. Gras P, Knuth S, Börner K, Marescot L, Benhaiem S, Aue A, et al. Landscape structures affect risk of canine distemper in urban wildlife. Frontiers in Ecology and Evolution. 2018;6:136.
19. Himsworth CG, Bidulka J, Parsons KL, Feng AYT, Tang P, Jardine CM, et al. Ecology of *Leptospira interrogans* in Norway rats (*Rattus norvegicus*) in

an inner-city neighborhood of Vancouver, Canada. PLoS Neglected Tropical Diseases. 2013;7(6):e2270.

20. Costa F, Hagan JE, Calcagno J, Kane M, Torgerson P, Martinez-Silveira MS, et al. Global morbidity and mortality of leptospirosis: a systematic review. PLoS Neglected Tropical Diseases. 2015;9(9):e0003898.
21. Byers KA, Booker TR, Combs M, Himsworth CG, Munshi-South J, Patrick DM, et al. Using genetic relatedness to understand heterogeneous distributions of urban rat-associated pathogens. Evolutionary Applications. 2021;14:198–209.
22. Oke TR. The energetic basis of the urban heat island. Quarterly Journal of the Royal Meteorological Society. 1982;108(455):1–24.
23. Heavyside C, Macintyre H, Vardoulakis S. The urban heat island: implications for health in a changing environment. Current Environmental Health Reports. 2017;4(3):296–305.
24. Chadwick Jr JG, Nislow KH, McCormick SD. Thermal onset of cellular and endocrine stress responses correspond to ecological limits in brook trout, an iconic cold-water fish. Conservation Physiology. 2015;3(1):cov017.
25. Hall JM, Warner DA. Thermal spikes from the urban heat island increase mortality and alter physiology of lizard embryos. Journal of Experimental Biology. 2018;221(14):jeb181552.
26. Megía-Palma R, Barja I, Barrientos R. Fecal glucocorticoid metabolites and ectoparasites as biomarkers of heat stress close to roads in a Mediterranean lizard. Science of the Total Environment. 2022;802:149919.
27. Mills JN, Gage KL, Khan AS. Potential influence of climate change on vector-borne and zoonotic diseases: a review and proposed research plan. Environmental Health Perspectives. 2010;118(11): 1507–14.
28. Wilke ABB, Vasquez C, Carvajal A, Medina J, Chase C, Cardenas G, et al. Proliferation of *Aedes aegypti* in urban environments mediated by the availability of key aquatic habitats. Nature. 2020;10:12925.
29. Lindsay SW, Wilson A, Golding N, Scott TW, Takken W. Improving the built environment in urban areas to control *Aedes aegypti*-borne diseases. Bulletin of the World Health Organization. 2017;95(8):607–8.
30. Sepp T, Ujvari B, Ewald PW, Thomas F, Giraudeau M. Urban environment and cancer in wildlife: available evidence and future research avenues. Proceedings of the Royal Society B: Biological Sciences. 2019;286(1894):20182434.
31. Chepesiuk R. Missing the dark: health effects of light pollution. Environmental Health Perspectives. 2009;117(1):A20–2.
32. Falcón J, Torriglia A, Attia D, Viénot F, Gronfier C, Behar-Cohen F, et al. Exposure to artificial light at night and the consequences for flora, fauna, and ecosystems. Frontiers in Neuroscience. 2020;14:602796.
33. Urbano T, Vinceti M, Wise LA, Filippini T. Light at night and risk of breast cancer: a systematic review and dose-response meta-analysis. International Journal of Health and Geography. 2021;20:44.
34. Ouyang JQ, de Jong M, van Grunsven RHA, Matson KD, Haussmann MF, Meerlo P. Restless roosts: light pollution affects behavior, sleep, and physiology in a free-living songbird. Global Change Biology. 2017;23(11):4987–94.
35. Kernbach ME, Cassone VM, Unnasch TR, Martin LB. Broad-spectrum light pollution suppresses melatonin and increases West Nile virus-induced mortality in house sparrows (Passer domesticus). Condor. 2020;122(3):1–13.
36. Francis CD, Barber JR. A framework for understanding noise impacts on wildlife: an urgent conservation priority. Frontiers in Ecology and the Environment. 2013;11(6):305–13.
37. Troïanowski M, Mondy N, Dumet A, Arcanjo C, Lengagne T. Effects of traffic noise on tree frog stress levels, immunity and color signaling. Conservation Biology. 2017;31(5)1132–40.
38. Demetillo MAG, Harkins C, McDonald BC, Chodrow PS, Sun K, Pusede SE. Space-based observational constraints on NO_2 air pollution inequality from diesel traffic in major US cities. Geophysical Research Letters. 2021;48(17):e2021GL094333.
39. Hilker N, Wang JM, Jeong C-H, Healy RM, Sofowote U, Debosz J, et al. Traffic-related air pollution near roadways: discerning local impacts from background. Atmospheric Measurement Techniques. 2019;12(10):5247–61.
40. Nardone A, Casey JA, Morello-Frosch R, Mujahid M, Balmes JR, Thakur N. Associations between historical residential redlining and current age-adjusted rates of emergency department visits due to asthma across eight cities in California: an ecological study. Lancet Planetary Health. 2020;4(1): e24–31.
41. Mohai P, Saha R. Racial inequality in the distribution of hazardous waste: a national-level reassessment. Social Problems. 2014;54(3):343–70.
42. Cushing L, Morello-Frosch R, Wander M, Pastor M. The haves, the have-nots, and the health of everyone: the relationship between social inequality and environmental quality. Annual Reviews in Public Health. 2015;36:193–209.

43. Nardone A, Chiang J, Corburn J. Historic redlining and urban health today in U.S. cities. Environmental Justice. 2020;13(4):109–19.
44. Murray MH, Sánchez CA, Becker DJ, Byers KA, Worsley-Tonks KEL, Craft ME. City sicker? A meta-analysis of wildlife health and urbanization. Frontiers in Ecology and the Environment. 2019;17(10):575–83.
45. Rashed MN, Soltan ME. Animal hair as biological indicator for heavy metal pollution in urban and rural areas. Environmental Monitoring and Assessment. 2015;110:41–53.
46. Jacobs DE, Clickner RP, Zhou JY, Viet SM, Marker DA, Rogers JW, et al. The prevalence of lead-based paint hazards in U.S. housing. Environmental Health Perspectives. 2002;110(10):A599–606.
47. Levallois P, St-Laurent J, Gauvin D, Courteau M, Prévost M, Campagna C, et al. The impact of drinking water, indoor dust and paint on blood lead levels of children aged 1–5 years in Montréal (Québec, Canada). Nature. 2014;24(2):185–91.
48. Snakin VV, Prisyazhnaya AA. Lead contamination of the environment in Russia. Science of the Total Environment. 2000;256(2–3):95–101.
49. Rifai N, Cohen G, Wolf M, Cohen L, Fraser C, Savory J, et al. Incidence of lead poisoning in young children from inner-city, suburban, and rural communities. Therapeutic Drug Monitoring. 1993;15(2):71–4.
50. Kalani TJ, South A, Talmadge C, Leibler J, Whittier C, Rosenbaum M. One map: using geospatial analysis to understand lead exposure across humans, animals, and the environment in an urban US city. One Health. 2021;13:100341.
51. Finkelstein ME, Gwiazda RH, Smith DR. Lead poisoning of seabirds: environmental risks from leaded paint at a decommissioned military base. Environmental Science and Technology. 2003;37(15):3256–60.
52. Roux KE, Marra PP. The presence and impact of environmental lead in passerine birds along an urban to rural land use gradient. Archives of Environmental Contamination and Toxicology. 2007;53(2):261–8.
53. Hariono B, Ng J, Sutton RH. Lead concentrations in tissues of fruit bats (*Pteropus* sp.) in urban and non-urban locations. Wildlife Research. 1993;20(3):315–19.
54. Provencher JF, Forbes MR, Hennin HL, Love OP, Braune BM, Mallory ML, et al. Implications of mercury and lead concentrations on breeding physiology and phenology in an Arctic bird. Environmental Pollution. 2016;218:1014–22.
55. Lemaire J, Bustamante P, Mangione R, Marquis O, Churlaud C, Brault-Favrou M, et al. Lead, mercury, and selenium alter physiological functions in wild caimans (*Caiman crocodilus*). Environmental Pollution. 2021;286:117549.
56. Andrew SC, Taylor MP, Lundregan S, Lien S, Jensen H, Griffith SC. Signs of adaptation to trace metal contamination in a common urban bird. Science of the Total Environment. 2019;650:679–86.
57. Poessel SA, Breck SW, Fox KA, Gese EM. Anticoagulant rodenticide exposure and toxicosis in coyotes (*Canis latrans*) in the Denver metropolitan area. Journal of Wildlife Disease. 2015;51(1):265–8.
58. Rudd JL, McMillin SC, Kenyon Jr MW, Clifford DL, Poppenga RH. Prevalence of first and second-generation anticoagulant rodenticide exposure in California mountain lions (*Puma concolor*). In: Woods DM (ed.) Proceedings of the 28th Vertebrate Pest Conference. Davis, CA: University of California Davis; 2018. p. 254–7.
59. Thornton GL, Stevens B, French SK, Shirose LJ, Reggeti F, Schrier N, et al. Anticoagulant rodenticide exposure in raptors from Ontario, Canada. Environmental Science and Pollution Research. 2022;29(49):34137–46.
60. Serieys LEK, Lea AJ, Epeldegui M, Armenta TC, Moriarty J, VandeWoude S, et al. Urbanization and anticoagulant poisons promote immune dysfunction in bobcats. Proceedings of the Royal Society B: Biological Sciences. 2018;285(1871):20172533.
61. Murray MH, Sánchez CA. Urban rat exposure to anticoagulant rodenticides and zoonotic infection risk. Biology Letters. 2021;17(8):20210311.
62. Raptors Are the Solution [Internet]. Berkeley, CA: Raptors Are the Solution; 2022 [cited 2022 Oct 13]. Available from: https://www.raptorsarethesolution.org/
63. da Rocha MEB, Freire FdCO, Maia FEF, Guedes MIF, Rondina D. Mycotoxins and their effects on human and animal health. Food Control. 2014;36(1):159–65.
64. Henke SE, Gallardo VC, Martinez B, Balley R. Survey of aflatoxin concentrations in wild bird seed purchased in Texas. Journal of Wildlife Disease. 2001;37(4):831–5.
65. Teuten EL, Saquing JM, Knappe DR, Barlaz MA, Jonsson S, Björn A, et al. Transport and release of chemicals from plastics to the environment and to wildlife. Philosophical Transactions of the Royal Society B: Biological Sciences. 2009;364(1526):2027–45.
66. de Souza Machado AA, Kloas W, Zarfl C, Hempel S, Rillig MC. Microplastics as an emerging threat to terrestrial ecosystems. Global Change Biology. 2018;24(4):1405–16.
67. US Food and Drug Administration (FDA). Food loss and waste [Internet]. Silver Spring, MD: US FDA;

2022 [cited 2022 Oct 13]. Available from: https://www.fda.gov/food/consumers/food-loss-and-waste

68. Statistics Canada. Composting practices of Canadian households [Internet]. Ottawa, Canada: Statistics Canada; 2021 Jun 1 [cited 2022 Oct 13]. Available from: https://www150.statcan.gc.ca/t1/tbl1/en/tv.action?pid=3810012801
69. Clark DN, Jones DN, Reynolds SJ. Exploring the motivations for garden bird feeding in south-east England. Ecology and Society. 2019;24(1):26.
70. Galbraith JA, Jones DN, Beggs JR, Parry K, Stanley MC. Urban bird feeders dominated by a few species and individuals. Frontiers in Ecology and the Environment. 2017;5:81.
71. Murray MH, Kidd AD, Curry SE, Hepinstall-Cymerman J, Yabsley MJ, Adams HC, et al. From wetland specialist to hand-fed generalist: shifts in diet and condition with provisioning for a recently urbanized wading bird. Philosophical Transactions of the Royal Society B: Biological Sciences. 2018;373(1745):20170100.
72. Gámez S, Potts A, Mills KL, Allen AA, Holman A, Randon PM, et al. Downtown diet: a global meta-analysis of increased urbanization on the diets of vertebrate predators. Proceedings of the Royal Society B: Biological Sciences. 2020;289(1970):20212487.
73. Murray MH, Becker DJ, Hall RJ, Hernandez SM. Wildlife health and supplemental feeding: a review and management recommendations. Biological Conservation. 2016;204:163–74.
74. Robb GN, McDonald RA, Chamberlain DE, Bearhop S. Food for thought: supplementary feeding as a driver of ecological change in avian populations. Frontiers in Ecology and the Environment. 2008;6(9):476–84.
75. Lin MJ, Chang SC, Lin TY, Cheng YS, Lee YP, Fan YK. Factors affecting the incidence of angel wing in white roman geese: stocking density and genetic selection. Asian-Australasian Journal of Animal Science. 2016;29(6):901–7.
76. Catto S, Sumasgutner P, Amar A, Thomson RL, Cunningham SJ. Pulses of anthropogenic food availability appear to benefit parents, but compromise nestling growth in urban red-winged starlings. Oecologia. 2021;197:565–76.
77. Schulte-Hostedde AI, Mazal Z, Jardine CM. Enhanced access to anthropogenic food waste is related to hyperglycemia in raccoons (*Procyon lotor*). Conservation Physiology. 2018;6(1):1–6.
78. Townsend AK, Staab HA, Barker CM. Urbanization and elevated cholesterol in American crows. Condor. 2019;121(3):duz040.
79. Zhen C. Food deserts: myth or reality? Annual Reviews in Resource Economics. 2021;13:109–29.
80. Murray MH, Lankau EW, Kidd AD, Welch CN, Ellison T, Adams HC, et al. Gut microbiome shifts with urbanization and potentially facilitates a zoonotic pathogen in a wading bird. PLoS ONE. 2020;15(3):e0220926.
81. Sugden S, Sanderson D, Ford K, Stein LY, St. Clair CC. An altered microbiome in urban coyotes mediates relationships between anthropogenic diet and poor health. Scientific Reports. 2020;10:22207.
82. Teyssier A, Rouffaer LO, Saleh Hudin N, Strubbe D, Matthysen E, Lens L, et al. Inside the guts of the city: urban-induced alterations of the gut microbiota in a wild passerine. Science of the Total Environment. 2018;612:1276–86.
83. Tyakht AV, Kostryukova ES, Popenko AS, Belenikin MS, Pavlenko AV, Larin AK, et al. Human gut microbiota community structures in urban and rural populations in Russia. Nature Communications. 2013;4:2469.
84. Vuong HE, Yano JM, Fung TC, Hsiao EY. The microbiome and host behavior. Annual Reviews in Neuroscience. 2017;40:21–49.
85. Brittingham MC, Temple SA. Impacts of supplemental feeding on survival rates of black-capped chickadees. Ecology. 1988;69(3):581–9.
86. Peneaux C, Grainger R, Lermite F, Machovsky-Capuska GE, Gaston T, Griffin AS. Detrimental effects of urbanization on the diet, health, and signal coloration of an ecologically successful alien bird. Science of the Total Environment. 2021;796:148828.
87. Murray MH, Hill J, Whyte P, St. Clair CC. Urban compost attracts coyotes, contains toxins, and may promote disease in urban-adapted wildlife. EcoHealth. 2016;13(2):285–92.
88. Hotchkiss ER, Davis A, Cherry JJ, Alitzer S. Mycoplasmal conjunctivitis and the behavior of wild house finches (*Carpodacus mexicanus*) at bird feeders. Bird Behavior. 2005;17(1):1–8.
89. Murray MH, Hernandez SM, Rozier RS, Kidd AD, Hepinstall-Cymerman J, Curry SE, et al. Site fidelity is associated with food provisioning and *Salmonella* in an urban wading bird. EcoHealth. 2021;18(3):345–58.
90. Ducatez S, Sayol F, Sol D, Lefebvre L. Are urban vertebrates city specialists, artificial habitat exploiters, or environmental generalists? Integrative and Comparative Biology. 2018;58(5):929–38.
91. Teitelbaum CS, Altizer S, Hall RJ. Habitat specialization by wildlife reduces pathogen spread in urbanizing landscapes. American Naturalist. 2022;199(2):238–51.

92. Reed JH, Bonter DN. Supplementing non-target taxa: bird feeding alters the local distribution of mammals. Ecological Applications. 2018;28(3):761–70.
93. Galbraith JA, Beggs JR, Jones DN, Stanley MC. Supplementary feeding restructures urban bird communities. Proceedings of the National Academy of Sciences of the United States of America. 2015;112(20):E2648–57.
94. Leibler JH, Zakhour CM, Gadhoke P, Gaeta JM. Zoonotic and vector-borne infections among urban homeless and marginalized people in the United States and Europe, 1990–2014. Vector-Borne Zoonotic Diseases. 2016;16(7):435–44.
95. McKinney ML. Urbanization as a major cause of biotic homogenization. Biological Conservation. 2006;127(3):247–60.
96. Keesing F, Holt RD, Ostfeld RS. Effects of species diversity on disease risk. Ecology Letters. 2006;9(4):485–98.
97. Johnson BJ, Munafo K, Shappell L, Tsipoura N, Robson M, Ehrenfeld J, et al. The roles of mosquito and bird communities on the prevalence of West Nile virus in urban wetland and residential habitats. Urban Ecosystems. 2012;15(3):513–31.
98. Huang ZYX, van Langevelde F, Estrada-Peña A, Suzán G, De Boer WF. The diversity–disease relationship: evidence for and criticisms of the dilution effect. Parasitology. 2016;143(9):1075–86.
99. Halliday FW, Rohr JR, Laine AL. Biodiversity loss underlies the dilution effect of biodiversity. Ecology Letters. 2020;23(11):1611–22.
100. Hope D, Gries C, Zhu WX, Fagan WF, Redman CL, Grimm NB, et al. Socioeconomics drive urban plant diversity. Proceedings of the National Academy of Sciences of the United States of America. 2003;100(15):8788–92.
101. Aerts R, Honnay O, van Nieuwenhuyse A. Biodiversity and human health: mechanisms and evidence of the positive health effects of diversity in nature and green spaces. British Medical Bulletin. 2018;127(1):5–22.
102. Methorst J, Rehdanz K, Mueller T, Hansjürgens B, Bonn A, Böhning-Gaese K. The importance of species diversity for human well-being in Europe. Ecological Economics. 2021;181:106917.
103. German D, Latkin CA. Exposure to urban rats as a community stressor among low-income urban residents. Journal of Community Psychology. 2016;44(2):249–62.
104. Byers KA, Cox SM, Lam R, Himsworth CG. "They're always there": resident experiences of living with rats in a disadvantaged urban neighbourhood. BMC Public Health. 2019;19(1):853.
105. Murray MH, Byers KA, Buckley J, Magle SB, Maffei D, Waite P, et al. "I don't feel safe sitting in my own yard": Chicago resident experiences with urban rats during a COVID-19 stay-at-home order. BMC Public Health. 2021;21(1):1008.
106. Susser SR, Perron S, Fournier M, Jacques L, Denis G, Tessier F, et al. Mental health effects from urban bed bug infestation (*Cimex lectularius* L.): a cross-sectional study. BMJ Open. 2012;2:e000838.
107. Gibb R, Redding DW, Chin KQ, Donnelly CA, Blackburn TM, Newbold T, et al. Zoonotic host diversity increases in human-dominated ecosystems. Nature. 2020;584(7821):398–402.
108. Britton AP, Trapp M, Sabaiduc S, Hsiao W, Joseph T, Schwantje H. Probable reverse zoonosis of influenza A(H1N1)pdm 09 in a striped skunk (*Mephitis mephitis*). Zoonoses and Public Health. 2019;66(4):422–7.
109. Cunningham AA, Daszak P, Rodriguez JP. Pathogen pollution: defining a parasitological threat to biodiversity conservation. Journal of Parasitology. 2003;89(Suppl):S78–S83.
110. Barros M, Cabezón O, Dubey JP, Almería S, Ribas MP, Escobar LE, et al. *Toxoplasma gondii* infection in wild mustelids and cats across an urban-rural gradient. PLoS One. 2018;13(6):e0199085.
111. Burgess TL, Tinker MT, Miller MA, Bodkin JL, Murray MJ, Saarinen JA, et al. Defining the risk landscape in the context of pathogen pollution: *Toxoplasma gondii* in sea otters along the Pacific Rim. Royal Society Open Science. 2018;5(7):171178.
112. Dunbar M, Lu S, Chin B, Huh L, Dobson S, Al-Rawahi GN, et al. Baylisascariasis: a young boy with neural larva migrans due to the emerging raccoon roundworm. Annals of Clinical Translational Neurology. 2019;6(2):397–400.
113. Rogers KH, Girard YA, Woods L, Johnson CK. Avian trichomonosis in spotted owls (*Strix occidentalis*): indication of opportunistic spillover from prey. International Journal of Parasitology: Parasites and Wildlife. 2016;5(3):305–11.
114. Dwyer JF, Hindmarch S, Kratz GE. Raptor mortality in urban landscapes. In: Boal CW, Dykstra CR (eds.) Urban Raptors. Washington, DC: Island Press; 2018. p. 199–213.
115. Real J, Manosa S, Munoz E. Trichomoniasis in a Bonelli's eagle population in Spain. Journal of Wildlife Diseases. 2000;36(1):64–70.
116. Landry V, Asselin H, Lévesque C. Link to the land and mino-pimatisiwin (comprehensive health) of Indigenous people living in urban areas in eastern Canada. International Journal of Environmental Research and Public Health. 2019;16(23):4782.

117. Kimmerer R. Braiding Sweetgrass: Indigenous wisdom, scientific knowledge and the teachings of plants. Minneapolis, MN: Milkweed Editions; 2013. 408 p.
118. Bartlett C, Marshall M, Marshall A. Two-eyed seeing and other lessons learned within a co-learning journey of bringing together Indigenous and mainstream knowledges and ways of knowing. Journal of Environmental Studies and Sciences. 2012;2(4):331–40.
119. Tomiak J, McCreary T, Hugill D, Henry R, Dorries H. Introduction. In: Dorries H, Henry R, Hugill D, McCreary T, Tomiak J (eds.) Settler City Limits: Indigenous resurgence and colonial violence in the urban prairie west. Winnipeg, Canada: University of Manitoba Press; 2019. p. 1–21.
120. Simpson LB. As We Have Always Done: Indigenous freedom through radical resistance. Minneapolis, MN: University of Minnesota Press; 2017.
121. City of Vancouver. Rain City Strategy [Internet]. Vancouver, Canada: City of Vancouver; 2019 [cited 2022 Oct 13]. Available from https://vancouver.ca/files/cov/rain-city-strategy.pdf
122. Kumar S, Preetha GS. Health promotion: an effective tool for global health. Indian Journal of Community Medicine. 2012;37(1):5.
123. Harm Reduction International. What is harm reduction? [Internet]. London: Harm Reduction International; 2022 [cited 2022 Oct 13]. Available from: https://www.hri.global/what-is-harm-reduction
124. Milaney K, Haines-Saah R, Farkas B, Egunsola O, Mastikhina L, Brown S, et al. A scoping review of opioid harm reduction interventions for equity-deserving populations. Lancet Regional Health. 2022;12:100271.
125. Gallagher CA Keehner JR, Hervé-Claude LP, Stephen C. Health promotion and harm reduction attributes in One Health literature: a scoping review. One Health. 2021;13:100284.
126. Kondo MC, Fluehr JM, McKeon T, Branas CC. Urban green space and its impact on human health. International Journal of Environmental Research and Public Health. 2018;15(3):445.
127. Dooling S. Ecological gentrification: a research agenda exploring justice in the city. International Journal of Urban and Regional Research. 2009;33(3):621–39.
128. Checker M. Wiped out by the "greenwave": environmental gentrification and the paradoxical politics of urban sustainability. City and Society. 2011;23(2):210–29.
129. Gould KA, Lewis TL. The environmental injustice of green gentrification: the case of Brooklyn's Prospect Park. In: DeSena JN, Shortell T (eds.) The World in Brooklyn: gentrification, immigration, and ethnic politics in a global city. Lanham, MD: Lexington Books; 2012. p. 113–46.
130. Svendsen E, Campbell LK. Urban ecological stewardship: understanding the structure, function and network of community-based urban land management. Cities and Environment. 2008;1(1):4.
131. Degeling C, Brookes V, Lea T, Ward M. Rabies response, One Health and more-than-human considerations in Indigenous communities in northern Australia. Social Science and Medicine. 2018;212: 60–7.
132. Health Effects Institute. State of Global Air 2019: air pollution a significant risk factor worldwide. Special Report. Boston, MA: Health Effects Institute; 2019. Available from: https://www.healtheffects.org/announcements/state-global-air-2019-air-pollution-significant-risk-factor-worldwide
133. Caminade C, McIntyre KM, Jones AE. Impact of recent and future climate change on vector-borne diseases. Annals of the New York Academy of Sciences. 2019;1436(1):157–73.
134. Mormile JE, Hill CM. Living with urban baboons: exploring attitudes and their implications for local baboon conservation and management in Knysna, South Africa. Human Dimensions of Wildlife. 2017;22(2):99–109.
135. Aplin KP, Chesser T, ten Have J. Evolutionary biology of the genus *Rattus*: profile of an archetypal rodent pest. In: Singleton GR, Hinds LA, Krebs CJ, Spratt DM (eds.) Rats, Mice and People: rodent biology and management. Canberra, Australia: Australian Centre for International Agricultural Research; 2003. p. 487–98.
136. Feng ATY, Himsworth CG. The secret life of the city rat: a review of the ecology of urban Norway and black rats (*Rattus norvegicus* and *Rattus rattus*). Urban Ecosystems. 2013;17(1):149–62.
137. Himsworth CG, Parsons KL, Feng AYT, Kerr T, Jardine CM, Patrick DM. A mixed methods approach to exploring the relationship between Norway rat (*Rattus norvegicus*) abundance and features of the urban environment in an inner-city neighborhood of Vancouver, Canada. PLoS One. 2014;9(5): e97776.
138. Byers KA, Lee MJ, Patrick DM, Himsworth CG. Rats about town: a systematic review of rat movement in urban ecosystems. Frontiers in Ecology and Evolution. 2019;7:13.
139. Ayral F, Artois J, Zilber AL, Widén F, Pounder KC, Aubert D, et al. The relationship between socioeconomic indices and potentially zoonotic pathogens carried by wild Norway rats: a survey in Rhône,

France (2010–2012). Epidemiology and Infection. 2015;143(3):586–99.

140. Rael RC, Peterson AC, Ghersi BM, Childs J, Blum MJ. Disturbance, reassembly, and disease risk in socioecological systems. Ecohealth. 2016;13(3): 450–5.
141. Leibler JH, Robb K, Joh E, Gaeta JM, Rosenbaum M. Self-reported animal and ectoparasite exposure among urban homeless people. Journal of Health Care for the Poor and Underserved. 2018;29(2): 664–75.
142. Himsworth CG, Parsons KL, Jardine C, Patrick DM. Rats, cities, people, and pathogens: a systematic review and narrative synthesis of literature regarding the ecology of rat-associated zoonoses in urban centers. Vector-Borne and Zoonotic Diseases. 2013;13(6):349–59.
143. Himsworth CG, Zabek E, Desruisseau A, Parmley EJ, Reid-Smith R, Jardine CM, et al. Prevalence and characteristics of *Escherichia coli* and *Salmonella* spp. in the feces of wild urban Norway and black rats (*Rattus norvegicus* and *Rattus rattus*) from an inner-city neighborhood of Vancouver, Canada. Journal of Wildlife Diseases. 2015;51(3):589–600.
144. Himsworth CG, Patrick DM, Mak S, Jardine CM, Tang P, Weese JS. Carriage of *Clostridium difficile* by wild urban Norway rats (*Rattus norvegicus*) and black rats (*Rattus rattus*). Applied Environmental Microbiology. 2014;80(4):1299–305.
145. Murray MH, Fidino M, Fyffe R, Byers KA, Pettengill JB, Sondgeroth KS, et al. City sanitation and socioeconomics predict rat zoonotic infection across diverse neighbourhoods. Zoonoses and Public Health. 2020;67(6):673–83.
146. Lee MJ, Byers KA, Cox SM, Stephen C, Patrick DM, Himsworth CG. Stakeholder perspectives on the development and implementation of approaches to municipal rat management. Journal of Urban Ecology. 2021;7(1):juab013.
147. Lee MJ, Byers KA, Cox SM, Stephen C, Patrick DM, Corrigan R, et al. Municipal urban rat management policies and programming in seven cities in the United States of America. Journal of Urban Affairs. 2022 Sep 20:1–5. Available from: https://doi.org/10.1080/07352166.2022.2091995
148. Lee MJ, Byers KA, Donovan CM, Bidulka JJ, Stephen C, Patrick DM, et al. Effects of culling on *Leptospira interrogans* carriage by rats. Emerging Infectious Diseases. 2018;24(2):356–60.
149. Byers KA, Lee MJ, Hill JE, Fernando C, Speerin L, Donovan CM, et al. Culling of urban Norway rats and carriage of *Bartonella* spp. bacteria, Vancouver, British Columbia, Canada. Emerging Infectious Diseases. 2022;28(8):1659–63.

CHAPTER 13

Developing a Toolbox for Urban Biodiversity Conservation

Erica Spotswood[‡,§], Max R. Lambert[¶], Selena Pang[§], Jonathan Young[#], and Lewis Stringer[#]

Introduction

Conserving biodiversity in every type of landscape involves people—people making decisions to take action to protect land, or people convincing each other to prioritize some things over others to protect species. It also requires considering the implications of conservation actions on local communities. The need to work together makes conservation challenging in any context. Yet in cities, the potential challenges are magnified and more acute. The intensity of land use, the density of buildings and other infrastructure, and the number of independently owned parcels all lead to a unique concentration of parties with different agendas. Nowhere is the need to reconcile competing priorities, engage communities, and align biodiversity actions with other social goals more urgent than in our thriving urban centers.

While this need for compromise may present challenges for urban conservation, it can also create opportunities found nowhere else. Urban biodiversity conservation can leverage many existing activities that were not originally intended for biodiversity. Aligning conservation with other social priorities creates novel pathways for financing and projects that are less possible in nonurban contexts. The proximity to and intensity of human activity also offer opportunities to make conservation happen with fewer resources. Harnessing volunteers for stewardship, for example, can be simpler because people can participate just outside their doorsteps or in nearby parks and open spaces. Cities also offer potential refuge from conflicts with human activities occurring in other land uses; for example, some species are thriving in cities because they can escape from threats they face in intensive agricultural landscapes (1,2).

Beyond direct benefits to conservation locally within cities, urban biodiversity conservation also has the potential to advance regional conservation efforts. Potential outcomes include a reduction in large-scale fragmentation and improved habitat connectivity which could facilitate wildlife movement across landscapes. More available habitat in cities could enable larger population sizes for many species, and support unique habitat types and rare organisms that may be isolated within or restricted to cities (3–5).

While cities have great potential to contribute to local and regional conservation efforts, very little work has been done to identify the most effective types of activities to achieve these objectives. Outside of cities, conservation scientists and practitioners have assembled and refined a set of tools and approaches used to enhance habitat and biodiversity. These tools are not currently being applied widely in urban landscapes, and

[*‡] Second Nature Ecology and Design, USA
[†§] San Francisco Estuary Institute, USA
[‡¶] Department of Environmental Science, Policy, and Management, UC Berkeley, USA
[§#] The Presidio Trust, USA

Erica Spotswood et al., *Developing a Toolbox for Urban Biodiversity Conservation.* In: *Urban Biodiversity and Equity.* Edited by: Max R. Lambert & Christopher J. Schell, Oxford University Press. DOI: 10.1093/oso/9780198877271.003.0013 many conservation maps leave urban areas entirely "grayed out," implying that cities are places where conservation should not be a priority, does not occur, or is not worth the effort (Figure 13.1). To our knowledge, a standard toolbox—accepted by the conservation community, refined, and widely applied—does not exist for urban biodiversity. This gap leaves questions about which tools from traditional conservation will be effective in cities, which may be less useful, and which may need to be reimagined for application to the unique urban context.

Here we begin assembling a toolbox for urban biodiversity conservation that could be used to inspire and guide actions and projects (Figure 13.2). Our starting place was to consider the tools used in traditional conservation contexts, many of which can be easily applied in urban settings (e.g., Geographic Information Systems mapping, camera trapping, invasive species removal, and species monitoring), though these tools may be used slightly differently to accommodate human activity, actions like vandalism, and socioeconomic conditions. Some tools used commonly in conservation planning (e.g., spatial prioritization and decision support approaches) would need to be substantially modified to account for unique and complex social-ecological urban contexts. Other tools such as species translocations are infrequently used in nonurban landscapes but may be more useful and widely applied in cities than in nonurban contexts and may benefit from being applied to a more diverse array of species than are typically considered. Lastly, a few tools included in this toolbox (e.g., synergies with urban greening and novel financing) harness unique opportunities that are mostly restricted to cities.

We draw on diverse examples, many from California's San Francisco Bay Area, a highly urbanized region where some of these tools are already being deployed. This toolbox is not necessarily comprehensive; there are things we have chosen not to include, and probably far more that we haven't yet considered. As urban biodiversity conservation becomes a more active and intentional practice, we hope that conservation can build upon and improve the ideas suggested here, as the toolbox expands.

Toolbox for urban biodiversity conservation

Spatial prioritization & decision support

Spatial prioritization and other decision support approaches are core tools for conservation in nonurban landscapes. One of the most widely used is systematic conservation planning, a tool for designing networks of conservation reserves (6). Key to applying this framework are to consider whether a reserve network contains all habitat types and species in need of protection (e.g., complementarity and representation), and the use of optimization to identify reserve networks that protect the most species at the least cost (e.g., efficiency) (7). Optimization typically uses current conditions (biodiversity or habitat data) as input, resulting in solutions that prioritize protecting the least degraded areas that currently have the best habitat (8).

Applying this approach in urban landscapes would lead to urban conservation actions focused mainly in parks and neighborhoods that are already the most green and biodiverse (8). In many cities, this approach would run counter to other societal goals, including the urgent need to redress racial and class inequity in the distribution of urban greenness and park access (9–11), or to expand climate adaptation services such as heat island mitigation and flood control (12,13).

A related issue is that species recolonization of degraded urban sites can take a long time; even years or decades after restoration many habitat patches may not necessarily contain all or even most species that could be supported (14–17). Further, although urban conservation is increasingly informed by citizen/community science data, these data still have substantial racial and geographic disparities.[1] Therefore, existing biodiversity data may not accurately represent the potential for a site to support species and thus confound spatial prioritization efforts.

Spatial prioritization tools could still be usefully applied in cities, particularly if tailored to incorporate societal goals alongside biodiversity

[1] Perkins et al.'s chapter details the inequities in citizen science data and how those inequities can mislead conservation.

Figure 13.1 Conservation mapping of the San Francisco Bay Area, California, USA, showing urbanized areas (purple) and lands identified as priorities for conservation (green). Urbanized areas are almost entirely excluded from lands to be considered for conservation planning.
Maps are from Conservation Lands Network 2.0, a regional conservation strategy created by Together Bay Area, a coalition of non-profits, local agencies, and communities working together to implement conservation action across the region (https://www.bayarealands.org/).

goals. For example, instead of using the number of existing species protected, least-cost algorithms could use the potential number of species gained through biodiversity conservation interventions as input (e.g., ecological gain). Combining predictive modeling with cost data, optimization algorithms would likely suggest that greening the least green areas using inexpensive techniques such as tree planting would result in substantial gains in species richness (an approach conceptually similar to that proposed by Callaghan et al. (4)). The number of people—particularly in underserved communities—potentially able to benefit from biodiversity projects could also be added along with other community goals, tailoring optimization for synergistic outcomes that weight biodiversity alongside other social priorities such as equity. For example, Kroeger et al. (13) used a

TOOL	EXAMPLES
SPATIAL PRIORITIZATION & DECISION SUPPORT	+ Optimizing biodiversity though integrating social goals alongside conservation + Cost-benefit optimization tools to identify projects with largest biodiversity benefit
SPECIES PRIORITIZATION	+ Threatened and endangered species protection + Identification of likely species to benefit from conservation
ASSISTED REINTRODUCTION	+ Reintroduction of locally extirpated species + Local translocation to recolonize areas after restoration
ECOLOGICAL RESTORATION	+ Grazing to mimic natural disturbance + Invasive species management, early detection, and rapid response + Habitat improvement + Monitoring and adaptive management + Environmental engineering (e.g., creek daylighting)
INFRASTRUCTURE & ACCIDENTAL HABITAT	+ Wildlife friendly lighting and windows + Wildlife crossings, bridges, culverts, and underpasses + Peregrine scrapes on skyscrapers and nest boxes + Analogous features (e.g., cliff swallows nesting on bridges) + Seasonal road closures (e.g., migrating California and rough-skinned newts) + Seasonal bird nest surveys pre disturbance + Enhancing the artificial habitats
SYNERGIES WITH URBAN GREENING	+ Tree planting + Stormwater detention ponds and bioretention basins + Park management and creation + Species conservation and protection + Corporate campus landscaping and management
COMMUNITY ENGAGEMENT IN SCIENCE & STEWARDSHIP	+ Milkweed planting + Native gardens + National Wildlife Federation wildlife-friendly garden certification + Citized science data collection + Community-led design and engagement in decision making + Outreach, certification, and collective action programs
PLANNING & POLICY	+ Biodiversity strategy planning and policy + Urban park and forest master plans + Climate adaptation planning + Land use policies (e.g., Transfer of Development Rights, zoning, from-based codes) + Regulatory policy (e.g., creek setbacks, open space requirements, tree ordinances) + General, specific, parks & open space plans + Biodiversity net gain + CEQA/NEPA - regulatory mandates that drive mitigation
NOVEL FINANCING	+ Infrastructure funding + Carbon trade funding + Capital & operations funding for nature-based solutions and mitigation banking + Private development funding, public/private partnerships and developer impact fees + Parks and open space provisions in city regulatory requirements + Green benefit districts and neighborhood tax districts

Figure 13.2 Toolbox for urban biodiversity conservation, including nine tools and examples that can be used to achieve both within-city and regional conservation objectives.

return-on-investment model to guide where to invest in tree planting for heat island mitigation, using as input variation in population density to capture where tree planting will reach the highest number of people. Prioritization of urban areas that are less green and biodiverse to align with other social priorities would likely lead to outcomes that favor conservation in smaller patches over larger ones, particularly if the smaller green spaces are located in marginalized communities.

Species prioritization

Selecting species to prioritize in traditional conservation relies on a variety of approaches, including ecological concepts such as umbrella species, ecosystem engineers, ecological function, and conservation status (18–22). Although the concepts of umbrella species or ecosystem engineers are not commonly used in urban conservation, there have been occasional applications such as with lorises in India and beavers in Seattle (23).[2] In the US, the Endangered Species Act is also a strong driver for species conservation, and priority for threatened and endangered species often precludes selecting other species (24).

Traditional conservation is increasingly recognizing the value of "keeping common species common" and there are growing efforts to ensure species that are not currently threatened remain that way (25,26). A focus on common species may be particularly helpful in cities, and species prioritization could expand species selection criteria in urban contexts by selecting common species based on particular characteristics such as their ecological function or public appeal. Greater efforts to prioritize conservation efforts around urban populations would also likely yield substantial benefits for threatened and endangered species, which have a higher presence in urban areas than one would expect (3,5).

Species prioritization may also need to be expanded in cities to more fully reflect social contexts. Diverse cultural values can and do overlap with biodiversity goals, but not always, and community input must be prioritized in order to build and maintain public trust. Including public input, as well as other social drivers, may lead to prioritization of species that would not have been selected if species diversity were the only priority. For example, in the San Francisco Bay Area, Indigenous communities place high priority on accessible local sources for harvesting of plants such as soap plant (*Chlorogalum pomeridianum*), blue dicks (*Dipterostemon capitatus*), and willow (*Salix* spp.) (27). In another example from San Francisco, the National Historic Preservation Act protects in perpetuity designated historic non-native invasive eucalyptus groves (*Eucalyptus* spp.), which makes removal of this species culturally challenging to accomplish (28).

[2] Kar Gupta et al.'s chapter outlines how umbrella, flagship, and other proxy species can be used in urban conservation.

Traditional conservation has a strong focus on protecting native species and minimizing human impacts. While these goals can be more challenging to apply in the urban environment, they often remain relevant. Invasive species management to protect a site for native species, for example, is a high priority in urban biodiversity management. In San Francisco, invasive cape ivy (*Delairea odorata*) is widespread, and while citywide eradication is considered impossible, site-specific targeted management is feasible. The ivy forms dense mats along riparian areas, stifling other vegetation and impairing aquatic ecosystems (29). Prioritizing ivy eradication within the city's few remaining riparian sites (priority habitat for avian diversity) makes sense from a biodiversity conservation perspective.

The novel ecosystem paradigm and no-analog community concept can also be useful and more realistic frameworks for urban biodiversity management and prioritization (17,30). Acknowledging that some novel habitats and non-native species can and do benefit native biodiversity—and that not all native species should be left unmanaged—is an important component of the prioritization process. For example, Western Monarch butterflies (*Danaus plexippus*) have come to rely almost exclusively on the non-native invasive eucalyptus (*Eucalyptus globulus*) for overwintering roosts on the highly urbanized California coast, therefore protecting eucalyptus groves is likely necessary in the near term to maintain overwintering sites (31). Similarly, in San Francisco, shallow open water wetlands are a rare valuable urban habitat that can be quickly and completely lost to native species such as willows (*Salix* spp.) and cattails (*Typha* spp.); preserving this habitat type requires management of aggressive native species.

Translocation

Translocation is the intentional movement of organisms from one site for release in another, and can be a valuable conservation tool for overcoming barriers to dispersal (32). Translocation includes both reintroduction—the reestablishment of a viable population at a site inside its native range from which it has disappeared—and assisted migration—the intentional release outside the species' native range. Translocation planning, methods, and monitoring are species-specific, and several frameworks have been developed to guide implementation, typically in nonurban habitats (33–35). Although unintended translocation regularly occurs in urban areas, intentional translocation for conservation has rarely been considered for the urban conservation toolbox.

Translocation can sometimes be the only option in highly fragmented landscapes. Many species of wildlife may not necessarily recolonize sites where they have been locally extirpated even once ecological restoration occurs and habitat conditions have improved. Wildlife in particular can face strong barriers to dispersal and may not be able to cross the highly fragmented urban matrix in order to colonize a site where restoration has occurred (15,35,36). Translocations can therefore be a logical next step and an important component of holistic restoration.

As attention begins to focus more on urban biodiversity, translocations are beginning to gain traction for a variety of species including birds, herpetofauna, and invertebrates (36–38). Examples of urban translocations are rare, and there is much to learn about species selection, implementation, and what factors predict success. In the Presidio in San Francisco, ongoing experimentation with translocations coupled with research are helping to fill the information gap. Over the last decade, several species, including the Pacific tree frog (*Pseudacris regilla*), the Variable checkerspot butterfly (*Euphydryas chalcedona*), the California ringlet butterfly (*Coenonymph tullia*), the Rough-skinned newt (*Taricha granulosa*), and several aquatic species (see the section on Mountain Lake), have been reintroduced following intensive ecological restoration and habitat improvement. Most of these species are continuing to thrive in the Presidio, providing much needed information to guide future translocation efforts. The Presidio has also considered translocation of California quail (*Callipepla californica*), a charismatic species with large public appeal. Research conducted prior to translocation has focused on whether the Presidio has the required habitat available to support a population of quail (15). This and other similar research can help to guide decision making before translocation takes place by evaluating whether a given area is likely to be able to support a species of interest.

Urban translocations also offer unique opportunities to increase public awareness, provide contact with native species, and advance the field of urban ecology. As with all urban biodiversity conservation tools, it is necessary to incorporate people throughout the process as public attitudes and behaviors vary and can play a significant role in success or failure. As successes occur, public trust and awareness will likely increase. However, translocation failures can also be an opportunity to learn if coupled with appropriate public interpretation. Urban translocations can have significant indirect benefits through bringing conservation action directly to an urban audience. Indeed, urban translocations of non-listed species may be an opportunity to enhance wildlife opportunities for communities in urban areas with limited biodiversity, producing a win-win situation by enhancing local urban biodiversity and providing benefits to marginalized communities.

Ecological restoration

Habitat restoration is an important and widely used tool in traditional conservation contexts with a long history of both success and failure (39). In urban ecosystems, restoration can increase biodiversity while improving ecological functions that benefit humans and wildlife (40). Highly degraded urban ecosystems can be restored to provide important, self-regulating ecological values and conserve rare species sometimes found only in cities (5,41). A several-acre daylighted creek for example, if designed and implemented well, has the potential to reduce downstream flood water damage, improve water quality, and provide valuable wildlife habitat, all while creating aesthetic enjoyment for urban residents (42–44). An urban grassland that is home to an endangered wildflower can be restored to increase biodiversity and prevent extinction (45).

Successful implementation of urban ecological restoration can require a diversity of disciplines beyond restoration ecology such as historical ecology, landscape architecture, hydrology, project planning, construction and project management, communications, compliance, utility management, and regulatory permitting. Sometimes novel combinations of these disciplines are required to stitch nature back into urban landscapes. In Singapore, soil bioengineering techniques created a habitable substrate for tropical riparian plants to grow over an old concrete canal, improving ecosystem services along the banks of the Kallang River. As a result, residents enjoy new access to a vegetated river designed to accommodate 40% more water during floods than the canal that preceded it (46). Similarly, "living shorelines" are being created in coastal cities throughout the world to build better resilience to sea-level rise (47). In these new urban edges, hardscapes become softscapes where natural materials are used to create a malleable scaffolding on which organisms like shellfish can reinforce the material function of the substrate, increase biodiversity, and hopefully regrow when perturbation inevitably comes (48,49). Eco-engineered seawall habitats also allowed migrating Pacific salmon—including federally threatened species—to be more successful in urban areas like Seattle (50).

The fact that cities are full of people is an advantage to urban restoration practitioners eager to engage the public. Community-based restoration programs can thrive in urban areas by providing meaningful volunteer advocacy and implementation opportunities for residents, while creating a dedicated source of labor for restoration activities such as invasive species control, seed collection, plant propagation, and monitoring. For example, ecological restoration efforts to remove non-native weeds and trees and to restore wetland and coastal dune habitats in the Presidio rely heavily on volunteers. These volunteers provide labor while also learning, acting as local stewards, and becoming ambassadors of conservation action for their communities (Figure 13.3). Long-term maintenance or ecological stewardship is essential to the success of restoration initiatives. Organizations that support community-based restoration programs are often well positioned to sustain this necessary stewardship over the long term. Like urban gardening, community-based restoration can help build a network of cultural interactions that enriches the community as well as the ecology of cities.

Infrastructure and accidental habitat

Urban infrastructure has many negative impacts on biodiversity ranging from bird collisions with windows to impenetrable barriers to movement. However, infrastructure also represents "accidental" habitat that can be built upon with conservation actions (Figure 13.5). Infrastructure projects such as road building and maintenance as well as development projects can incorporate bird-friendly window design and wildlife-friendly lighting directly into design, or as retrofits on existing buildings and outdoor spaces. Road construction and other transportation infrastructure projects can also include features that facilitate connectivity for wildlife such as culverts and underpasses, as well as larger wildlife crossings over freeways. Biodiversity-supporting elements such as nest boxes and water features can also be included in many types of residential, commercial, and public spaces.

Use of urban infrastructure by wildlife and plants often occurs spontaneously without planning or intervention from people. Sometimes infrastructure can serve as analogous habitats for organisms in nonurban landscapes. For example, anthropogenic structures such as buildings, bridges, and metal pipes are used by many species of birds including swifts, swallows, and starlings, and skyscrapers are used by peregrine falcons for nesting and roosting (54,55). These accidental habitats can be passively accepted, or in some cases augmented and improved through conservation action. For example, boxes with gravel placed on window ledges can improve nesting habitat for peregrine falcons, which already show a high affinity for tall buildings (54). Stormwater detention ponds are also sources of accidental biodiversity support, and can be colonized by a wide variety of aquatic species, amphibians, and plants (56,57). In fact, a diversity of sensitive amphibian species are as common (or more common) in constructed stormwater ponds as they are in relictual urban wetlands, despite the lack of habitat considerations in stormwater pond design (58,59). These basins can be reengineered to better support amphibians, and native habitat can be improved through intentional planting to mimic native habitats. More broadly acknowledging and studying these accidental habitats will help us to

Figure 13.3 Ecological restoration at the Presidio of San Francisco in 2013 was conducted with the goals of creating healthy wetland habitat for wildlife and capturing stormwater. A series of detention basins was constructed after removal of invasive plant species, primarily blue gum eucalyptus trees (*Eucalyptus globulus*) and English ivy (*Hedera helix*). The basins are located within the historic footprint of Mountain Lake, which was reduced in size by the creation of Highway 1 (see the Mountain Lake case study). Restoration at these sites revives some ecological functions by creating seasonal wetlands to capture and process nutrient-rich runoff from a nearby golf course and provide groundwater recharge. Community volunteers planted the basins with native wetland and riparian vegetation (primarily *Juncus* spp. and *Scirpus* spp.). After vegetation establishment, several species of aquatic wildlife were translocated to the site, including Sierran chorus frogs (*Pseudacris sierra*) and rough-skinned newts (*Taricha granulosa*). The site currently supports a high diversity of local flora and fauna in a city with few remaining wetlands and captures stormwater runoff.
Photos courtesy of the Presidio Trust.

Case study San Francisco's Mountain Lake

Mountain Lake is located adjacent to Highway 1 within the Presidio of San Francisco, California, an urban national park managed by The Presidio Trust, a US Federal agency. Following over a century of degradation, the lake has undergone a dramatic recovery in water quality and biodiversity in the 21st century resulting from a combination of ecological restoration, infrastructure installation, invasive species removal, and the translocation of multiple species including a species of special concern. The story of the lake's recovery highlights how applying the tools described here in collaboration with the local community can lead to large changes in ecosystem health with synergistic benefits for people (Figure 13.4).

For thousands of years, the lake was home to plants and animals and provided resources for the Indigenous Ohlone people. In 1776, the first Spanish colonizers established a base at the Presidio and established farming and livestock grazing (51). The US military took control of the land in the 1840s as the city was growing. A golf course built in the 1890s caused runoff and nutrient input into the lake, and the construction of a highway along its western edge reduced the lake's surface area by nearly 40%. Non-native species accumulated as generations of people released unwanted aquatic pets and exotic recreational fishing species into the lake.

When the Presidio became a park in the late 1990s, much of the lake's biodiversity had been extirpated. No native fish remained and the lake was overrun by common carp (*Cyprinus carpio*), an invasive species that degrades water quality (52). The lake had also become highly eutrophic and seasonal toxic algae blooms (*Microcystis* sp.) often resulted in massive fish die-offs (53). Unmanaged issues with mosquitos and *E. coli* outbreaks were a public health concern, and the local community demanded the issues with the lake be addressed.

In 2000, a restoration plan was developed by the Presidio Trust that identified three main goals: improve water quality, increase native biodiversity, and raise public awareness. Terrestrial habitat restoration (Tool 4) began around the lake's riparian edge habitat with the removal of non-native eucalyptus trees (*Eucalyptus globulus*) and the planting of native species conducted with community volunteers. Non-native red slider turtles (*Trachemys scripta*) were trapped and the entire non-native fish community was eradicated using a pesticide application. Stormwater retention wetlands and bioswales were constructed at the golf course to control runoff and drainages were rerouted along the highway to reduce contaminated input into the lake. An aeration mixing system was installed in the lake and an unwanted pet safe-surrender dropbox was placed on-site.

Reintroductions of extirpated species (Tool 3) began with submerged aquatic vegetation (e.g., *Stuckenia pectinata*) in order to provide structural habitat, oxygenation, and water quality improvement. Wildlife reintroductions followed with Pacific chorus frogs (*Pseudacris sierra*), threespine stickleback fish (*Gasterosteus aculeatus*), western pond turtles (*Emys marmorata*, a species of special concern in California and which are undergoing review for listing under the U.S. Endangered Species Act), freshwater mussels (*Anodonta californiensis*), and rough-skinned newts (*Taricha granulosa*). Several plants were also reintroduced to the site including lesser pond weed (*Potamogeton pusillus*). Each species was identified for candidacy based on its conservation status and its functional value (Tool 2). For example, freshwater mussels were identified because their filter feeding can improve water quality, and the sticklebacks, the host fish for the freshwater mussels, were identified in order to enable population persistence of the mussel.

A public engagement campaign (Tool 7) was launched including volunteer stewardship, public walks and talks led by partner scientists, on-site education, multimedia coverage, comprehensive interpretative signage, school field trips, and community science monitoring programs. The western pond turtle was designated the "Ambassador Species," or official mascot for the project, and has helped bring the story of Mountain Lake's biodiversity restoration to thousands of people around the world. In 2021, the first on-site hatchling western pond turtle was discovered, showcasing the value of this urban restoration and translocation for a highly sensitive species. Despite many setbacks and lessons learned, initial efforts have generated tremendous excitement and helped build trust between managers and the public while also providing opportunities for the field of urban ecology to advance (e.g., Ismail et al. (53)). Based on identified metrics, the project has been a resounding success that merges multiple tools like habitat restoration, species prioritization, and translocations. However, the urban lake is a dynamic system with endless challenges ahead that will require ongoing monitoring, management, and public support.

Case study *Continued*

Figure 13.4 Mountain Lake (top left) sits at the edge of the Presidio between a golf course, Highway 1, and the dense urban landscape of San Francisco. Prior to restoration, the lake was dominated by invasive species, was severely contaminated, and suffered from frequent toxic algal blooms. Ecological restoration of the lake included non-native eucalyptus tree removal (top right) and riparian restoration (middle left), involving many volunteers through outreach and stewardship programs. Post-restoration species translocations of listed western pond turtles (middle right), Pacific chorus frogs (bottom left), and several other species helped to restore biodiversity and ecosystem function to the lake. Today, the restored lake supports a thriving community of waterbirds, aquatic organisms, and terrestrial wildlife around its edges (bottom right).
Photos courtesy of the Presidio Trust.

Figure 13.5 Infrastructure can support wildlife in cities both unintentionally and through construction of features meant to support wildlife. Bridges such as the Congress Avenue bridge in Austin, Texas (top) and the chimney at the Chapman Elementary School in Portland, Oregon (second from top) can support roosting bats and swifts. Analogous features such as nest boxes (second from bottom) can replace natural tree holes which are a rare limiting resource for cavity-nesting organisms in cities. Engineered or eco-engineered infrastructure such as living shorelines (bottom) can also create habitat that supports biodiversity like migrating salmon while delivering other benefits to people such as sea-level rise protection and flood mitigation.

Photos courtesy of: Daniel Spiess, Chase Elliott Clark, Phillip Shoffner, Jason Toft, respectively.

augment them more effectively, reducing reliance on (and negative impacts from) areas that may cause harm to urban species.

Synergies with urban greening

Just because an urban space is green or blue does not mean that it is biodiverse or advances conservation goals. Many urban greening activities are already occurring in urban landscapes for reasons that are most often not directly oriented toward biodiversity protection. Bioretention basins and stormwater ponds are built to manage flooding and water quality. Tree planting is often motivated by a desire to reduce urban heat islands, beautify neighborhoods, or provide access to trees in underserved communities. Even park creation and management is often focused on goals related to recreation and community amenities. In private spaces, landscaping on commercial property and gardening in residential yards are conducted largely for reasons related to beautification, personal enjoyment, or food production.

These activities can all help to achieve biodiversity conservation objectives, though their benefits to conservation are best maximized if the actions can be tailored to meet both human and biodiversity goals. Synergies can often be achieved through slight modifications to the design of existing projects. For example, bioretention basins, landscaping, backyard gardening, and tree planting programs can prioritize planting native trees and other vegetation. Backyard gardening programs such as those run by local organizations like the California Native Plant Society can incentivize and motivate residents to install native plants in their yards (Figure 13.6). In Mountain View, California, a public–private partnership between the city, Google, and several other corporations expanded a stormwater detention basin by removing non-native trees and parking spaces and installing native willows and cattails (Figure 13.6). The resulting basin serves stormwater and flood detention functions, has a boardwalk and publicly accessible walking path around its perimeter, and provides native habitat for many species of wildlife and plants.

Many urban tree programs also include heritage tree ordinances which protect large trees of particular species or historic significance on both public and private property, ensuring trees with outsized impacts for biodiversity are conserved (60). For example, the city of Palo Alto in California has a heritage tree ordinance that protects large native trees including several species of California oaks and coast redwoods (*Sequoia sempervirens*). Park design can combine amenity spaces with areas dedicated for habitat that can also serve as places for psychological restoration and contemplation.

Figure 13.6 Native gardening programs with plaques (left) that residents can place in their yards such as this program run by the California Native Plant Society (www.cnps.org) can incentivize private residents to install native plants in their yards. In Mountain View (right) a stormwater detention basin was expanded through removing parking spaces and restoring with native willows, cottonwoods, and cattails through a public–private partnership between the city, Google, and other local landowners.
Photos courtesy of Erica Spotswood (left) and Shira Bezalel (right).

Community engagement in science and stewardship

Cities offer truly unique opportunities for engaging large numbers of people in science, stewardship, and conservation. Community science is a rapidly growing and evolving field, and the data quality, types, and taxonomic groups represented are growing every year. Community science has rapidly become the largest source of biodiversity records, and data are often more comprehensive in cities than elsewhere. Community science data collection also offers a powerful avenue for community participation, learning, and engagement with potential benefits for both conservation (61) and human well-being (62). Even so, community science is not immune to biases and many urban biodiversity datasets have substantial racial-spatial biases that can interfere with their utility for conservation.

Community involvement in local stewardship activities such as backyard gardening offers significant potential for contributing to urban biodiversity conservation. Coordinated involvement through certification programs such as the National Wildlife Federation wildlife-friendly gardens certification, or through thematically led campaigns for particular species (examples include Monarch Watch & Pollinator pathways) also serves to bring people together around shared goals, creating a community of participants and building social resilience (63,64). Engagement in stewardship activities has also been linked to increases in pro-environmental behavior and willingness to invest in conservation (65). Thus, there is a huge and largely untapped potential to link urban to nonurban conservation through increasing the involvement of urban citizens in stewardship of the spaces where they live and work.

Planning and policy

Biodiversity conservation efforts in urban spaces stand to benefit greatly from building on existing urban planning and policy tools, which are typically robust and already address many environmental issues in the city. There are many opportunities for synergy between biodiversity goals and those for water resource management, human health, active lifestyles, landscaping, heat management, and other aspects that make for a comfortable and safe urban space (Table 13.1). Cities do not exist in a vacuum, however, and multiple levels of policy making from the city to regional, state, and national levels may contribute to advancing biodiversity goals in urban and urbanizing regions.[3] Even so, currently few agencies at higher echelons are meaningfully and actively engaged in urban biodiversity conservation planning and policy. Regional urban growth boundaries like that in the Portland Metro Area or in various regions of the San Francisco Bay Area are largely intended to constrain the impact of urbanization on surrounding hinterlands but may be avenues for also advancing biodiversity conservation in cities.

City fabrics are defined and upheld by a unique system of urban policies, plans, and rules. Therefore, to create a planning and policy space where urban biodiversity initiatives can succeed, cities must approach biodiversity planning on multiple levels. First, a city needs high-level planning and benchmarking for biodiversity goals, including target-setting (66).[4] High-level planning can be national in scale, such as the UK's recent adoption of Biodiversity Net Gain, wherein new development projects will be required to produce a 10% increase in biodiversity, maintained for 30 years (67). High-level planning and benchmarking can also happen at the city scale (68). This level of planning provides important guidance and a vision for how people and nature should coexist within urban space.

Plans and policies can also explicitly incorporate biodiversity conservation. Many existing plans and policies are facilitative of urban biodiversity, in that they promote the creation of new green space or identify and protect existing biodiversity resources. However, by adopting biodiversity as a goal, cities can make an explicit commitment to growing and protecting their urban biodiversity and create a road map to do so using city-specific information. Natural resource master plans (such as urban forest, park, and river master plans) can be easily adapted

[3] Larson's and Brown's chapter outlines how human motivations at multiple social scales shape biodiversity in cities.

[4] Stanford et al.'s chapter demonstrates how city planning can include conservation and prioritizing biodiversity.

Table 13.1 Types and examples of national, city, and local planning and regulation that can contribute to biodiversity conservation.

Planning type	Examples
Regional planning	Urban growth boundaries, growth management policies at regional scales
City & community planning	City general & specific plans, community master planning
Utility related	Green stormwater infrastructure, water resource protection, waterway management, floodplain protection, envision sustainable infrastructure certification targets, street tree selection tools (powerlines)
Resource master plans	Climate adaptation plans, urban park master plans, urban forest plans, biodiversity plans, green infrastructure master plans
Transit planning	Green streets, transit-oriented development, shade ordinances, green space networks
Health related	Pesticide bans, health impact assessments, brownfields and green remediation, urban heat mitigation through impervious cover reductions, smoking bans (butts)
Public facilities	Parks, privately owned public spaces, school and hospital campuses
Habitat related	Riparian habitat buffers, conservation easements, protections, e.g., wetland/ critical habitat designation/ protected species or host plants. Smaller-scale protective tree ordinances, heritage trees, creek setbacks
Landscape related	Ethnic planting palettes, historic or cultural landscape restoration, planting diversity requirements for aesthetics or biodiversity specifically
Environmental impact mitigation, permitting, and regulatory compliance	National Environmental Quality Act, National Endangered Species Act, Migratory Bird Treaty Act, national air and water quality regulation
Development regulation	Biodiversity Net Gain (UK), transfer of development rights, creek setback and open space requirements, Sustainable SITES certification, onsite stormwater mitigation (in lieu of fees), conservation subdivisions, green roof requirements
Zoning and code	Form-based codes, varied density, and transit-oriented development create more opportunities for greenway creation, multi-use communities, ecological and social connectivity

to become part of a city's biodiversity strategy. Resource master plans that conduct analyses at the property or parcel level could also provide high value to urban biodiversity conservation.

In order to realize biodiversity goals, cities need paths to implementation. This happens when biodiversity goals are interpreted and enforced by the city to actively increase and protect urban biodiversity. Implementation can ultimately result in measurable improvements to biodiversity resources, create accountability for set benchmarks, and create tangible benefits for urban residents. Minor revisions to existing landscaping requirements could increase baseline urban biodiversity, for example requiring a certain percentage of native or culturally important species, or setting a minimum number of species. Existing mechanisms that typically ensure compliance, such as city ordinances and permit processing, could thus be used to ensure implementation of biodiversity-focused policy tools and could also be incorporated into development, redevelopment, and infill as green space is created for setbacks, park and open space requirements, conservation easements, or through transfers of development rights. These policies and plans can also simultaneously account for environmental equity and justice goals. For instance, multiple agencies and organizations in the City of Portland are working under the city's plan to enhance the city's biodiversity while also addressing the region's long history of environmental racism.[5]

Novel financing

One key characteristic of cities is the diverse array of activities, projects, policies, and funding sources. Much of conservation planning assumes limited scarce resources, and a need to strongly prioritize conservation dollars to identify the highest-leverage places in which to invest (e.g., systematic

[5] Guderyahn's and Logalbo's chapter details these conservation organizations' work for biodiversity and equity.

conservation planning and other spatial prioritization frameworks). Within this framework, allocating finite conservation dollars to urban areas may be perceived as wasteful, and the degraded nature of urban habitat patches usually precludes them from being selected by conservation least-cost algorithms. However, within urban landscapes, there are opportunities to tap into novel sources of financing that can combine in synergistic ways without competing with conservation funds for nonurban landscapes.

Funding for biodiversity conservation can come from a diverse range of sources in urban landscapes, often through aligning with other activities that are either required or have a strong social priority. For example, California has the only state-led cap-and-trade program in the US. Funds generated from the program through carbon taxes partially finance urban tree planting through a state-led granting program that funds cities to inventory their city-owned trees, create management plans for trees, and plant trees that will sequester large amounts of carbon (see: https://www.fire.ca.gov/what-we-do/grants/urban-and-community-forestry-grants). This program, funded through sources that don't compete with conservation dollars and for reasons unrelated to biodiversity, offers an opportunity to achieve some conservation objectives through increasing urban tree canopy.

The movement toward valuing ecosystem services has also created opportunities for considering nature-based solutions as infrastructure, opening doors to new sources of funding through infrastructure dollars. In one particularly novel example, dollars to purchase open space just upstream from the city of San Jose, California, US, were included in a 2018 infrastructure bond. Voters approved the bond, and the city council unanimously voted to spend $46 million to purchase 937 acres to be managed as public open space (https://www.openspaceauthority.org/our-work/planning-coyote-valley/about-coyote-valley.html). The justification for using infrastructure funding was that, left undeveloped, the land provides a crucial service by reducing flooding downstream in the city of San Jose. If the land were to be developed, this service would be lost, exacerbating downstream flooding in the city (69).

Global perspectives: place-based needs to develop the urban conservation toolbox

There are many actions taking place around the world to support biodiversity. Some cities are creating biodiversity strategies (70). Others are adopting biodiversity indices to monitor biodiversity change. These efforts are important steps toward creating biodiversity objectives, and tracking whether objectives are being met.

Vast differences among cities in climate, geography, urban form, and socio-economic context all lead to large variability in the types of opportunities for urban biodiversity conservation. In Europe, proactive policies for urban biodiversity are ahead of those in the US, although multiple state and federal agencies in the US proactively established many large urban and urban-adjacent natural areas in the middle of the 20th century (68). In the Global South, large informal settlements in and around many of the world's cities present a different context and challenge for urban conservation. Each city has a slightly different set of wildlife to manage, and cities are surrounded by highly variable contexts, influencing what is possible to achieve in connecting conservation in and outside cities to achieve regional conservation goals.

While we expect that established and emerging urban conservation tools will be widely applicable across the world, we also anticipate that which tools are most important or easily deployed will vary from city to city depending on context. Ecological restoration, for example, can be conducted in any city in the world, though the ability to finance and implement restoration projects may be much greater in some cities than others. Similarly, planning and policy tools may be more important in cities with strong regulatory frameworks, whereas community-led projects may be more impactful in cities with fewer top-down governance structures available to influence green infrastructure planning. The use of planning to influence development may be more important in growing cities, and possibly in very high-density cities where the biggest opportunities lie in on-structure greening such as green roofs and living walls.

Innovation in urban conservation is likely to happen in diverse ways across the world and in ways that are informed by local culture and political structures. Continued support of biodiversity conservation in cities across all continents will help conservationists tailor actions and tools to their city's particular needs, history, and conditions. We anticipate that a global community of urban biodiversity conservationists will foster a richer dissemination of new tools and approaches for advancing conservation goals across scales from cities to regions, countries, and continents.

Would conservation in cities make a difference? If so, for whom?

We know from urban ecology research that greener areas, more habitat patches, larger habitat patches, and areas with greater connectivity in cities tend to be associated with more biodiversity (71). It is therefore reasonable to assume that deliberate efforts to green cities, keeping biodiversity-supporting elements in mind (66), would increase the number and abundance of species that can thrive in urban habitats. Currently, cities are considered in a conservation context primarily and solely for their negative impacts to surrounding nature, and not for the potential gains to biodiversity that could be derived from habitat improvements within the urban footprint. This represents a missed opportunity given urban areas have the highest potential proportional gain in habitat and biodiversity value to be derived from conservation actions, and thus the greatest chances for net gain relative to other land use types.

The outcomes of biodiversity conservation in cities for a particular species depend on whether the changes ultimately help the species persist across its broader range. This could occur if urban habitat improvement enabled larger population sizes, led to higher population growth rates, increased genetic diversity, or allowed individuals to move more easily across the urban landscape. All these results would yield positive outcomes for regional conservation. Larger numbers of individuals and greater genetic diversity can buffer populations from periods of stress, and the ability to move unhindered across the landscape can enable species and individuals to access more resources.

Species traits such as behavioral flexibility and a generalist diet have been associated with an ability to tolerate cities (2). However, these findings do not necessarily imply that the set of species that can use cities is fixed—patterns of species use of cities are fluid, and there is evidence that they can change over time and in response to shifting conditions. For example, in the San Francisco Bay Area, harbor porpoises returned to the Bay after nearly a century of being absent in response to improving water quality and habitat conditions (72). And coyotes returned to San Francisco after decades of absence after being hunted to local extinction in the mid-20th century (73). These examples highlight the dynamic nature of biodiversity in cities; removing pressure by improving habitat and eliminating threats such as hunting can lead species to bounce back after long periods of absence.

Species assemblages in cities are highly altered, and habitat improvement would shift current patterns. Many species currently avoid cities altogether, and some of these would likely begin to venture into them (Figure 13.7). Over time, a greater fraction of the regional species pool would be found in cities, and urban species compositions would be less different from those in adjacent wildlands. Species that currently move back and forth, or have populations both within and adjacent to cities, would likely use cities more frequently as new habitat became available (Figure 13.7). Species that are already common in cities could benefit, though some might lose out as unique urban habitats become less common or if they are negatively impacted by new interspecific interactions. Of the species found in cities, some are of conservation concern and are not common in surrounding landscapes. These "last chance" species deserve special conservation consideration, and stand to benefit from targeted protection in urban habitats on their behalf (5).

It is possible that improving habitat in cities could cause downsides that will themselves need to be managed within a conservation framework. For instance, habitat in cities could create or exacerbate existing ecological traps (habitat that attracts a species but which results in impaired fitness) where species are drawn into cities despite facing reduced survival, reproduction, or population viability (74).

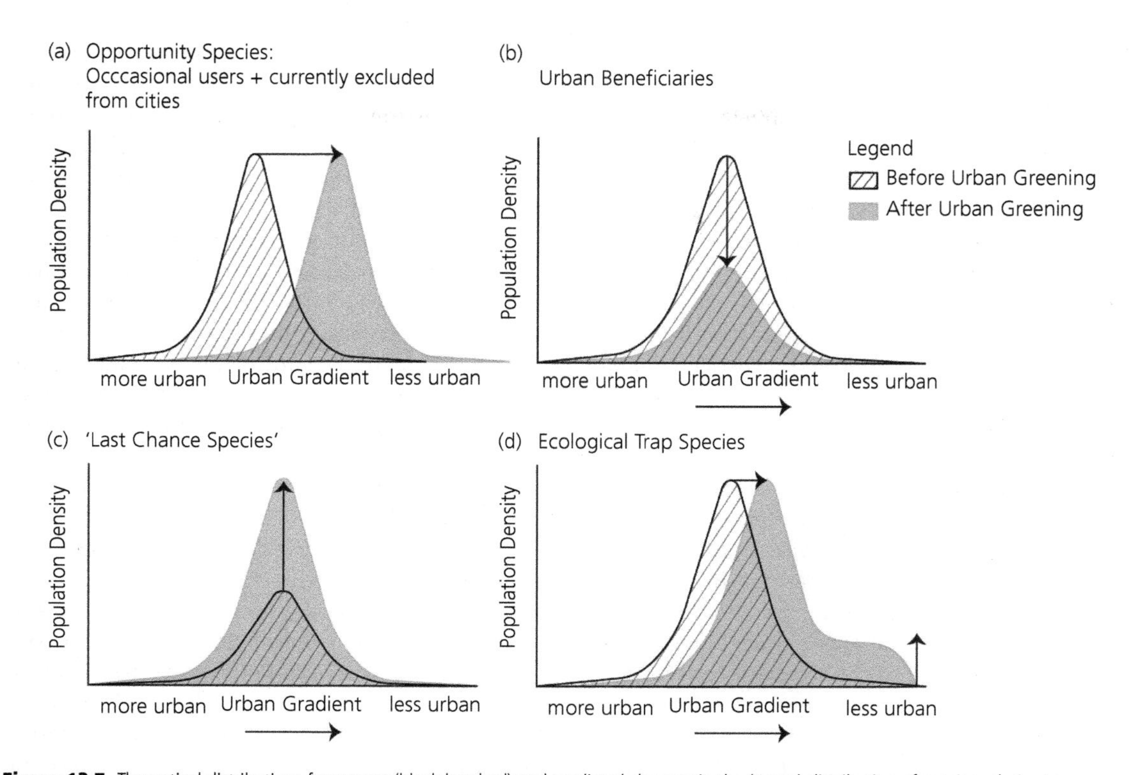

Figure 13.7 Theoretical distributions for current (black hatched) and predicted changes in the (green) distribution of species relative to urbanization gradients as urban conservation improves habitat within cities. Opportunity species (a) are those that are currently excluded from cities or venture into them only occasionally. As urban habitat improves, a greater fraction of regional populations would likely begin to use cities. Urban beneficiaries (b) are currently common and benefiting from habitat in cities, and may see populations in cities shrink if habitat improvement leads to a loss in novel resources available only in cities, or if new habitat leads to an increase in interspecific interactions such as predation that limit populations. Populations of "last chance" (c) species currently restricted to cities because cities contain their only remaining habitat, could see population sizes increase as urban biodiversity conservation actions improve the amount and quality of their habitat. Ecological trap species (d), or those for which cities contain attractive habitat but where mortality rates are high, could see their distributions shift towards greater use of cities, as well as more successful use, leading to larger population sizes.

There is some evidence that ecological traps can occur, though a recent meta-analysis concluded that most relevant studies have demonstrated that urban areas are functioning as safe sites for species studied rather than traps (75). It is also likely that improving habitat in cities could reduce the potential for ecological traps, as a greater fraction of the urban landscape transitions to better habitat.

Social-ecological applications: managing conflicts to achieve conservation objectives

Protecting and supporting biodiversity in cities requires reconciliation between human goals and biodiversity goals. This can be best achieved by seeking compromise, balancing among multiple priorities, and finding win-win solutions. Balancing among human needs and biodiversity is not all about overcoming challenges—there are many synergies, and numerous actions that can provide benefits to both people and ecosystems. Here, we offer a brief overview of the types of conflicts that can arise.

Many of the conflicts that arise in cities, while not unique to urban landscapes, can be most acutely felt in urban contexts because of the higher numbers of people, and a strong need to consider many competing priorities. Human–wildlife conflict is occurring across the globe and is not restricted to urban landscapes (76). Conflict between people and plants also occurs and can take many forms. For example, trees can cause conflicts with infrastructure, can cause sidewalks to heave, and can drop limbs or fall during storms, damaging property and endangering people (77). Poisonous plants can be a concern for domestic pets and people, and plants that produce pollen or attract bees and other stinging insects can pose hazards for people. Trees and shrubs that produce acorns or other hard seeds or fruit can drop them on cars and sidewalks, causing tripping hazards or messiness.

A movement toward more biodiversity conservation in cities may lead to additional and new forms of acute conflicts in cases where greening actions intensify the frequency of interaction between people and the source of the conflict. In other cases, urban biodiversity conservation may help reduce conflict. For example, increasing population sizes of medium-sized mammal carnivores in cities is likely to lead to an increase in the potential for human–wildlife conflict. On the other hand, more carnivores in cities could also help to control pests, leading to less conflict between people and rats and other undesirable pest species.

Education and public awareness related to sources of conflict is a high priority for any urban biodiversity conservation organization project. Finding solutions can take many forms and is often best achieved through use of a variety of approaches (e.g., digital tools like social media or traditional approaches like publication presentations and door hangers). Clear and transparent communication is essential in building and maintaining trust, articulating the issues, and clarifying goals and actions. Knowing the diversity of urban audiences is also essential for understanding cultural drivers of conflicts and identifying solutions. Essential to this work is meaningfully engaging communities to understand needs and to achieve equity goals.

Public education can be combined with more active interventions that help to directly reduce the probability of conflict. For example, signs and temporary fencing can help to keep people out of sensitive areas in parks, while education can help people understand what it means to live alongside wildlife (see the coyote case study). City arborists can help to maintain urban trees to resolve infrastructure conflicts alongside tree care education for private residents. For seasonal and predictable conflict, having established guidelines to implement and standardize conflict reporting and response protocols can help maintain consistency. Additionally, regular training can help all public-facing agency employees learn how to handle certain types of potential conflicts, particularly surrounding high-profile wildlife such as large carnivores. Training should include information regarding the official responsible agencies and departments, as well as where to direct for appropriate and up-to-date information.

Conclusion

Cities are similar to traditional conservation contexts in that resources limit what it is functionally possible for biodiversity managers to implement. Laws, culture, and values provide the social context within which we identify goals and apply a social-ecological understanding that guide decision making. The human influences in urban ecosystems interact with biodiversity in a multitude of ways creating feedback loops ranging from negative to benign to beneficial (17,76). Societal values and ecosystems trajectories are not static, and priorities should be firm yet adaptable as changes continue to unfold. We outline a series of tools and approaches to advance conservation in urban areas. We are eager to see this toolbox expand as urban biodiversity conservation becomes more common and practitioners and researchers around the world innovate new approaches.

A key theme identified here is that biodiversity conservation in cities requires compromise and can be best achieved through creating synergies between biodiversity objectives and other social goals. Creating projects with this structure can be complex, requiring coordination across sectors, and inclusion of diverse ideas, partners, and approaches. Crossing disciplinary boundaries can require overcoming language barriers that can occur when biologists talk with urban planners, landscape architects, and developers. More than in any other landscape, urban conservation needs to take place with and through engagement with diverse stakeholders, and the role of community

Case study coyote conflict management in San Francisco

After an absence of more than 75 years, coyotes (*Canis latrans*) reestablished a population in San Francisco, California, in the early 2000s. The presence of coyotes in San Francisco, as in other cities, has been a controversial topic, and public attitudes range from fear and hate to lack of awareness, admiration, and respect. Human–coyote conflict, both real and perceived, is an ongoing challenge for city wildlife authorities. As with many other North American cities, San Francisco has adopted a "coexistence" approach, which acknowledges that the species is here to stay and proactive management should aim to reduce conflict to the extent possible (73).

Over the first decade following coyote establishment, city authorities were inexperienced, and management was primarily reactive. Urban coyote information was limited, and management and messaging from the various agencies across the city were inconsistent. This changed when representatives from all agencies came together to develop a citywide approach, including a standardized incident response framework, which guides conflict assessment and response. This framework acts as a consistent guide to assess the circumstances of a conflict and decide on a course of action. This citywide working group has gained more experience with each new challenge, refining strategies and building public trust. Meanwhile a San Francisco Bay Area network including research institutions, scientists, and wildlife authorities is growing and sharing knowledge.

As the San Francisco coyote population begins its third decade, management strategies are becoming more agile, sophisticated, and prepared. For example, pupping season consistently results in conflict when domestic dogs get too close to an active den. Seasonally appropriate signage now informs dog walkers how to respond, and in some cases trail closures close to den sites also significantly reduce this conflict (Figure 13.8). Clear and regular communication, including how to report incidents, is deployed though a variety of outlets from local media to door hangers, websites, and brochures. One-on-one conversations with residents are conducted by agency staff, and while this is the most effective means for shifting perceptions of coyotes, it is also time-intensive. Neighborhood townhall meetings are also used, and can be a good forum for discussing concerns. Other standard coyote management tools such as vegetation removal, fencing, hazing, and trash management have yielded mixed results, and are most effectively deployed when communities are involved.

Figure 13.8 San Francisco coyotes (left) and example signage used as part of ongoing management (right). Individuals are collared and tracked to identify den locations, track movement outside of the Presidio, and monitor population sizes. Signage is used to reduce conflict with people and domestic dogs which occurs particularly during the breeding season when coyotes actively protect denning locations. When dens are identified close to trails, signage is used for temporary trail closure.
Photos courtesy of Daniel Ramirez (left) and the Presidio Trust (right).

Case study *Continued*

Collaring and tracking coyotes through collaborative research with outside institutions is also generating high-resolution information including movement patterns, pup dispersal, diet, and health. This information is feeding back into management and education and helping to address public concern about coyotes in the city. As San Francisco coyote management continues to learn and grow, public education and engagement remains the number one most important component. While managing coyote behavior is sometimes possible, urban coyote management is mostly about human management, placing it at the intersection of ecological and social sciences. Human–coyote conflict is impossible to eliminate completely, but it is possible to reduce. Scientists and managers still have a lot to learn about the animals, but arguably more to learn about society's relationship with nature and how to effectively influence attitudes and behaviors that promote coexistence.

engagement is both essential and beneficial to any urban conservation project.

Managing conflicts between people and biodiversity is also a defining feature of urban conservation projects, and is an ongoing activity that does not end once projects are implemented. We hope that some conflicts may become less acute as people adjust to having more wildlife and a greater diversity of plants in their cities, but in the meantime, ongoing public outreach, and direct intervention to reduce conflict will be needed to help people successfully live alongside thriving biodiversity.

Making cities greener for biodiversity will change who shows up, altering species assemblages, and creating opportunities for new species to step into cities for the first time. Evidence strongly suggests that actions to support conservation in cities will also benefit regional conservation efforts in nonurban areas. Being intentional about this goal will help conservation projects in cities better understand the implications of their efforts. Over time, we hope that through collaboration and communication, better connections between conservation actions inside and outside cities can help both efforts go farther.

References

1. Hall DM, Camilo GR, Tonietto RK, Ollerton J, Ahrné K, Arduser M, et al. The city as a refuge for insect pollinators. Conservation Biology. 2017;31(1):24–9.
2. Spotswood EN, Beller EE, Grossinger R, Grenier JL, Heller NE, Aronson MF. The biological deserts fallacy: cities in their landscapes contribute more than we think to regional biodiversity. BioScience. 2021;71(2):148–60.
3. Ives CD, Lentini PE, Threlfall CG, Ikin K, Shanahan DF, Garrard GE, et al. Cities are hotspots for threatened species: the importance of cities for threatened species. Global Ecology and Biogeography. 2016;25(1):117–26.
4. Callaghan CT, Bino G, Major RE, Martin JM, Lyons MB, Kingsford RT. Heterogeneous urban green areas are bird diversity hotspots: insights using continental-scale citizen science data. Landscape Ecology. 2019;34(6):1231–46.
5. Soanes K, Lentini PE. When cities are the last chance for saving species. Frontiers in Ecology and the Environment. 2019;17(4):225–31.
6. Margules CR, Pressey RL. Systematic conservation planning. Nature. 2000;405(6783):243–53.
7. Schwartz MW, Cook CN, Pressey RL, Pullin AS, Runge MC, Salafsky N, et al. Decision support frameworks and tools for conservation. Conservation Letters. 2018;11(2):e12385.
8. Jalkanen J, Vierikko K, Moilanen A. Spatial prioritization for urban biodiversity quality using biotope maps and expert opinion. Urban Forestry & Urban Greening. 2020;49:126586.
9. Nesbitt L, Meitner MJ, Girling C, Sheppard SR, Lu Y. Who has access to urban vegetation? A spatial analysis of distributional green equity in 10 US cities. Landscape and Urban Planning. 2019;181:51–79.
10. Venter ZS, Shackleton CM, Van Staden F, Selomane O, Masterson VA. Green Apartheid: urban green infrastructure remains unequally distributed across income and race geographies in South Africa. Landscape and Urban Planning. 2020;203:103889.
11. Spotswood EN, Benjamin M, Stoneburner L, Wheeler MM, Beller EE, Balk D, et al. Nature inequity and higher COVID-19 case rates in less-green

neighbourhoods in the United States. Nature Sustainability. 2021;4(12):1092–8.

12. McDonald RI, Kroeger T, Boucher T, Wang L, Salem R. Planting Healthy Air: a global analysis of the role of urban trees in addressing particulate matter pollution and extreme heat. Arlington, VA: The Nature Conservancy; 2016.
13. Kroeger T, McDonald RI, Boucher T, Zhang P, Wang L. Where the people are: current trends and future potential targeted investments in urban trees for PM_{10} and temperature mitigation in 27 U.S. cities. Landscape and Urban Planning. 2018;177:227–40.
14. Alagona PS. The Accidental Ecosystem: people and wildlife in American cities. Berkeley, CA: University of California Press; 2022.
15. Iknayan KJ, Wheeler MM, Safran SM, Young JS, Spotswood EN. What makes urban parks good for California quail? Evaluating park suitability, species persistence, and the potential for reintroduction into a large urban national park. Journal of Applied Ecology. 2022;59(1):199–209.
16. Watts K, Whytock RC, Park KJ, Fuentes-Montemayor E, Macgregor NA, Duffield S, et al. Ecological time lags and the journey towards conservation success. Nature Ecology and Evolution. 2020;4(3): 304–11.
17. Lambert MR, Donihue CM. Urban biodiversity management using evolutionary tools. Nature Ecology and Evolution. 2020;4(7):903–10.
18. Mills LS, Soulé ME, Doak DF. The keystone-species concept in ecology and conservation. BioScience. 1993;43(4):219–24.
19. Jones CG, Lawton JH, Shachak M. Organisms as ecosystem engineers. In: Samson FB, Knopf FL (eds.) Ecosystem Management. New York, NY: Springer; 1994. p. 130–47.
20. Roberge JM, Angelstam PER. Usefulness of the umbrella species concept as a conservation tool. Conservation Biology. 2004;18(1):76–85.
21. Miles L, Newton AC, DeFries RS, Ravilious C, May I, Blyth S, et al. A global overview of the conservation status of tropical dry forests. Journal of Biogeography. 2006;33(3):491–505.
22. Sekercioglu CH. Increasing awareness of avian ecological function. Trends in Ecology and Evolution. 2006;21(8):464–71.
23. Bailey DR, Dittbrenner BJ, Yocom KP. Reintegrating the North American beaver (*Castor canadensis*) in the urban landscape. WIREs Water. 2019;6(1):e1323. Available from: https://onlinelibrary.wiley.com/doi/10.1002/wat2.1323.
24. Noss RF, Cartwright JM, Estes D, Witsell T, Elliott G, Adams D, et al. Improving species status assessments under the U.S. Endangered Species Act and implications for multispecies conservation challenges worldwide. Conservation Biology. 2021;35(6):1715–24.
25. Association of Fish and Wildlife Agencies, Teaming With Wildlife Committee, State Wildlife Action Plan (SWAP) Best Practices Working Group. Best Practices for State Wildlife Action Plans: voluntary guidance to states for revision and implementation [Internet]. Washington, DC: Association of Fish and Wildlife Agencies; 2012. 80 p. Available from: https://www.fishwildlife.org/application/files/3215/1856/0300/SWAP_Best_Practices_Report_Nov_2012.pdf
26. Sterrett SC, Katz RA, Brand AB, Fields WR, Dietrich AE, Hocking DJ, et al. Proactive management of amphibians: challenges and opportunities. Biological Conservation. 2019;236:404–10.
27. Anderson K. Tending the Wild: Native American knowledge and the management of California's natural resources. Berkeley, CA: University of California Press; 2005. 570 p.
28. NPS & Presidio Trust. Presidio of San Fransciso: Vegetation Management Plan and Environmental Assessment [Internet]. San Francisco, CA: Golden Gate National Recreational Area, United States Department of the Interior / National Park Service, The Presidio Trust; 2001 [cited 2022 Oct 30]. Available from: https://www.presidio.gov/presidio-trust/planning-internal/Shared%20Documents/Planning%20Documents/PLN-344-VmpEa_200112.pdf
29. Robison R, DiTomaso JM. Distribution and community associations of Cape ivy (*Delairea odorata*) in California. Madroño. 2010;57(2):85–94.
30. Hobbs RJ, Higgs E, Harris JA. Novel ecosystems: implications for conservation and restoration. Trends in Ecology and Evolution. 2009;24(11):599–605.
31. Pelton EM, Schultz CB, Jepsen SJ, Black SH, Crone EE. Western monarch population plummets: status, probable causes, and recommended conservation actions. Frontiers in Ecology and Evolution. 2019;7:258.
32. Seddon PJ. From reintroduction to assisted colonization: moving along the conservation translocation spectrum. Restoration Ecology. 2010;18(6):796–802.
33. Batson WG, Gordon IJ, Fletcher DB, Manning AD. Translocation tactics: a framework to support the IUCN Guidelines for wildlife translocations and improve the quality of applied methods. Journal of Applied Ecology. 2015;52(6):1598–607.

34. Sampson L, Riley JV, Carpenter AI. Applying IUCN reintroduction guidelines: an effective medium for raising public support prior to conducting a reintroduction project. Journal of Nature Conservation. 2020;58:125914.
35. Clinton SM, Hartman J, Macneale KH, Roy AH. Stream macroinvertebrate reintroductions: a cautionary approach for restored urban streams. Freshwater Science. 2022;41(3):507–20.
36. van Heezik Y, Seddon PJ. Animal reintroductions in peopled landscapes: moving towards urban-based species restorations in New Zealand. Pacific Conservation Biology. 2018;24(4):349–59.
37. Nagy C. 2004. The Eastern Screech Owl Reintroduction Program in Central Park, New York City: habitat, survival, and reproduction [Internet]. [New York]: Fordham University; 2004 [cited 2022 Oct 31]. Available from: https://www.proquest.com/docview/2203352345
38. Cook RP. Potential and limitations of herpetofaunal restoration in an urban landscape. In: Mitchell JC, Jung RE, Bartholomew B (eds.) Urban Herpetology. Salt Lake City, UT: Society for the Study of Amphibians and Reptiles; 2008. p. 465–78.
39. Suding KN. Toward an era of restoration in ecology: successes, failures, and opportunities ahead. Annual Reviews in Ecology, Evolution, and Systematics. 2011;42:465–87.
40. Klaus VH, Kiehl K. A conceptual framework for urban ecological restoration and rehabilitation. Basic and Applied Ecology. 2021;52:82–94.
41. Soanes K, Sievers M, Chee YE, Williams NS, Bhardwaj M, Marshall AJ, et al. Correcting common misconceptions to inspire conservation action in urban environments. Conservation Biology. 2019;33(2):300–6.
42. Buchholz TA, Madary DA, Bork D, Younos T. Stream restoration in urban environments: concept, design principles, and case studies of stream daylighting. Sustainable Water Management in Urban Environments. 2016;121–65.
43. Baho DL, Arnott D, Myrstad KD, Schneider SC, Moe TF. Rapid colonization of aquatic communities in an urban stream after daylighting. Restoration Ecology. 2021;29(5):e13394. Available from: https://onlinelibrary.wiley.com/doi/10.1111/rec.13394
44. American Rivers. Daylighting streams: breathing life into urban streams and communities [Internet]. Washington, DC: American Rivers; 2016 [cited 2022 Aug 29]. Available from: https://www.americanrivers.org/wp-content/uploads/2016/05/AmericanRivers_daylighting-streams-report.pdf
45. Niederer C, Weiss SB, Stringer L. Identifying practical, small-scale disturbance to restore habitat for an endangered annual forb. California Fish and Game. 2014;100:61–78.
46. An Z, Chen Q, Li J. Ecological strategies of urban ecological parks – a case of Bishan Ang Mo Kio Park and Kallang River in Singapore. E3S Web of Conferences. 2020;194:05060.
47. Davis JL, Currin CA, O'Brien C, Raffenburg C, Davis A. Living shorelines: coastal resilience with a blue carbon benefit. PLoS One. 2015;10(11):e0142595.
48. Safak I, Norby PL, Dix N, Grizzle RE, Southwell M, Veenstra JJ, et al. Coupling breakwalls with oyster restoration structures enhances living shoreline performance along energetic shorelines. Ecological Engineering. 2020;158:106071.
49. Smith CS, Rudd ME, Gittman RK, Melvin EC, Patterson VS, Renzi JJ, et al. Coming to terms with living shorelines: a scoping review of novel restoration strategies for shoreline protection. Frontiers in Marine Science. 2020;7:434.
50. Sawyer AC, Toft JD, Cordell JR. Seawall as salmon habitat: eco-engineering improves the distribution and foraging of juvenile Pacific salmon. Ecological Engineering. 2020;151:105856.
51. Blind EB, Voss BL, Osborn SK, Barker LR. El Presidio de San Francisco: at the edge of empire. Historical Archaeology. 2004(3):135–49.
52. Roberts J, Chick A, Oswald L, Thompson P. Effect of carp, *Cyprinus carpio* L., an exotic benthivorous fish, on aquatic plants and water quality in experimental ponds. Marine and Freshwater Research. 1995;46(8):1171–80.
53. Ismail NS, Dodd H, Sassoubre LM, Horne AJ, Boehm AB, Luthy RG. Improvement of urban lake water quality by removal of *Escherichia coli* through the action of the bivalve *Anodonta californiensis*. Environmental Science and Technology. 2015;49(3):1664–72.
54. Boal CW, Dykstra CR. Urban Raptors: ecology and conservation of birds of prey in cities. Washington, DC: Island Press; 2018.
55. James Reynolds S, Ibáñez-Álamo JD, Sumasgutner P, Mainwaring MC. Urbanisation and nest building in birds: a review of threats and opportunities. Journal of Ornithology. 2019;160(3):841–60.
56. Filazzola A, Shrestha N, MacIvor JS. The contribution of constructed green infrastructure to urban biodiversity: a synthesis and meta-analysis. Journal of Applied Ecology. 2019;56(9):2131–43.
57. Hale R, Swearer SE, Sievers M, Coleman R. Balancing biodiversity outcomes and pollution management in urban stormwater treatment wetlands. Journal of Environmental Management. 2019;233:302–7.
58. Guderyahn LB, Smithers AP, Mims MC. Assessing habitat requirements of pond-breeding amphibians in

a highly urbanized landscape: implications for management. Urban Ecosystems. 2016;19(4):1801–21.

59. Ostergaard EC, Richter KO, West SD. Amphibian use of stormwater ponds in the Puget Lowlands of Washington, USA. In: Mitchell JC, Jung RE, Bartholomew B (eds.) Urban Herpetology. Salt Lake City, UT: Society for the Study of Amphibians and Reptiles; 2008. p. 259–70.
60. Stagoll K, Lindenmayer DB, Knight E, Fischer J, Manning AD. Large trees are keystone structures in urban parks. Conservation Letters. 2012;5(2):115–22.
61. McKinley DC, Miller-Rushing AJ, Ballard HL, Bonney R, Brown H, Cook-Patton SC, et al. Citizen science can improve conservation science, natural resource management, and environmental protection. Biological Conservation. 2017;208:15–28.
62. Williams CR, Burnell SM, Rogers M, Flies EJ, Baldock KL. Nature-based citizen science as a mechanism to improve human health in urban areas. International Journal of Environmental Research and Public Health. 2021;19(1):68.
63. McMillen H, Campbell LK, Svendsen ES, Reynolds R. Recognizing stewardship practices as indicators of social resilience: in living memorials and in a community garden. Sustainability. 2016;8(8):775.
64. Mumaw L. Transforming urban gardeners into land stewards. Journal of Environmental Psychology. 2017;52:92–103.
65. Jones MS, Teel TL, Solomon J, Weiss J. Evolving systems of pro-environmental behavior among wildscape gardeners. Landscape and Urban Planning. 2021;207:104018.
66. Spotswood E, Grossinger R, Hagerty S, Bazo M, Benjamin M, Beller E, et al. Making Nature's City: a science-based framework for building urban biodiversity. Richmond, CA: San Francisco Estuary Institute Publication #947; 2019.
67. Knight-Lenihan S. Achieving biodiversity net gain in a neoliberal economy: the case of England. Ambio. 2020;49(12):2052–60.
68. Pierce JR, Barton MA, Tan MMJ, Oertel G, Halder MD, Lopez-Guijosa PA, et al. Actions, indicators, and outputs in urban biodiversity plans: a multinational analysis of city practice. PloS One. 2020;15(7):e0235773.
69. Santa Clara Valley Open Space Authority, Conservation Biology Institute. Coyote Valley Landscape Linkage: a vision for a resilient, multi-benefit landscape. San Jose, CA: Santa Clara Valley Open Space Authority; 2017.
70. McDonald RI, Hamann M, Simkin R, Walsh B. Nature in the Urban Century: a global assessment of where and how to conserve nature for biodiversity and human wellbeing. Arlington, VA: The Nature Conservancy; 2018.
71. Beninde J, Veith M, Hochkirch A. Biodiversity in cities needs space: a meta-analysis of factors determining intra-urban biodiversity variation. Ecology Letters. 2015;18(6):581–92.
72. Pratt-Bergstrom B. When Mountain Lions are Neighbors: people and wildlife working it out in California. Berkeley, CA: Heyday; 2014.
73. Greer M. Coyote Management in San Francisco [Master's Thesis]. [San Francisco, CA]: University of San Francisco; 2021.
74. Battin J. When good animals love bad habitats: ecological traps and the conservation of animal populations. Conservation Biology. 2004;18(6):1482–91.
75. Zuñiga-Palacios J, Zuria I, Castellanos I, Lara C, Sánchez-Rojas G. What do we know (and need to know) about the role of urban habitats as ecological traps? Systematic review and meta-analysis. Science of the Total Environment. 2021;780: 146559.
76. Schell CJ, Stanton LA, Young JK, Angeloni LM, Lambert JE, Breck SW, et al. The evolutionary consequences of human–wildlife conflict in cities. Evolutionary Applications. 2021;14(1): 178–97.
77. Wolf KL, Lam ST, McKeen JK, Richardson GR, van den Bosch M, Bardekjian AC. Urban trees and human health: a scoping review. International Journal of Environmental Research and Public Health. 2020;17(12):4371.

CHAPTER 14

Making Nature's City

An Applied Science Framework to Guide Evaluation and Planning for Urban Biodiversity Conservation

Bronwen Stanford[‡], Kelly Ikyanan[‡], Robin Grossinger[‡,§], Erin Beller[¶], Matthew Benjamin[‡,#], J. Letitia Grenier[‡], Micaela Bazo[‡,§], Nicole Heller[||], Myla F. J. Aronson[‡‡], Alexander Felson[§§], Peter Groffman[¶¶,##], and Erica Spotswood[‡,§]

Introduction

In many ways, cities are entirely novel environments: they are filled with human structures, concrete, fast-moving cars, and highly managed plant communities and soils. Yet, cities are simply another type of landscape for plants, animals, and invertebrates to navigate, made up of areas rich in resources interspersed with inhospitable terrain. The field of urban ecology explores the ways that species respond to urban environments and applies ecological tools to understand how species interact with these highly modified landscapes. Using these tools, scientists can attempt to "see" cities as species do and use those understandings to improve the ability of species to survive or even thrive in urban spaces. In particular, urban ecologists have found that urban design strongly influences urban biodiversity: the size, quality, and connectedness of urban green spaces affect how biodiverse a city can be.

‡ San Francisco Estuary Institute, USA
§ Second Nature Ecology and Design, USA
¶ Google Inc., USA
# City of San Jose, USA
|| Carnegie Museum of Natural History, USA
‡‡ Rutgers University, USA
§§ University of Melbourne, Australia
¶¶ City University of New York, USA
Cary Institute of Ecosystem Studies, USA

We distilled key findings from urban ecological research into seven elements that can be used to identify opportunities and align planning efforts with the needs of species: patch size, connections, matrix quality, habitat diversity, native vegetation, special resources, and stewardship and management (Figure 14.1). The seven elements are presented with additional examples and illustrations in a companion report *Making Nature's City: a science-based framework for building biodiversity* (1). Together, these elements can help planners or anyone with the capacity to modify urban habitat understand the ways that species use urban spaces, so that design can proactively support ecological function within urban ecosystems. To successfully increase urban biodiversity, efforts should include a broad array of partners, including city and county staff, individuals, nonprofits, landscape architects, neighborhood associations, and backyard gardeners.

This urban biodiversity analysis should be only one piece of a larger assessment that incorporates human community needs, environmental justice concerns, and equity and vulnerability assessments. In addition to the many benefits for biodiversity of strategic urban greening, the addition of green space has a host of benefits for people, from mental health to climate resilience (2). An urban biodiversity plan needs to reflect local values and preferences both to ensure long-term success and

Bronwen Stanford et al., *Making Nature's City*. In: *Urban Biodiversity and Equity*. Edited by: Max R. Lambert & Christopher J. Schell, Oxford University Press.
 DOI: 10.1093/oso/9780198877271.003.0014

public participation, and to ensure that projects provide community benefits. When integrated with additional information about the human landscape, an understanding of the needs of biodiversity can help build an urban biodiversity conservation plan that supports both human and nonhuman residents.

Social-ecological applications

Human needs and benefits are not explicitly incorporated in this framework, although these can be a strong motivation for urban design supporting biodiversity. Nature can help limit flooding, mitigate water pollution, reduce urban heat, and capture carbon from the atmosphere, increasing urban resilience to climate change and extreme events (2). While some of these services can be achieved in biotically homogenized green spaces, biodiverse nature can substantially improve ecosystem service provision (3). In addition, integration of biodiversity into cities brings people closer to nature, enhancing their sense of and connection to place.

However, in some cases design for native biodiversity can conflict with design for human needs. For example, trees provide shade and cooling that benefit people even in arid climates that would naturally have few trees. In some areas, a native-based planting palette may only include a few dozen tree species, which can raise concerns about susceptibility to disease. Residents may prefer monocultures and exotic species such as turf grass, or may insist on manicured spaces and insect control that conflict with biodiversity goals (4).[1]

Societal needs and a focus on equity should help inform and guide priorities for urban biodiversity design. For example, many historically redlined communities have very low percentages of tree canopy (5), and prioritizing efforts to increase connectivity or introduce patches of green space in these communities could have large societal and justice benefits. By contrast, an effort to expand a large, high-quality green space in an affluent neighborhood may have high biodiversity benefits, but serve to increase social inequities. A failure to consider the characteristics of human communities in planning and prioritization can lead to projects that reinforce existing inequities (6).

Due to the many benefits of urban greening, an increase in green space can result in gentrification and displacement if not done strategically and in collaboration with the local community. Improvements pursued without local engagement may also fail to meet the needs of local residents and be disused. Because cities are first and foremost places for people, any plan to manage cities for biodiversity needs to consider potential impacts on human communities through inclusive engagement with residents.

Seven elements to support urban biodiversity

We have identified seven elements to highlight distinct aspects of the urban environment that city planning and design can influence (Figure 14.1). The elements interact with and support each other: for example, improved connections between large patches can make cities more liveable for species that are primarily reliant on these patches of habitat. The elements work together at both landscape and site scales to meet the needs of a broad variety of species (Figure 14.2): three of the elements focus on the landscape scale (patch size, connections, and matrix quality), three address conditions within a site (native vegetation, special resources, and management), and one includes considerations at both landscape and site scales (habitat diversity). We emphasize the importance of explicitly considering biodiversity needs at both site and landscape scales and focusing on features that will be broadly protective of many species, rather than tailored to individual species.[2]

Several of the elements distinguish habitat patches from the urban matrix. We define habitat patches as larger green spaces (>1 ha) that provide resources for species, such as natural habitat

[1] See Larson's and Brown's chapter which explores human motivations for urban nature.

[2] Spotswood et al.'s and Avilés-Rodriguez et al.'s chapters explore species-specific approaches.

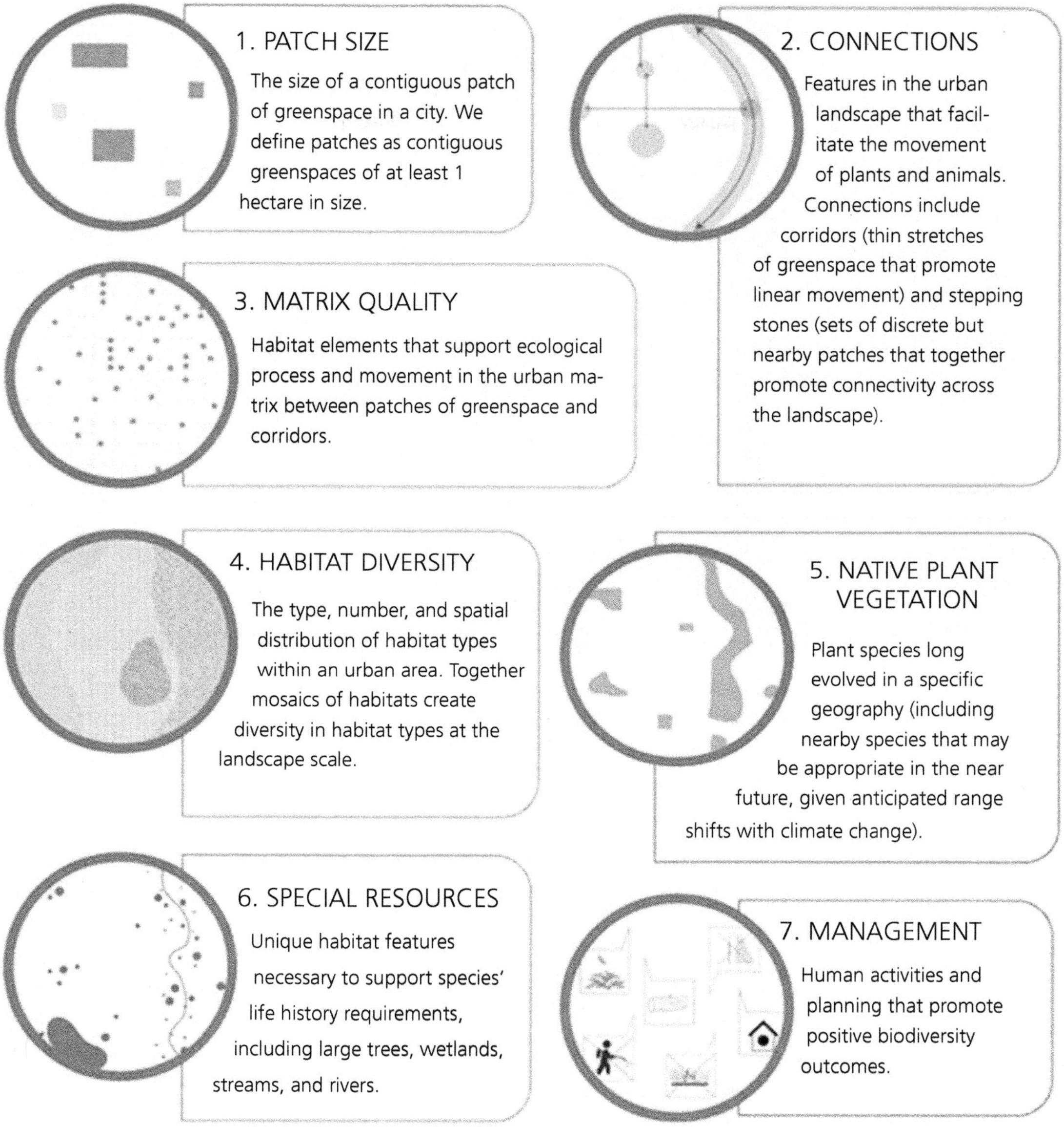

Figure 14.1 Summary of the seven elements to support urban biodiversity. These seven elements seek to distill broadly understood ecological concepts into practical guidance for managers.

remnants, parks, and gardens. These spaces are typically covered with either vegetation or bare ground. Habitat patches are critical to the survival of many urban species. Everything else surrounding habitat patches, including both hardscape areas and smaller areas of green space, is considered the urban matrix. Many species travel through the matrix to reach patches of habitat, and may use features such as street trees, green roofs, or pocket parks to navigate within the city.

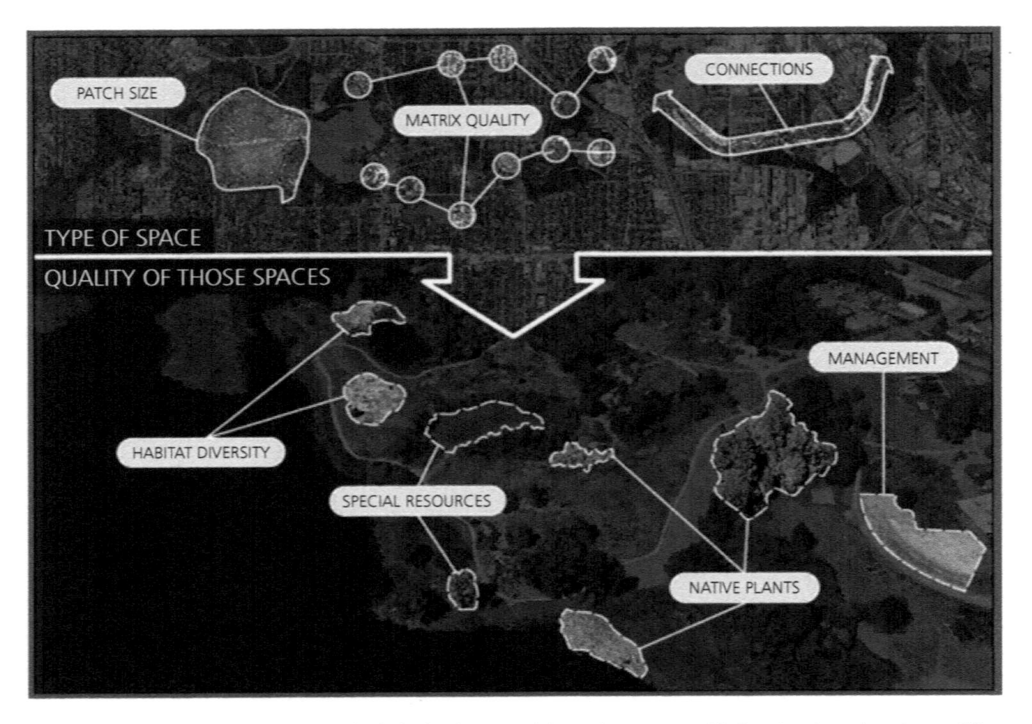

Figure 14.2 The seven elements work together at both the landscape and site scales to support biodiversity, shown here in two different landscapes. The top panel features San Antonio, TX, and highlights a large, vegetated area as a patch, a long linear riparian corridor as a connection, and portions of the cityscape as examples of the urban matrix. The bottom panel is zoomed in on Lake Washington in Seattle, WA, and contrasts grassland with shrubland to show habitat diversity, a large tree and small pond as special resources, varied vegetation types for native plants, and a mown field for management.
Imagery from Google Earth.

Patch size

One of the most predictable patterns in ecology is the relationship between the size of a patch and the number of species found in it (7). Larger patches tend to have more microhabitats, creating variation that broadens the number of species that can find adequate resources for survival, and larger patches are also able to buffer species from the area surrounding the patch (8). Position and shape also matter: similarly sized patches can support more biodiversity if, in addition to being large, they are less isolated (9), have less edge compared to core area (10), and if the overall habitat quality is higher (11).

Patches of different sizes are likely to perform different functions in the landscape. Small patches are commonly defined in the literature as a contiguous green space greater than 1 hectare (ha) in area, but patches smaller than 4.4 ha typically have limited diversity of even urban-adapted species (12). Large patches (>53 ha) can function as biodiversity hubs, supporting species with lower tolerance for urbanization.

Most cities will be extremely constrained in their ability to create new large habitat patches, but may have more opportunities to create or improve smaller habitat patches. Even small patches can provide benefits and improve the ability of animals to move through the urban landscape, so we recommend that cities create patches of a variety of different sizes to support species with varied tolerance of urbanization (13). Due to the often harsh contrast between conditions inside and outside of habitat patches in urban areas, it may be important to include a buffer zone around core habitat features, particularly for urban-avoiding species.

Connections

Movement across the landscape can be an essential part of daily, seasonal, or lifetime cycles for many species. For example, animals may need to move between where they roost and where they feed within a day, move seasonally to track food supplies, or move to new areas to establish a territory and reproduce successfully. Connections between patches, and between critical habitat features within patches, enable individuals to move more freely.

Continuous corridors of native habitat between large habitat patches are the most effective way of enhancing connectivity in the urban landscape (12). Corridors can facilitate movement within a city and connect fragmented habitats to larger open spaces (14). Urban landscapes with high connectivity can support higher numbers of species (14), including those species that need connections to tolerate urbanization (15). In general, the wider and more continuous a vegetated corridor is, the more ecological function and species richness it will support.

Communities seeking to enhance connectivity in cities must often rely on features that already connect the landscape, such as riparian corridors, repurposed infrastructure corridors, or urban rights-of-way. In particular, protecting and restoring riparian corridors can be a strategy for maintaining natural networks of movement across the landscape, protecting aquatic habitat, and boosting climate resilience (16). In areas without remnant natural connections, the edges of streets and repurposed transit corridors can be planted with native species to provide connectivity for small mammals, insects, and birds (17).

Cities may also consider the creation of "stepping stones"—patches of green space positioned so that species can use them to move between larger habitat patches, effectively enhancing connectivity across the landscape. Stepping stones can include vacant lots, backyards, green roofs, and campuses, and are particularly valuable for species that can fly (18). Although generally less effective than a continuous corridor (12), these small features can improve connectivity where larger patches or continuous corridors are not possible.

Connectivity can also be enhanced through the removal or softening of barriers that impede movement. Common examples of barriers in urban areas include roads and other impervious surfaces (19), buildings (20), outdoor lighting (21), and culverted stream reaches or gaps in canopy (22,23). These barriers could be softened by introducing small green spaces within large areas of impervious cover, including vegetated medians and green walls or roofs. The impact of outdoor lighting on migrating birds can be minimized by turning off unnecessary lights and pointing lights downwards.

Matrix quality

In cities, most wildlife species navigate both habitat patches and the mix of buildings, trees, streets, yards, roads, and medians that make up the urban matrix. The contrast in habitat quality between habitat patches and the matrix can be stark, serving as a barrier to movement. In the absence of a corridor, improving the habitat quality of the matrix can increase connectivity by enabling organisms to move more easily from one patch to another (24). Higher matrix quality is also linked to higher species richness in the matrix itself (25,26).

Land managers and stewards can improve habitat quality in the matrix by adding trees and other vegetation along streets and in private yards to provide cover and food sources for wildlife (27,28). Matrix quality can also be improved by adding structural complexity, such as shrubs and small trees that create vegetation layers between trees and herbaceous cover (25,26). (See "Habitat diversity.")

Land managers can also consider creating small green spaces within the matrix, such as pocket parks or landscaping that incorporates locally appropriate species and habitat structure. However, very small patches can also expose species to high risk of predation and other threats, and the best size and configuration to ensure species benefit is still not well understood (18). (See more discussion of ecological traps under "Special resources.")

Habitat diversity

Habitat diversity represents the variation in both the types of habitat patches across a landscape and the heterogeneity within a habitat patch. Natural habitats are defined by assemblages of plants that tend to occur together, forming predictable physical structures and resources that are used by wildlife. At regional or landscape scales, different assemblages of plants and animals form distinct habitat types, typically following topographic and hydrologic gradients. This regional habitat diversity results in a greater variety of resources, boosting ecological resilience as species can access resources in multiple types of habitats as they move across the landscape (29,30).

In urban landscapes, much of this natural spatial variation in habitat patch types and available resources has been lost due to development and may need to be intentionally reintroduced. To organize urban habitats around existing and historical gradients, one approach can be to first establish habitat zones. Habitat zones are a spatially explicit planning tool that divide a region into areas that would be most appropriate for different vegetation assemblages based on existing gradients and constraints (31). These zones can be constructed using historical ecology information as a template and then modified as needed to accommodate both the urban landscape and climate change projections (30). For example, habitat zones might differentiate locations that are suitable for wetlands or bioswales from oak savanna habitats (see Putting Theory to Practice below). Managing by habitat zone can help increase the diversity of habitats at the city scale, while also providing coherence within habitat types (31).

Habitat complexity within a patch also affects habitat quality. In urban areas, habitat attributes associated with greater wildlife diversity include the variety of plants, the presence of woody debris and leaf litter, and a multilayered canopy that includes combinations of trees, shrubs, and herbaceous vegetation (12,32). For many invertebrates, or species in more arid environments, bare ground can also be an important type of "cover" (33). Designing urban habitats around naturally occurring species assemblages can help wildlife access needed resources.

Native vegetation

Within a site, one easy way to boost native urban biodiversity is to plant a diverse and locally appropriate mix of native plants, defined as plant species that have evolved over time in the local geography. Urban vegetation is often intensively managed, so cities can actively select plant species, including native species that may be locally rare or threatened (34). Urban green spaces with a greater diversity of native plants can support a greater diversity of native wildlife species (35,36). Many insects have strongly specialized relationships with host plants used for reproduction and larval life-stages, such as Cynipid gall-forming wasps and California native oak trees (37). These insects can form the basis of food webs that cascade upwards, enhancing resources for other native wildlife (38). Deep coevolutionary relationships between local wildlife and native plants often mean that non-native plants do not necessarily provide the same resources to animals (39).

In urban areas, a purely native planting palette is often impractical. For example, in desert climates, native trees may be naturally rare or largely absent, and urban trees (likely non-native) can be desirable for their many human benefits. Exotic trees can also benefit native wildlife: exotic species can provide food or other resources (40), and, in some cases, exotic plant species may be an important component of urban vegetation, due to tolerance of stressful conditions, ease of maintenance, or the abundance of resources they provide (36,41). However, a reliance on exotics requires detailed understanding of individual species needs and interactions that is often lacking. In many cases, supporting native plant assemblages is a proxy for supporting important species interactions that may not yet be well understood (36). Use of native plants also avoids the homogenizing effect of widespread planting of non-native plants, and supports the expression of multiple distinct habitat types across urban landscapes (42,43).

Any amount of native vegetation may help native species, but higher percentages are often more beneficial. Some research has shown that 48% native plant cover in landscaping leads to higher native bee abundance (44), whereas others found wildlife benefits from 30% native species or 30% native tree cover in landscaping (35,45). In other research, landscaping with over 70% native biomass (shrubs and trees) was found to sustain native bird and insect populations, maintaining intact food webs (38). Higher proportions of native plants better reflect natural habitat composition and are likely to provide more value to native wildlife.

Special resources

Large trees, water features, and riparian areas support species that may not be able to exist elsewhere in the urban landscape. To represent their importance, we group these features together as "special resources," another site-scale consideration. Large trees often serve as the foundation for a local ecosystem on a small scale, providing large canopies and woody debris for shade and cover; cavities for nesting; flowers and fruits for forage; and litter that fertilizes nearby soil (46). Even small water features can provide habitat for insects and amphibians (47), as well as birds and mammals. Vegetated riparian corridors can simultaneously improve connectivity for terrestrial species (48), support a unique assemblage of more hydrophilic species, and buffer aquatic species from some of the stresses of urbanization (49,50). Where special resources are not present, man-made analog features can sometimes play a similar role (51). For example, nest boxes can effectively serve as substitutes for tree cavities, fulfilling a specialized habitat requirement for cavity-nesting birds (52). Human-created water features like ponds or bioswales can improve connectivity between natural aquatic habitats and provide water access.

Wildlife may require the presence of one or all of these features to survive in an urban landscape. For example, access to a reliable water source and an appropriate tree cavity for nesting limits the suitability of a site for some species of breeding birds. As a result, both the siting of patches and design within a patch should consider these special resources, and seek to enhance access to these resources for native species.

Stewardship of either natural or man-made resources in urban areas should work to ensure that these features do not function as ecological traps, in which animals prefer habitats that reduce fitness. Ecological traps occur when cues animals are using for habitat selection become decoupled from habitat quality, a pattern that can be common in highly altered urban environments (53). Recent research has documented several cases of ecological traps in urban areas, including nesting birds that preferred large artificial cavities even though these resulted in lower fledging success (53), and the inability of amphibians to select against stormwater wetlands with high contaminant loads, even though the contaminants lower survival (54). In many cases, reduction of the risk that a habitat feature will function as an ecological trap may require active monitoring and efforts to reduce the attractiveness of features that function as traps, for example by removing vegetation along roads and highly polluted sites. Efforts to improve the quality of habitats that function as ecological traps can also be beneficial, such as reducing pollution loads or managing predators (55).

Management

The final site-scale element, management or stewardship, incorporates many different types of actions that people can take to restore ecological function and support biodiversity, some of which have been mentioned as part of the six prior elements. Some of these actions require reducing management intensity. For example, replacing lawns with more drought-tolerant native vegetation that requires fewer inputs can help support biodiversity while reducing other impacts, such as water demand and pesticide use (56). Similarly, reducing the frequency of mowing and pruning, and leaving fallen leaves, logs, and fruits on the ground can increase available food and cover for local wildlife (42). Dead trees and branches support cavity-nesting animals in the urban environment (57), and these resources are particularly important given their rarity in urban landscapes.

Other changes may require more active management of invasive and exotic species. Urban-tolerant invasive species may have negative impacts on native biodiversity through predation (58), competition (59), and disease exposure (60), requiring active management of these populations. For example, domestic cats can be prolific predators that limit bird populations (61). Management actions to control pests, pathogens, and invasive species can help reduce pressures on native species.

Finally, people can change their behaviors to reduce abiotic stressors such as chemical pollution, light pollution, and noise. Limiting or eliminating use of pesticides, in particular, can benefit biodiversity (32). Street and building lighting can be reduced in areas important to wildlife species (62), and building windows can be designed, or films added, to reduce bird collisions (63). Directly reducing the effects of road barriers through overpasses, underpasses, and culverts can reduce wildlife mortality and improve population connectivity (64), as can reducing vehicle speed and planting trees near roadways (65).

Global perspectives

The seven elements for biodiversity support presented here were developed through an extensive global literature review. As a result, the principles and metrics described here can be adapted to cities worldwide. Tropical areas and the Global South are underrepresented in our review due to lack of accessible published research on urban ecology in those geographies, so targets and specific examples need to be developed for a broader array of climates and social contexts. Our experience is largely from the US, and particularly California.

Cities will need to adapt each of these elements to better suit their local ecology, climate, and cultural context. For example, management priorities will depend on the local species, building density and ownership structures, and current conditions in the city (e.g., pesticide use, feral animal populations, light and noise pollution). High-density cities may have few opportunities to enhance ground cover or create new parks, and may instead explore opportunities for on-structure greening, whereas sprawling low-density cities may work with residents to promote biodiversity in private backyards.

Variation in natural conditions also changes the appropriate targets for cities in different bioregions. Naturally occurring diversity of habitat and vegetation structures in a desert, temperate rainforest, and tropical climate look dramatically different (Figure 14.3). Patch sizes and corridor widths that sustain healthy populations vary by species, and studies of focal species needs could be used to calibrate these metrics to meet local needs.

Developing appropriate examples that are representative of distinct cultural and ecological contexts that cities can adapt to their needs will be critical to successful implementation of these concepts. However, the ecological principles of connections, patches, diversity of habitat, special resources, and the importance of native vegetation hold true worldwide.

Putting theory to practice

These seven elements can help cities identify opportunities to improve support for biodiversity in the urban landscape from the neighborhood to the city scale. Landscape analysis can help planners identify

Figure 14.3 Contrasting images of native vegetation structure in Saguaro National Park and Olympic National Park, USA, and the Great Rift Valley, Kenya.

Images courtesy of Christoph Von Gellhorn and author Bronwen Stanford.

opportunities to protect and enhance biodiversity at the site or neighborhood scale, develop plans for these areas, and prioritize interventions. For example, land acquisition can create new parks in park-poor areas and can be used to fill gaps in corridors. Existing parks can be enhanced by identifying and cultivating appropriate habitat types, changing management practices, and protecting special resources. Corridors can be restored and widened to minimize disturbance and increase connectivity. Integrating across all seven elements of the framework ensures that the needs of multiple species across scales will be addressed. The online companion Making Nature's City Toolkit provides examples of analyses that can be performed to evaluate each of the seven elements, along with suggested datasets and case study examples (https://www.makingnaturescity.org/).

Real-world implementation of this framework also needs to engage local residents. The human landscape should inform prioritization of efforts: for example, many marginalized human communities lack green spaces, so the addition of new green spaces in those communities should be prioritized for their human benefits, even if there would be limited benefits for biodiversity. To provide a real benefit to communities, meaningful engagement to determine local needs and priorities is essential. In addition to identifying needs and opportunities, practitioners will need to build institutional capacity and public support to implement change on the ground. Through this engagement, communities can also provide information on the type of intervention that would be most beneficial. In most cases, a mix of city-led actions and public–private partnerships, incentive programs, and public outreach and engagement will be required to meet urban biodiversity goals.

To illustrate how these seven elements can be used to assess biodiversity needs, we provide an example analysis in Sunnyvale, CA, USA. Sunnyvale is at the southern end of the San Francisco Bay and occupies a key transitional zone from salt marshes at its northern end to the foothills of the Santa Cruz Mountains in the south. Sunnyvale is largely urbanized, with extensive office parks and suburban development. Our analysis focused on Moffett Park, a 5 sq km planning district in Sunnyvale adjacent to the San Francisco Bay that is dominated by office parks and impervious surfaces, with surface parking covering ~33% of the area. Within the project area habitat cover was historically tidal marsh and wet meadow, with patches of oak savanna (66). As an update to the Specific Plan, the City of Sunnyvale is creating a long-range vision for Moffett Park as an ecological innovation district. We assessed existing and potential features for biodiversity support in Moffett Park to help develop metrics, goals, and policies for the updated Specific Plan.

To perform the analysis, we assembled spatial datasets describing historical habitat cover, present-day location of green spaces, city trees, and stream and wetland locations (Figure 14.4 and Case Study). Using these datasets, we performed a series of simple analyses to summarize current conditions, including assessing the distribution of patch sizes, landscape fragmentation and connectivity, percent tree cover and native tree cover, and percent of streams with vegetated buffers (Table 14.1). To extend this analysis, socioeconomic data could be layered on top of the biophysical datasets. For example, a map of human communities most vulnerable to heat stress could be overlaid with a map of the urban canopy so that areas with high vulnerability and low tree canopy could be prioritized for urban tree planting programs.

Through this approach, we were able to generate a list of priority actions that consider both landscape- and site-scale needs and should support many different species. Cities can then use these suggested actions and identified gaps to develop goals that will benefit biodiversity. In this way, the seven elements can provide a science-based foundation that informs city planning and can help select coordinated suites of actions.

Once this type of planning document is approved by a city, a variety of biodiversity conservation tools[3] can be used to improve conditions for urban species. For example, ecological restoration, community engagement, and stewardship activities can select plant palettes based on habitat zones, and novel financing pathways can be used to fund

[3] Spotswood et al.'s chapter starts assembling a toolbox of approaches to biodiversity conservation in cities.

Table 14.1 Example of an analysis of the seven elements of this urban biodiversity framework in Sunnyvale, CA (for more details see (66)). The analysis is a recommendation and has not been finalized or approved by the city council.

Urban biodiversity framework element	Data source	Scale	Example analysis	Ecological assets	Identified gap	Suggested actions
Patch size	Protected Areas Database, aerial imagery	District	Assess number of patches by area	Extensive bayland habitat on northern edge forms regional habitat hub	Few additional patches > 4.4 ha	Prioritize creation of larger patches to support a broader suite of species. Create >18 ha additional high value green space, including some patches >4.4 ha in size
Connections	Aerial imagery	District	Stream cover analysis; landscape connectivity analysis	Presence of several streams with surface channels, regionally connected baylands	Low-quality riparian habitat, lack of connectivity from baylands to uplands	Widen vegetated riparian corridors to >50 m to improve water filtration and support wildlife movement
Matrix quality	Canopy cover	District	Spatial assessment of % canopy cover by neighborhood	Existing tree cover 9% overall in study area, higher in east	Increase tree cover to match regional targets	Plant street trees to match locally appropriate canopy cover of >25%
Habitat diversity	Historical habitat mapping, soils maps	District	Identify habitat zones, assess which habitat types are missing from the landscape	Historical ecology mapping exists, saltwater marshes on northern edge form important and distinct habitat	Freshwater wetland habitats have been almost entirely removed	Match restoration and revegetation to the underlying environmental gradients to create a system with a variety of habitat types; seek opportunities to restore freshwater wetland habitats
Native vegetation	Local vegetation inventories	Site scale	Assess % native vegetation		City trees are 97.7% non-native, and only 3 native street tree species are present	Encourage private landowners to prioritize native species, plant appropriate native street trees (or exotics with similar functions), select plant palettes with >80% native species
Special resources	Municipal tree inventory, national hydrography dataset	District	Map locations of large trees, streams, and wetlands	Row of large trees along northern edge of study area, streams, baylands	26 large trees (>32 in. diameter) in municipal inventory within study area; lack of riparian vegetation limits value of streams	Protect large trees; design patches around large trees and stream/wetland features
Management		District and site scale	Timing of bird migrations			Wildlife-friendly building and lighting, particularly during migration periods, raptor perch deterrents on structures near wetlands

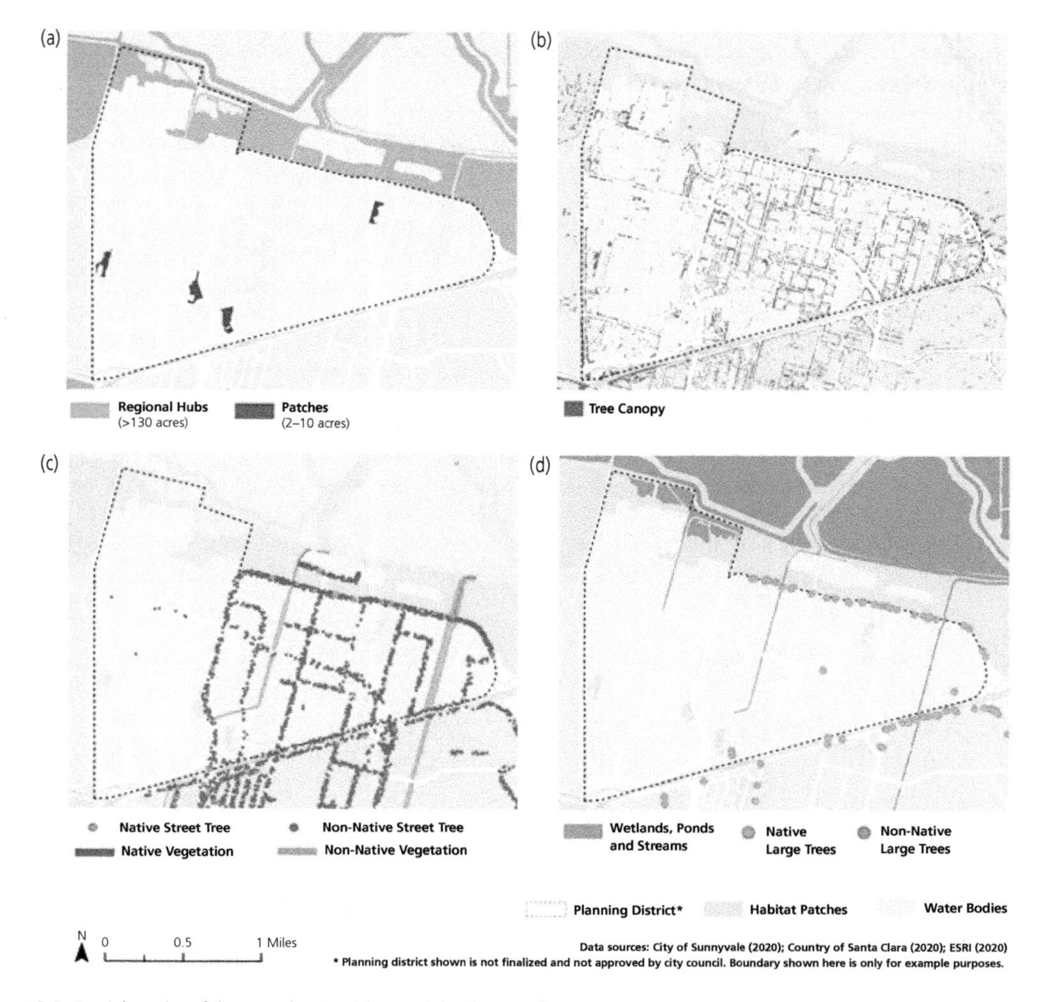

Figure 14.4 Spatial overlay of datasets showing (a) spatial distribution of patches; (b) tree canopy cover; (c) street trees, both native and non-native; and (d) location of special resources, in this case large trees, creeks, wetlands, and ponds.
Figure adapted from our report: Iknayan et al. (66).

projects to improve connectivity around areas identified for new green spaces. By evaluating each of the seven elements and thinking about the city as a type of ecological space, managers can improve coordination of actions and more effectively implement biodiversity conservation tools.

Conclusions

Drawing on research from cities around the world, this urban biodiversity framework provides an accessible science-based foundation for biodiversity planning and design. Planning that considers each of these elements can help align the actions that shape our cities to be more supportive of local ecosystems and species. We envision these seven elements informing actions at multiple scales, from regional open space or climate adaptation plans to backyard landscape design.

This approach is intended as a unifying scientific guide for city planners, corporations, engineers, designers, concerned citizens, and others to maximize the value of their collective efforts to create healthy, resilient, and ecologically vibrant cities.

Case study Habitat zones, Sunnyvale, CA

To improve habitat diversity at the landscape scale, we identified distinct habitat zones using a combination of historical vegetation patterns, present-day native species, and present-day land use (Figure 14.5). These zones are each best suited to support a different assemblage of native species and ecological functions. Historically, wet meadows of short flowering plants dominated the Moffett Park inland area, with patches of willow grove and tidal marshes occupying its shoreline. However, land development, including grading, soil compaction, fill, and some reduction of groundwater levels (from formerly artesian conditions), has profoundly altered the landscape, making it now more suitable for oak woodland habitats, valley oak groves, willow groves, riparian habitat, or freshwater marsh (67).

In this altered landscape, habitat zones can guide plant selection, help focus attention on unique features, and maintain a diversity of habitat types across the region. For this area, the zones highlight a need to maintain or enhance willow groves, freshwater wetlands, and riparian habitats, which are relatively rare but can support plants and animals otherwise absent from the landscape. The zones also distinguish three dryland habitats: coastal grassland, valley oak mix, and coast live oak mix. Moffett Park retains relatively high groundwater, so it has the potential to support healthy and fast-growing communities of native trees of high ecological, cultural, and aesthetic value, such as arroyo willow, valley oak, Fremont cottonwood, white alder, box elder, and western sycamore. Since, not all areas are likely to support all of these species, the habitat zones can be used to guide planting decisions. By matching planting to underlying physical gradients, city managers can reintroduce local heterogeneity while also increasing plant viability.

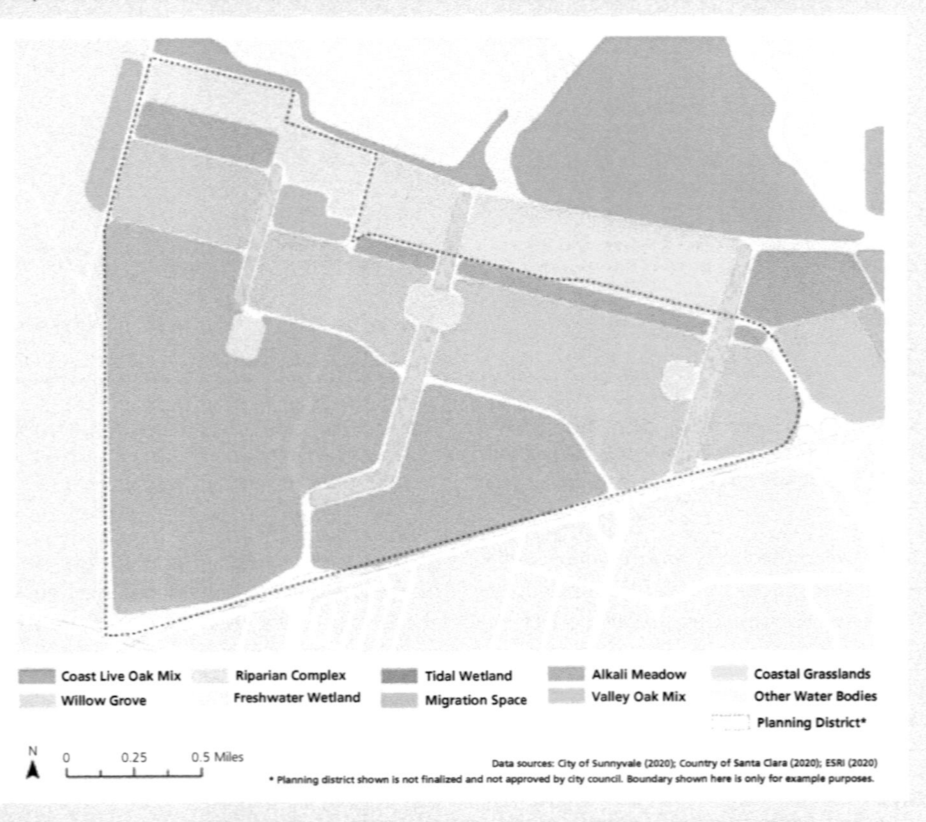

Figure 14.5 Habitat zones in Moffett Park. Habitat zone types have been selected based on a combination of historical and contemporary data and future projections, representing the species most likely to succeed and support local ecology. Each of these zones could be used to guide selection of a plant palette and management strategy to both match underlying physical gradients and support heterogeneity of resources and habitat types at the landscape scale.

Figure adapted from our report: Iknayan et al. (66).

As the body of research in urban environments continues to grow, further work must bring the results of this research into practice through synthesis and translation efforts. Our vision is that urban biodiversity planning can unite people, creating a shared agenda where each actor can have a role in shaping healthier, wildlife-supporting cities.

Acknowledgments

This material is based on work supported by the San Francisco Bay Water Quality Improvement Fund of the US Environmental Protection Agency Region IX, the Google Ecology Program, the Peninsula Open Space Trust, and the City of Sunnyvale. Many thanks to Brandon Herman and Cate Jaffe for images developed for this work.

References

1. Spotswood E, Grossinger R, Hagerty S, Bazo M, Benjamin M, Beller E, et al. Making Nature's City: a science-based framework for building urban biodiversity. Richmond, CA: San Francisco Estuary Institute; 2019. (A product of the Healthy Watershed, Resilient Baylands project. Funded by the San Francisco Bay Water Quality Improvement Fund, EPA Region IX.) SFEI Publication #947.
2. World Bank. A Catalogue of Nature-Based Solutions for Urban Resilience [Internet]. Washington, DC: World Bank; 2021 [cited 2022 Sep 26]. 240 p. Available from: https://openknowledge.worldbank.org/handle/10986/36507
3. Sandifer PA, Sutton-Grier AE, Ward BP. Exploring connections among nature, biodiversity, ecosystem services, and human health and well-being: opportunities to enhance health and biodiversity conservation. Ecosystem Services. 2015;12(Apr):1–15.
4. Larson KL, Nelson KC, Samples SR, Hall SJ, Bettez N, Cavender-Bares J, et al. Ecosystem services in managing residential landscapes: priorities, value dimensions, and cross-regional patterns. Urban Ecosystems. 2016;19(1):95–113.
5. Locke DH, Hall B, Grove JM, Pickett STA, Ogden LA, Aoki C, et al. Residential housing segregation and urban tree canopy in 37 US cities. npj Urban Sustainability. 2021;1(1):15.
6. Stanford B, Zavaleta E, Millard-Ball A. Where and why does restoration happen? Ecological and sociopolitical influences on stream restoration in coastal California. Biological Conservation. 2018;221:219–27.
7. Connor EF, McCoy ED. The statistics and biology of the species-area relationship. American Naturalist. 1979;113(6):791–833.
8. Haddad NM, Gonzalez A, Brudvig LA, Burt MA, Levey DJ, Damschen EI. Experimental evidence does not support the habitat amount hypothesis. Ecography. 2017;40(1):48–55.
9. Nielsen AB, van den Bosch M, Maruthaveeran S, van den Bosch CK. Species richness in urban parks and its drivers: a review of empirical evidence. Urban Ecosystems. 2014;17(1):305–27.
10. Soga M, Yamaura Y, Koike S, Gaston KJ. Woodland remnants as an urban wildlife refuge: a cross-taxonomic assessment. Biodiversity Conservation. 2014;23(3):649–59.
11. Angold PG, Sadler JP, Hill MO, Pullin A, Rushton S, Austin K, et al. Biodiversity in urban habitat patches. Science of the Total Environment. 2006;360(1–3):196–204.
12. Beninde J, Veith M, Hochkirch A. Biodiversity in cities needs space: a meta-analysis of factors determining intra-urban biodiversity variation. Ecology Letters. 2015;18(6):581–92.
13. Spotswood EN, Beller EE, Grossinger R, Grenier JL, Heller NE, Aronson MFJ. The biological deserts fallacy: cities in their landscapes contribute more than we think to regional biodiversity. BioScience. 2021;71(2):148–60.
14. Shanahan DF, Miller C, Possingham HP, Fuller RA. The influence of patch area and connectivity on avian communities in urban revegetation. Biological Conservation. 2011;144(2):722–9.
15. Matsuba M, Nishijima S, Katoh K. Effectiveness of corridor vegetation depends on urbanization tolerance of forest birds in central Tokyo, Japan. Urban Forestry and Urban Greening. 2016;18(Aug):173–81.
16. Fremier AK, Kiparsky M, Gmur S, Aycrigg J, Craig RK, Svancara LK, et al. A riparian conservation network for ecological resilience. Biological Conservation. 2015;191:29–37.
17. Leston L, Koper N. Urban rights-of-way as extensive butterfly habitats: a case study from Winnipeg, Canada. Landscape and Urban Planning. 2017;157:56–62.
18. Lynch AJ. Creating effective urban greenways and stepping-stones: four critical gaps in habitat connectivity planning research. Journal of Planning Literature. 2019;34(2):131–55.

19. Rondinini C, Doncaster CP. Roads as barriers to movement for hedgehogs. Functional Ecology. 2002;16(4):504–9.
20. Beninde J, Feldmeier S, Werner M, Peroverde D, Schulte U, Hochkirch A, et al. Cityscape genetics: structural vs. functional connectivity of an urban lizard population. Molecular Ecology. 2016;25(20):4984–5000.
21. Hale JD, Fairbrass AJ, Matthews TJ, Davies G, Sadler JP. The ecological impact of city lighting scenarios: exploring gap crossing thresholds for urban bats. Global Change Biology. 2015;21(7):2467–78.
22. Tremblay MA, St. Clair CC. Permeability of a heterogeneous urban landscape to the movements of forest songbirds: songbird movements in urban landscapes. Journal of Applied Ecology. 2011;48(3): 679–88.
23. Favaro C, Moore JW. Fish assemblages and barriers in an urban stream network. Freshwater Science. 2015;34(3):991–1005.
24. Baum KA, Haynes KJ, Dillemuth FP, Cronin JT. The matrix enhances the effectiveness of corridors and stepping stones. Ecology. 2004;85(10):2671–6.
25. Fernandez-Juricic E. Avifaunal use of wooded streets in an urban landscape. Conservation Biology. 2000;14(2):513–21.
26. Belaire JA, Whelan CJ, Minor ES. Having our yards and sharing them too: the collective effects of yards on native bird species in an urban landscape. Ecological Applications. 2014;24(8):2132–43.
27. Carbó-Ramírez P, Zuria I. The value of small urban greenspaces for birds in a Mexican city. Landscape and Urban Planning. 2011;100(3):213–22.
28. Bateman PW, Fleming PA. Big city life: carnivores in urban environments. Journal of Zoology. 2012;287(1):1–23.
29. Tscharntke T, Tylianakis JM, Rand TA, Didham RK, Fahrig L, Batáry P, et al. Landscape moderation of biodiversity patterns and processes - eight hypotheses. Biological Reviews. 2012;87(3):661–85.
30. Beller EE, Spotswood EN, Robinson AH, Anderson MG, Higgs ES, Hobbs RJ, et al. Building ecological resilience in highly modified landscapes. BioScience. 2019;69(1):80–92.
31. Goddard MA, Dougill AJ, Benton TG. Scaling up from gardens: biodiversity conservation in urban environments. Trends in Ecology and Evolution. 2010;25(2):90–8.
32. Shwartz A, Muratet A, Simon L, Julliard R. Local and management variables outweigh landscape effects in enhancing the diversity of different taxa in a big metropolis. Biological Conservation. 2013;157: 285–92.
33. Knisley CB. Anthropogenic disturbances and rare tiger beetle habitats: benefits, risks, and implications for conservation. Terrestrial Arthropod Reviews. 2011;4(1):41–61.
34. Blackmore S. Cities: the final frontier for endangerd plants? Sibbaldia. International Journal of Botanic Garden Horticulture. 2019;17:3–10.
35. Threlfall CG, Mata L, Mackie JA, Hahs AK, Stork NE, Williams NSG, et al. Increasing biodiversity in urban green spaces through simple vegetation interventions. Journal of Applied Ecology. 2017;54(6):1874–83.
36. Berthon K, Thomas F, Bekessy S. The role of "nativeness" in urban greening to support animal biodiversity. Landscape and Urban Planning. 2021;205:103959.
37. Spotswood E, Grossinger R, Hagerty S, Beller E, Grenier JL, Askevold R. Re-oaking Silicon Valley: building vibrant cities with nature. SFEI Contribution No. 825. Richmond, CA: San Francisco Estuary Institute; 2017.
38. Narango DL, Tallamy DW, Marra PP. Nonnative plants reduce population growth of an insectivorous bird. Proceedings of the National Academy of Sciences of the United States of America. 2018;115(45):11549–54.
39. Strauss SY, Lau JA, Carroll SP. Evolutionary responses of natives to introduced species: what do introductions tell us about natural communities? Ecology Letters. 2006;9(3):357–74.
40. Leuzinger S, Rewald B. The who or the how? Species vs. ecosystem function priorities in conservation ecology. Frontiers in Plant Science. 2021;12:758413. Available from: https://www.frontiersin.org/article/10.3389/fpls.2021.758413
41. Gleditsch JM, Carlo TA. Fruit quantity of invasive shrubs predicts the abundance of common native avian frugivores in central Pennsylvania. Diversity and Distributions. 2011;17(2):244–53.
42. McKinney ML. Urbanization as a major cause of biotic homogenization. Biological Conservation. 2006;127(3):247–60.
43. Schwartz MW, Thorne JH, Viers JH. Biotic homogenization of the California flora in urban and urbanizing regions. Biological Conservation. 2006;127(3): 282–91.
44. Pardee GL, Philpott SM. Native plants are the bee's knees: local and landscape predictors of bee richness and abundance in backyard gardens. Urban Ecosystems. 2014;17(3):641–59.
45. Ikin K, Knight E, Lindenmayer DB, Fischer J, Manning AD. The influence of native versus exotic streetscape vegetation on the spatial distribution of birds in suburbs and reserves. Diversity and Distributions. 2013;19(3):294–306.

46. Stagoll K, Lindenmayer DB, Knight E, Fischer J, Manning AD. Large trees are keystone structures in urban parks. Conservation Letters. 2012;5(2): 115–22.
47. Gaston KJ, Smith RM, Thompson K, Warren PH. Urban domestic gardens (II): experimental tests of methods for increasing biodiversity. Biodiversity Conservation. 2005;14(2):395–413.
48. Hilty JA, Merenlender AM. Use of riparian corridors and vineyards by mammalian predators in northern California. Conservation Biology. 2004;18(1):126–35.
49. Walsh CJ, Roy AH, Feminella JW, Cottingham PD, Groffman PM, Morgan RP. The urban stream syndrome: current knowledge and the search for a cure. Journal of the North American Benthological Society. 2005;24(3):706–23.
50. Sweeney BW, Newbold JD. Streamside forest buffer width needed to protect stream water quality, habitat, and organisms: a literature review. JAWRA Journal of the American Water Resources Association. 2014;50(3):560–84.
51. Lundholm JT, Richardson PJ. Habitat analogues for reconciliation ecology in urban and industrial environments. Journal of Applied Ecology. 2010;47(5): 966–75.
52. Milligan MC, Dickinson JL. Habitat quality and nest-box occupancy by five species of oak woodland birds. The Auk. 2016;133(3):429–38.
53. Demeyrier V, Lambrechts MM, Perret P, Grégoire A. Experimental demonstration of an ecological trap for a wild bird in a human-transformed environment. Animal Behavior. 2016;118:181–90.
54. Sievers M, Parris KM, Swearer SE, Hale R. Stormwater wetlands can function as ecological traps for urban frogs. Ecological Applications. 2018;28(4):1106–15.
55. Hale R, Coleman R, Pettigrove V, Swearer SE. Identifying, preventing and mitigating ecological traps to improve the management of urban aquatic ecosystems. Journal of Applied Ecology. 2015;52(4): 928–39.
56. Donofrio J, Kuhn Y, McWalter K, Winsor M. Research article: Water-sensitive urban design: an emerging model in sustainable design and comprehensive water-cycle management. Environmental Practice. 2009;11(3):179–89.
57. Sandström UG, Angelstam P, Khakee A. Urban comprehensive planning – identifying barriers for the maintenance of functional habitat networks. Landscape and Urban Planning. 2006;75(1–2):43–57.
58. Rodewald AD, Gehrt SD. Wildlife population dynamics in urban landscapes. In: McCleery RA, Moorman CE, Peterson MN (eds.) Urban Wildlife Conservation: theory and practice [Internet]. Boston, MA: Springer US; 2014. p. 117–47. Available from: https://link.springer.com/10.1007/978-1-4899-7500-3_8
59. Faeth SH, Warren PS, Shochat E, Marussich WA. Trophic dynamics in urban communities. BioScience. 2005;55(5):399–407.
60. McCleery RA, Moorman CE, Peterson MN (eds.) Urban Wildlife Conservation: theory and practice. Boston, MA: Springer US; 2014. 408 p.
61. Loss SR, Will T, Marra PP. The impact of free-ranging domestic cats on wildlife of the United States. Nature Communications. 2013;4(1):1396.
62. Longcore T, Rich C. Ecological light pollution. Frontiers in Ecology and the Environment. 2004;2(4): 191–8.
63. Erickson WP, Johnson GD, Young Jr DP. A summary and comparison of bird mortality from anthropogenic causes with an emphasis on collisions. In: Ralph JC, Rich TD (eds.) Bird Conservation Implementation and Integration in the Americas: proceedings of the third International Partners in Flight conference. 2002 Mar 20–24; Asilomar, CA, Volume 2 Gen. Tech. Rep. PSW-GTR-191. Albany, CA: US Department of Agriculture, Forest Service, Pacific Southwest Research Station; 2005. p. 1029–42.
64. Mata C, Hervás I, Herranz J, Suárez F, Malo JE. Are motorway wildlife passages worth building? Vertebrate use of road-crossing structures on a Spanish motorway. Journal of Environmental Management. 2008;88(3):407–15.
65. Hobday AJ, Minstrell ML. Distribution and abundance of roadkill on Tasmanian highways: human management options. Wildlife Research. 2008;35(7):712–26.
66. Iknayan K, Bazo M, Benjamin M, Ndayishimiye E, Grossinger R. Moffett Park Specific Plan Urban Ecology. Report No.: 985. Richmond, CA: San Francisco Estuary Institute; 2020. 54 p. Available from: https://www.sfei.org/sites/default/files/biblio_files/MPSP_Ecology_20_04.pdf
67. Beller E, Salomon M, Grossinger R. Historical Vegetation and Drainage Patterns of Western Santa Clara Valley: a technical memorandum describing landscape ecology in Lower Peninsula, West Valley, and Guadalupe Watershed management areas. Richmond, CA: San Francisco Estuary Institute; 2010. 71 p.

CHAPTER 15

Conclusion—Biodiversity for the People

Future Directions for Urban Biodiversity Conservation

Christopher J. Schell[‡], Max R. Lambert[‡], Simone Des Roches[§], Travis Gallo[¶], and Nyeema C. Harris[#]

Conservationists have long acknowledged that their field is values-based (1). There is nothing "objective" about prioritizing management actions to increase hunting or fishing opportunities, to increase the diversity of species present, or because a given species is listed as endangered. We manage species for consumptive purposes, to change their listing status, or for any other reasons that align with societal values. Doing urban conservation is a scientific necessity to achieve broader conservation goals, and environmental justice is a necessary tool to ensure the success of those goals. However, a superficial or performative implementation of environmental justice principles by conservationists solely to combat the biodiversity crisis will inevitably fuel the same exclusionary practices that have perpetuated environmental harms on minoritized communities for centuries (2). Exercising legitimate environmental justice action relies on authentically embedding just values into how the discipline operates, from the processes that govern what we decide to protect, to the ways in which we build and engage with community. Consequently, conservationists across the globe must acknowledge that justice is the central tenet to their discipline.

[‡] Environmental Science, Policy, and Management, UC Berkeley, USA
[§] University of Washington, USA
[¶] University of Maryland, USA
[#] Applied Wildlife Ecology (AWE) lab, Yale University, USA

Our collective challenge rests in our ability to produce malleable, multipurpose, adaptive strategies and tools that benefit both human and nonhuman entities. Such an agenda will require uncomfortable conversations about recalcitrant systems of power and oppressive authorities that limit successful outcomes of biodiversity conservation writ large (2). We must redefine what biodiversity is and who has the right to make such assertions, the conscious and unconscious biases that dictate the conservation actions we elevate over others, and what types of knowledge we accept as objective or superior (2–4). Moreover, conservation scientists and practitioners, even in cities, have been reluctant in past decades to infuse political and scholar-activism into research and decision-making processes, for fear of appearing as partisan, unobjective actors that use state-appropriated funds for personal political agendas.[1] Such fears implicitly justify impediments to environmental justice and equity discourses that explicitly interrogate the role of White supremacy, sexism, xenophobia, homophobia, and ableism in perpetuating environmental harms and inequities.

There is a wide multiverse of definitions of, experiences of, and approaches to biodiversity conservation. Cities, more so than any other

[1] Yale School of the Environment editorial on political will for climate change action (https://environment.yale.edu/news/article/building-public-and-political-will-for-climate-change-action).

Christopher J. Schell et al., *Conclusion—Biodiversity for the People.* In: *Urban Biodiversity and Equity.* Edited by: Max R. Lambert & Christopher J. Schell, Oxford University Press. © Oxford University Press (2023). DOI: 10.1093/oso/9780198877271.003.0015 human-dominated landscape, are perfectly situated to assume a leading role in developing the figurative cookbook for how we elevate multiple knowledges, epistemologies, cultures, and relationships with biodiversity and nature (5). And, despite what feels like a Sisyphean exercise in combating the inequitable development and societal practices that hinder effective biodiversity conservation, there is cause for considerable hope. Emerging discourses in the last decade alone have begun to interrogate who benefits from urban conservation actions (3), how to broaden urban greening efforts to include anti-racist and emancipatory approaches (6,7), spotlight case studies for authentic and intentional coproduction with impacted communities of color (8), and redefine what biodiversity is to equitably distribute the benefits and ecosystem services provided therein (3). Literature addressing the urgent need to embed environmental justice principles into sustainability practice (9), science disciplines such as ecology (10), urban planning and infrastructure (11), and nature-based solutions (4,12,13) has also accelerated. Taken together, these relevant perspectives and empirical works suggest we are at the precipice of a global paradigm shift in how we perceive our natural world, biodiversity, and our role in this entirely complex tapestry.

The next critical step to a transformative, justice-infused urban conservation science will require a radical reimagining of Western societies' relationship with biodiversity and nature. Such a transformation will require revolutionary interventions in decision-making, policy, and governance structures that are both robust enough to withstand drastic environmental changes while simultaneously elastic enough to mutate as conditions change (14,15). This transformation must equally include a more diffuse power structure and inclusive enterprise. Conservation as traditionally constructed is ill-equipped to achieve such a Herculean task, despite the equally superhuman efforts of conservationists to conserve sensitive habitats, ecosystems, and species (16). What feels like fighting a losing battle, however, is only because conservation science is still struggling with creating robust strategies that comprehensively account for the most consequential ecosystem engineer on this planet: people.

The emergence and continued maturation of urban biodiversity conservation may provide the much-needed antidote to our contemporary dilemma.[2] Since its inception, urban conservation has been compelled to consider how human–nature relationships shape fundamental properties of the landscape. Conservation strategies and implementation legislation enacted by cities across the globe emphasize this fact, from Atlanta, GA, USA to Singapore (8,17,18). Though the relative successes and barriers of urban conservation programs vary as a function of sociopolitical systems, local and national ordinances, cultural views on conservation efforts, and economic institutions, such programs are unified in the ideology that: (1) society and the environment are both interdependent and interconnected; (2) decision makers and practitioners must work with local communities to effect positive change; and (3) implemented strategies need to be hyper-agile to morph and grow with society in mind. Still, even the most financially supported and politically backed programs are grappling with effectively meeting the needs of the most marginalized people in society. Hence, the central quandary of urban conservation resurfaces: how do we promote biodiversity conservation in cities that also elevates justice and equity for those most impacted by historical, current, and future environmental harms?

One of the hard truths that the entire conservation community must contend with is that we, too, are but one group among the many diverse, informed, and justified communities that have a tremendous stake in saving our natural world (2). Too often has knowledge from scientific institutions been elevated over other ways of knowing, often perpetuating environmental harms, and sabotaging the long-term conservation success of implemented recommendations. These include traditional ecological knowledge, Indigenous ways of knowing, and experiential and place-based knowledge, all of which stress that the multitude of diverse experiences with nature create a much richer tapestry

[2] The Nature Conservancy on elevating cities as the solution to biodiversity loss (https://www.nature.org/en-us/newsroom/urban-expansion-impacts-for-biodiversity-planning-yale/).

for being in community with biodiversity. Thus, in this closing chapter we further emphasize that decentering narrowly-focused conservation plans that rest almost exclusively in the scientific objectivity of the elite and privileged few are destined to fail.

We end here by providing a set of recommendations and spotlight what will be required of us collectively to truly build ecological resilience in our cities, suburbs, towns, and megapolitan areas. Importantly, the "us" in this context is both expansive to include the collective of diverse and interconnected communities across urban environments, as well as specific to local communities of practice and place-based efforts. First, we discuss the need to center pluralism in our conceptions of biodiversity, and how doing so allows for greater equity and inclusivity in conservation efforts. Second, we call attention to the role that other disciplines of research and practice can play in facilitating urban biodiversity conservation, emphasizing that breaking disciplinary silos is required for transcending myopic solutions. Third, we address how creativity and imagination are critical to envisioning just future cities, spotlighting how equitable practices and structural design are necessary for implementing nature-based solutions. In parallel, we underscore how the pervasive impacts of capitalism, have generated the environmental ills we currently face noting that our efforts to move toward an environmentally just city will require that we dispel racial capitalism (i.e., the concept that race is the primary factor in structuring social and labor hierarchies in capitalist societies) (19). Finally, we conclude with critical questions that will guide future urban conservation discourses, emphasizing that collective action will be required for saving the natural world, even if current generations may be unlikely to experience the benefits of our labor.

Centering a pluralistic view of biodiversity

How and why do we define biodiversity? This is an elementary, yet profound question situated at the nexus of all the solutions and policy recommendations addressed in the previous chapters. At its core, biodiversity is a unit of measurement we use to practice conservation; hence, having a defined entity provides a target to achieve. Consider for a moment the push to restore sensitive habitat or establish wildlife crossings to increase landscape connectivity. In those examples we assume that "biodiversity" will increase because restoring habitat creates more ecological niches for a wider array of species to survive, or wildlife crossings will bolster landscape connectivity enabling greater permeability and movement across the urban matrix. In both instances, the implicit, perhaps unconscious assumption, is that biodiversity equates to more species (i.e., greater species richness). This definition is not necessarily wrong or misplaced: alpha diversity metrics like species richness have been a useful and robust statistic for measuring community and ecosystem function (20). Ironically, species richness is simply one definition of biodiversity. Within the ecological sciences alone, there are multiple ways of calculating biodiversity across scales—including alpha, beta, and gamma diversity. The biological sciences broadly consider biodiversity at three distinct scales as well: genetic, species, and ecosystem diversity (21). Case in point: the natural sciences already have multiple definitions at various scales for what biodiversity is.

These details are not trivial and have economic, conservation decision-making, and governance implications. Thus, the specific interpretation of biodiversity that is elevated in the decision-making process matters considerably. For example, if an endangered species is discovered on a small collection of urban green spaces, conservation professionals may support legislation that effectively reduces human activity in those spaces, using their selected definition as justification to bolster such policies. The enacted policies to protect the sensitive habitat, however, may violently displace residents or reduce their access to nature's benefits in urban environments, leading to a negative feedback loop that further annexes the most marginalized communities from experiencing the beneficial ecosystem services provided by urban nature. In this instance, the interpretation of biodiversity that gives greater weight to species that are endangered superseded the communities' well-being, perpetuating insidious environmental harms. Sadly, this hypothetical scenario highlights the very real and repeated injustice of conservation-induced displacement (22) that is intrinsic to colonial and postcolonial conservation efforts.

Singular, narrow definitions of biodiversity not only have the potential to perpetuate systemic injustices faced by minoritized communities, but also severely compromise collaborations and community engagement efforts to support conservation agendas (3). Narrow definitions are frequently used to undercut the legitimacy of other culturally or experientially situated definitions of biodiversity, invalidating diverse perspectives that are no less valid than academic definitions. As a result, conservation and management organizations run the risk of further alienating local communities from the decision-making process, limiting (a) the power of those communities to effect change in their own neighborhoods; (b) the long-term success of conservation actions; and (c) the overall legitimacy of biodiversity conservation policies for society (3).

Throughout this volume, we and other contributors have stressed the significance of infusing justice and equity into conservation practices and policies; both because it is morally just and because it is the necessary catalyst for effective urban conservation. Without it, such agendas are doomed to fail. As researchers in the conservation space ourselves, it is imperative to call out the biases that assert academic knowledge as superior to other ways of knowing. A significant part of biodiversity conservation's reckoning thus rests in working to restore trust with disaffected communities, including Black, Indigenous, and other persons of color. Such work can only be initiated by authentically validating differing world views, perspectives, and definitions of biodiversity.

Centering a pluralistic view in urban biodiversity conservation opens space for diverse, complex, and culturally meaningful conceptions of a biodiverse world and our place in it. Rather than considering what singular definition to use as the defining factor, multiple approaches—including social, cultural, spiritual, and quantitative—can be used to identify common nodes of collaboration among conservationists and community members (23). The coproduction of shared common goals thus emerges from placing less emphasis on the "how?" and "what?," and more emphasis on the "why?" Relaxing rigidity around how biodiversity is defined clears the way for answering the question of why we choose to conserve certain species, and whom those actions are for (4)? This is akin to a poignant line in the film "Black Panther: Wakanda Forever," where both the antagonist Namor and the previous film's antagonist Killmonger state that "How is never as important as why."[3] Moreover, a pluralistic perspective effectively democratizes the decision-making process and decenters certain forms of knowledge. As a result, conservationists, practitioners, and academics must cede any perceived moral or intellectual authority (i.e., power) in laying claim to how people should perceive biodiversity and nature (3).

Pluralism also supports more effective scaling of biodiversity conservation policies. Urban conservationists are frequently considering how policies work at multiple scales (21,24) and whether those policies are equipped to deal with the interconnections of societal and ecological systems (25). Policies built on a pluralist foundation confer legitimacy to the varied intersectional identities of the communities vested in those policies. An urban conservation plan that validates and works to abolish the oppressive constraints of sexism, homophobia, ableism, classism, xenophobia, and racism in society is equally buoyed by the reciprocal support of a diverse coalition of stakeholders. This poetic positive feedback is reminiscent of biodiversity itself, in which greater diversity fortifies ecosystem resilience and bolsters function and health (26). Finally, pluralism allows humans as a species to simultaneously celebrate human–biodiversity relationships and consider the value of urban ecosystems beyond their utility to society. Equally holding the valuation of human and nonhuman organisms as interconnected beings echoes both the 17 foundational principles of Environmental Justice (27), as well as Indigenous practices and rituals,[4] which

[3] Rolling Stone perspective on the movie "Black Panther: Wakanda Forever" (https://www.rollingstone.com/tv-movies/tv-movie-features/black-panther-wakanda-forever-mcu-colonialism-1234628690/).

[4] Website from the Nature Conservancy addressing Indigenous conservation practices (https://www.nature.org/en-us/about-us/who-we-are/how-we-work/

deserve substantial deference and elevation. Necessarily, Indigenous-led urban conservation should assume a leading role, as generations of Indigenous peoples globally have refined conservation strategies that colonialism and urban development have forcefully pushed to the margins of society (28,29).

In sum, expanding the definition of biodiversity to include the multitude of variants paves the way for embedding the practice of coproduction into conservation programs (7).

Expanding the disciplinary table

This volume covers a wide breadth of interdisciplinary fields, making a concerted effort to illustrate the interconnectedness of these previously siloed disciplines. Despite the broad expertise represented by the contributors, there are invariably disciplines and synergies that were not captured in prior chapters. It is our hope that future works and literature will continue to explicate the various disciplines that can provide support and insight in building biodiverse, resilient cities. Breaking silos within academic disciplines will be necessary to improve urban conservation efforts, and such a task may be facilitated by building transdisciplinary coalitions that produce unorthodox solutions. These considerations notwithstanding, some of the most impactful changes will occur outside of the ivory tower of academia or scientific practice more broadly, demanding that we intentionally create space for community members and laypersons with less conferred privileges and capital to join a more expansive and inclusive table. Manifesting such welcoming spaces will further require interrogating the structures that prevent other professionals from contributing to urban conservation discourses, as well as deep reflexivity in the systemic constraints that hinder academic and governmental professionals from biodiversity conservation action.

Centering environmental education and opportunities for experiential learning is arguably one of the highest-priority areas. Urban environments are the ideal landscapes for cross-generational learning, providing opportunities to freely observe, explore, and hypothesize their social-ecological reality (30). Lesson plans coproduced among urban practitioners, researchers, and educators can subsequently serve as the scaffolding for a student's learning. For instance, Wildlife Neighbors[5] is an informal education program developed by the Applied Wildlife Ecology (AWE) lab at Yale School of the Environment and the Detroit Zoological Society funded by the National Science Foundation. The program leverages an extensive wildlife camera survey to build science capacities and identities, promote environmental stewardship, and enhance sense of place in urban youth of metro Detroit.

Educators trained in effective science pedagogy from preschool to adult education, in formal and informal settings, will have extraordinary insight into the most effective ways of demystifying the interconnectedness of urban nature, society, and environmental balance (31). Because of its malleability in strategies and tools, urban conservation may be the perfect medium for a lesson plan in conservation more broadly that transcends age, class, gender, country of origin, and mobility. Moreover, the development of teaching modules that center nature experiences of students will give them agency in their learning, bolstering their intrinsic valuing of urban nature and thus their willingness to conserve and protect it. Urban biodiversity and nature experiences can also reduce physiological stress markers in children, highlighting the public health benefits of integrating urban conservation practice with K-12 pedagogy (32,33).

Importantly, this learning exercise is a two-way street, as urban conservation researchers and practitioners can deepen their relationships with local communities by listening to students' experiences of nature. Youth living in various cities across the globe express exceptional interest in their natural world and make profound ecological observations that are not constrained by academic ways of knowing. Fresh perspectives born out of genuine curiosity on the environmental mechanisms

community-led-conservation/); National Geographic editorial on Indigenous people's role in conservation (https://www.nationalgeographic.com/environment/article/can-indigenous-land-stewardship-protect-biodiversity-).

[5] National Science Foundation (NSF) funded award for the Wildlife Neighbors program (https://www.nsf.gov/awardsearch/showAward?AWD_ID=2005812).

that govern their realities subsequently serve as the foundation for groundbreaking ideas that can greatly benefit biodiversity conservation. Adolescent imagination thus provides a critical fountain of innovation, warranting full representation at the conservation decision-making table. Certainly, the fact that young children will also inherit our planet elevates their perspectives as central to these conversations, further conferring their right to these expanding discourses. Whether through art, open-ended play, or guided lesson plans, we must be open to children's perspectives as a wellspring of ideas that remind us how to honor and value biodiversity. Through open exploration and valuing the common species that call our urban systems home, younger generations of conservationists and environmental stewards can be substantial reminders of what we are collectively fighting for.

Certainly, there are other disciplines within the biological sciences alone that deserve mention. The emergence and proliferation of microbiome studies may provide us critical insight into the role small biological universes play in shaping species resilience toward global environmental change. Similarly, an in-depth investigation into ecophysiological processes may spotlight the regulatory mechanisms that are most robust to change, and those that are most directly linked to reduced health and fitness. Further still, this volume did not have a detailed explication on ecosystem ecology and nutrient cycling, broad properties that are certainly dictated by social-ecological function and socially driven urban heterogeneity. Data science and smart technologies, environmental economics, political ecology, religious studies, and philosophy will also have roles to play in this narrative. However, these integrations will be neither possible nor worthwhile if academic institutions are unwilling to grapple with the incentive structures that govern how applied, community-oriented, place-based research is valued and rewarded.

Community organizations and research institutions have fundamentally misaligned goals, which can present genuine challenges to enacting local, place-based conservation efforts. Researchers working in an academic model are often urged to prioritize peer-reviewed scientific publications as a marker of career success. These publications are conferred extraordinary weight compared to other aspects of the profession, with activities like community engagement and diversity, equity, inclusion, and justice (DEIJ) efforts minoritized and devalued. State and federal granting agencies are further entrenched in this system: even though many of these agencies require statements explicating the broader impacts of proposed research plans, it has been notoriously difficult to quantify the true impact of proposed activities. Taken together, this results in extractive and exploitative research practices that take resources away from communities often without reciprocity. Local communities are therefore left with a written document that may detail environmental patterns relevant to their issue(s), but no mechanism nor financial backing to adjudicate those issues and find robust solutions. Ironically, academic journals often have costly paywalls that prevent access to local communities outside of academia, meaning that those individuals may not even get the opportunity to read the research conducted on their environmental issue. To effectively do urban biodiversity conservation, incentive structures within scientific research institutions must be reoriented toward effective positive community change, rather than simply being a third-party scribe disassociated from local environmental issues. Deprioritizing peer-reviewed publications and conferring greater significance to applied community-based research is thus a substantial move in the right direction.

Beyond fields of research, urban conservation inherently will rely on integrative transdisciplinarity among practitioner disciplines as well. City engineers, social workers, public works, arborists, utility companies, and countless others who play a role in managing urban landscapes and organizing people are all part of the conservation community. As communities of practice, professionals in the public sector dedicate considerable time, energy, and resources to the betterment of our neighborhoods and cities. Moreover, these public sector professionals frequently serve as the "boots on the ground," hearing directly from community members about their various environmental and societal concerns. As such, those individuals are responsible for holding the community's concerns, discussing potential implementation strategies to address those concerns, then eventually being tasked with developing effective tools that translate to beneficial legislation for the impacted communities. Conservation action is therefore impossible without the expertise and buy-in of public sector experts.

Understanding the legal frameworks for urban conservation is also critical. For instance, agencies at different scales from local to regional, state, and national are charged with different authorities that may constrain when, where, and how conservation actions take place in cities and suburbs. However, these same authorities may also present opportunities for new conservation actions beyond traditional conservation agencies including those which focus on housing, transportation, energy, and education. Urban conservation may therefore have different legal or policy constraints and incentives than nonurban conservation. Consider the urban forest which inevitably is managed by diverse authorities: city foresters, transportation officials, parks departments, and private landowners may all have a vested interest in forest health, but utility companies may be compelled to trim or remove trees when they encroach on power lines. Consequently, natural resource agencies are likely to have little authority but will instead consult with stormwater or transportation agencies who have legal authorities to maintain and modify those components of the built environment. Making sure all entities have the time, resources, and latitude to create environmental action plans that equally value input across sectors will be critical, especially in the management of our environmental commons.

As biodiversity conservation in cities continues to transcend into mainstream discourses from academia to the public, it is our hope that these disciplines use their respective expertise to collectively innovate strategies for protecting urban biodiversity.

Back to the (just) future

It is uncertain how recognizable our environmental landscape will be 100–150 years from the present. Continued landscape conversion, development, and shifts in the global climate paint a grim picture of future cities. In obstinate defiance of this doomsday, environmental scientists and thinkers have worked to envision alternative futures and pathways that humanity may take (34). Importantly, this envisioning exercise liberates singular-minded narratives that are restricted to contemporary systems of oppression to imagine possible future world states (34). As a result, these thought exercises facilitate modern-day time travel that activates one's ability to concurrently hold historical and contemporary processes while imagining how to make positive changes to the system. Gaining such openness, however, requires interrogating the figurative skeleton in the closet: racial capitalism.

The rise of racial capitalism saw the concept of nature transform from a shared resource owned by no one, to a commodity to be exploited and monetized (35). The commodification of nature, wilderness, and biodiversity is born out of structural processes inherent in racial capitalism, tying many conceptions of nature in the West to the institutions of slavery and land dispossession (35,36). In the US, the transition of nature to a limited, often privatized resource festered over the centuries to fuel both de jure and de facto atrocities, that serve as the foundation of our cities today. Envisioning a more just future mandates reconciliation with historical injustices, and this is especially pertinent for cities, as the geographies of neighborhoods, highways, and municipal services are birthed out of those past inequities with fear limiting progress (2).

Envisioning a future system resilient to extreme climatic events and biodiversity loss means uncovering in detail how past transgressions shape our present ecological reality (Figure 15.1). Our cities, if nothing else, are epic storytellers: the concrete, the buildings, the siting of pollutants, and even the location of trees are due to past events decades prior to our current time. All the social and ecological elements of our cities were informed by generations prior, and the cadre of recent studies typify how pervasive policies enforcing residential segregation (e.g., redlining) shape the ecology and evolution of nonhuman organisms (37–40), as well as contemporary health outcomes for people (41–43). The critical next step is creating solutions that both demonstrate critical learning from past

Figure 15.1 Sewage discharged by wastewater treatment plants, paired with prolonged heat waves and warm waters due to climate change, led to increased nutrient loads that accelerated harmful algal bloom growth in urban water bodies of the San Francisco Bay Area. The proliferation of this red tide (i.e., when accumulating algae plant colonies, often rust-colored, overpopulate an aquatic system) led to massive fish die-offs, some of which were especially pronounced in areas like Lake Merritt, Oakland, CA pictured here. Fish carcasses quickly began to accumulate on the shores of Lake Merritt and elsewhere as a result. This example further demonstrates how inextricably linked society and ecology are in cities, with negative feedbacks among these realms leading to ecosystem collapse. Radically reimagining just urban futures necessarily means deconstructing the infrastructural and municipal histories of urban landscapes to mitigate social-ecological calamities and prevent future ones from occurring.
Photos courtesy of Chris Martin.

inequities and embed justice mechanisms in planning, procedure, and practice that eliminate the possibility of repeating history. This may manifest as urban land reparations for individuals with enslaved African ancestors or serious advances in Land Back discourses.[6] Emancipatory and abolitionist movements may also need greater prominence in conservation and development processes, as a radical reimagining becomes far more likely when all peoples are free.

[6] City of Oakland and Sogorea Te' Land Trust announce plan to return land for Indigenous stewardship (https://www.oaklandca.gov/news/2022/sogorea-te-land-trust-and-city-of-oakland-announce-plan-to-return-land-to-indigenous-stewardship).

Fortifying cities to become ecologically resilient to the intensification of global climate-related weather events (e.g., drought, wildfires, flooding, hurricanes) can also benefit tremendously by learning about how past inequities shape our contemporary urban ecosystems. Fortunately, urban ecologists, planners, and climate scientists have developed a breadth of knowledge on this very topic over the past decade. At its core, ecological resilience describes the overall capacity of a region, site, or habitat to flexibly adjust with disturbances that can jeopardize ecosystem balance and function (44–46). Equally well-developed have been accompanying critical discourses on who benefits most from resilience initiatives and why (47).

The formative question of who benefits and why is a recurrent theme, as the implementation of proposed solutions for building resilience, if done outside of an environmental justice and equity lens, can completely neglect or detrimentally impact the very marginalized communities that the proposed activities were intended to benefit (4). For instance, take the emergence of Nature-Based Solutions (NBS) and their growing acceptability as a prominent tool to mitigate climate change (48,49). Improving green and blue infrastructure via improved stormwater infrastructure, urban greening, and transitioning to green energy resources are offered as multifunctional solutions that (a) safeguard cities against environmental catastrophes and (b) address a variety of societal challenges like increasing access to nature (48). Because of these considerations, it is automatically assumed that NBS are socially just; however, the implementation of these solutions can exacerbate segregation, displacement, and dispossession, further widening social inequities born from historical injustices (13).

Urban greening efforts stand as one of the most prominent examples of NBS exacerbating social inequities. Tree planting efforts in low-income communities and communities of color in the US, for example, are often perceived as a net benefit because these activities increase access to nature for the most disenfranchised. Though greening efforts can subsequently increase local environmental health and aesthetics, such activities also contribute to increasing property values, taxes, and rental costs. Hence, well-meaning efforts to increase access to nature and biodiversity for the most disenfranchised consequently backfire, leading to the displacement of residents to the fringes of urban environments. This form of "green gentrification" has been studied extensively (11,47,48), emphasizing how officials in the past have weaponized climate-mitigating greening efforts against the most vulnerable communities' needs (49). However, environmental justice and NBS do not need to be mutually exclusive. Justice principles can inform which sites are prioritized, provide a blueprint for effective implementation of proposed strategies, and require clear assessment plans that measure implementation success (49). These principles can also inform future housing developments,[7] tree planting efforts,[8] and reconciliation[9] that helps us heal from the past to build healthier cities.[10] Such an effort will require a recognition of environmental injustices as an equally concerning threat to urban resilience as the climate crisis, which is inherently a political, social, and cultural task.

In building future just cities, we must accept that cities in proceeding generations will look markedly different than our past or present but are inextricably linked to both. Moreover, we cannot begin the building process without acknowledging that status quo approaches under racial capitalism are inadequate for addressing our collective global struggle. Accepting that our previously distorted relationship with nature has been destructive and counterproductive is the only viable path forward. In so doing, we can build proactive strategies that simultaneously address environmental justice and biodiversity loss concerns. As previously stated by Dr. Martha Munoz, Evolutionary Biologist and Assistant Professor at Harvard University, we must "Dismantle by building differently."[11] If we genuinely want ecological resilience, it will require that we dismantle racial capitalism and its antagonistic relationship with nature.[12]

Conclusion

Urban biodiversity conservation is an emerging discipline with extraordinary promise for providing all peoples agency in conserving our natural world. We have attempted to bring together transdisciplinary expert researchers and practitioners to detail the

[7] Pejchar and Reed's chapter details housing development in suburban systems.

[8] Locke et al.'s chapter discusses justice in urban greening and tree planting.

[9] Hoover and Scarlett's chapter addresses environmental justice and reconciliation.

[10] Byers et al.'s chapter details a One Health approach and role in urban biodiversity.

[11] Quote from Dr. Martha Munoz, cited by Dr. Ambika Kamath (https://ambikamath.com/).

[12] Editorial by Vox on the relationship between capitalism and nature (https://www.vox.com/down-to-earth/23518769/cop15-un-biodiversity-conference-montreal-biodiversity-wwf).

state-of-the-science for urban biodiversity conservation. In doing so, we acknowledge that many disciplines and voices were invariably left out of this conversation and that urban conservation will continue to grow and expand. Accordingly, we challenge readers to constantly consider these questions so that the science and practice of urban conservation are intentional and inclusive:

- Who has the legal authority to do urban conservation where you are? What about the moral authority?
- Who is included in decision making and setting priorities for urban conservation? Who is being forgotten or left out of the decision-making process?
- Who are we calling a conservationist? Are our definitions inclusive of city engineers, transportation agencies and utilities, neighbors, and community groups?
- How are we defining biodiversity? Which species are we prioritizing and why are we doing it?

The recommendations we put forth come with a caveat: current professionals may never see the proverbial fruits of their labor. It is quite likely that our time on this plane of existence will not afford us the opportunity to experience a just future, rife with biodiverse and wildlife-friendly cities that function in equilibrium with society, rather than assume an adversarial role. Boomers, Gen X, Millennials, and perhaps even Gen Z persons are thus tasked with creating the conditions necessary for leveraging cities as hubs of biodiversity, knowing they may never reap the subsequent benefits in their lifetimes. Our responsibility is to disrupt the status quo, to spotlight the role that justice and equity play as the figurative foundation upon which we can conserve urban species. It is our task to ensure that future generations and changemakers have the tools needed to persist in a chaotic future, and to remind our progenitors that the power to innovate exists within them. The lessons from our ancestors compel us now to collectively organize and construct a blueprint for saving our natural world. This exercise alone—the act of preparation, activism, and sustainability—gives us hope.

As John Lewis, prominent civil rights activist and US representative serving for over 30 years, once stated, "Freedom is not a state; it is an act... Freedom is the continuous action we all must take, and each generation must do its part to create an even more fair, more just society." Let us hope that, if we can strive for justice every day, we can build an urban conservation practice that looks radically different once we have transcended this reality.

References

1. Soulé ME. What is conservation biology? Bioscience. 1985;35(11):727–34.
2. Oke C, Bekessy SA, Frantzeskaki N, Bush J, Fitzsimons JA, Garrard GE, et al. Cities should respond to the biodiversity extinction crisis. npj Urban Sustainability. 2021;1(1):9–12.
3. Pascual U, Adams WM, Díaz S, Lele S, Mace GM, Turnhout E. Biodiversity and the challenge of pluralism. Nature Sustainability. 2021;4(7):567–72.
4. Anguelovski I, Brand AL, Connolly JJT, Corbera E, Kotsila P, Steil J, et al. Expanding the boundaries of justice in urban greening scholarship: toward an emancipatory, antisubordination, intersectional, and relational approach. Annals of the American Association of Geographers. 2020;110(6):1743–69.
5. Mullenbach LE, Breyer B, Cutts BB, Rivers L, Larson LR. An antiracist, anticolonial agenda for urban greening and conservation. Conservation Letters. 2022;15(4):1–12.
6. Montambault JR, Dormer M, Campbell J, Rana N, Gottlieb S, Legge J, et al. Social equity and urban nature conservation. Conservation Letters. 2018;11(3):e12423.
7. Kellogg S. Urban Ecosystem Justice: strategies for equitable sustainability and ecological literacy in the city. Abingdon, UK: Routledge; 2021. 260 p.
8. Hoover FA, Meerow S, Grabowski ZJ, McPhearson T. Environmental justice implications of siting criteria in urban green infrastructure planning. Journal of Environmental Policy & Planning. 2021;23(5):665–82.
9. Langemeyer J, Connolly JJT. Weaving notions of justice into urban ecosystem services research and practice. Environmental Science & Policy. 2020;109:1–14.
10. Pineda-Pinto M, Frantzeskaki N, Nygaard CA. The potential of nature-based solutions to deliver ecologically just cities: lessons for research and urban planning from a systematic literature review. Ambio. 2022;51(1):167–82.
11. Tozer L, Hörschelmann K, Anguelovski I, Bulkeley H, Lazova Y. Whose city? Whose nature? Towards inclusive nature-based solution governance. Cities. 2020;107(Dec):102892.

12. Morrison TH, Adger WN, Agrawal A, Brown K, Hornsey MJ, Hughes TP, et al. Radical interventions for climate-impacted systems. Nature Climate Change. 2022;12(Dec):1100–6.
13. Staudinger MD, Carter SL, Cross MS, Dubois NS, Duffy JE, Enquist C, et al. Biodiversity in a changing climate: a synthesis of current and projected trends in the US. Frontiers in Ecology and the Environment. 2013;11(9):465–73.
14. Massarella K, Nygren A, Fletcher R, Büscher B, Kiwango WA, Komi S, et al. Transformation beyond conservation: how critical social science can contribute to a radical new agenda in biodiversity conservation. Current Opinion in Environmental Sustainability. 2021;49:79–87.
15. Lysaght T, Capps B, Bailey M, Bickford D, Coker R, Lederman Z, et al. Justice is the missing link in One Health: results of a mixed methods study in an urban city state. PLoS One. 2017;12(1):e0170967.
16. Puppim de Oliveira JA, Balaban O, Doll CNH, Moreno-Peñaranda R, Gasparatos A, Iossifova D, et al. Cities and biodiversity: perspectives and governance challenges for implementing the convention on biological diversity (CBD) at the city level. Biological Conservation. 2011;144(5):1302–13.
17. Swan CM, Pickett STA, Szlavecz K, Warren P, Willey KT. Biodiversity and community composition in urban ecosystems: coupled human, spatial, and metacommunity processes. In: Niemelä J (ed.) Urban Ecology: patterns, processes, and applications. Oxford: Oxford University Press; 2013. p. 179–86.
18. Uchida K, Blakey RV, Burger JR, Cooper DS, Niesner CA, Blumstein DT. Urban biodiversity and the importance of scale. Trends in Ecology & Evolution. 2021;36(2):123–31.
19. Agrawal A, Redford K. Conservation and displacement: an overview. Conservation and Society. 2009;7(1):1.
20. Woelfle-Erskine CA. Underflows: queer trans ecologies and river justice. Seattle, WA: University of Washington Press; 2022.
21. Goddard MA, Dougill AJ, Benton TG. Scaling up from gardens: biodiversity conservation in urban environments. Trends in Ecology & Evolution. 2010;25(2): 90–8.
22. Aronson MFJ, Lepczyk CA, Evans KL, Goddard MA, Lerman SB, MacIvor JS, et al. Biodiversity in the city: key challenges for urban green space management. Frontiers in Ecology and the Environment. 2017;15(4):189–96.
23. Schell CJ, Guy C, Shelton DS, Campbell-Staton SC, Sealey BA, Lee DN, et al. Recreating Wakanda by promoting Black excellence in ecology and evolution. Nature Ecology and Evolution. 2020;4(10): 1285–7.
24. Reese G, Jacob L. Principles of environmental justice and pro-environmental action: a two-step process model of moral anger and responsibility to act. Environmental Science and Policy. 2015;51:88–94.
25. Fletcher M-S, Hamilton R, Dressler W, Palmer L. Indigenous knowledge and the shackles of wilderness. Proceedings of the National Academy of Sciences of the U. S. A. 2021;118(40):e2022218118.
26. Hernandez, V. Indigenizing restoration: Indigenous lands before urban parks. Human Biology. 2020;92(1):37–44.
27. Krasny ME, Lundholm C, Shava S, Lee E, Kobori H. Urban landscapes as learning arenas for biodiversity and ecosystem services management. In: Elmqvist E, Fragkias M, Goodness J, Güneralp B, Marcotullio PJ, McDonald RI, et al. (eds.) Urbanization, Biodiversity and Ecosystem Services: challenges and opportunities. Dordrecht: Springer Netherlands; 2013. p. 629–64.
28. Muvengwi J, Kwenda A, Mbiba M, Mpindu T. The role of urban schools in biodiversity conservation across an urban landscape. Urban Forest & Urban Greening. 2019;43(Jul):126370.
29. Hunter MR, Gillespie BW, Chen SY-P. Urban nature experiences reduce stress in the context of daily life based on salivary biomarkers. Frontiers in Psychology. 2019;10:722.
30. Corraliza JA, Collado S, Bethelmy L. Nature as a moderator of stress in urban children. Procedia Social and Behavioral Sciences. 2012;38:253–63.
31. Moore ML, Milkoreit M. Imagination and transformations to sustainable and just futures. Elementa. 2020;8(1):1–17.
32. Desmond M. Capitalism. In: Roper C, Silverman I, Silverstein J (eds.) The 1619 Project: a new origin story. London: One World; 2021. p. 165–85.
33. Taylor DE. The Rise of the American Conservation Movement: power, privilege, and environmental protection. Durham, NC: Duke University Press Books; 2016. 496 p.
34. Schmidt C, Garroway CJ. Systemic racism alters wildlife genetic diversity. Proceedings of the National Academy of Sciences of the U. S. A. 2022;119(43):e2102860119.
35. Schell CJ, Dyson K, Fuentes TL, Des Roches S, Harris NC, Miller DS, et al. The ecological and evolutionary consequences of systemic racism in urban environments. Science (80-.). 2020;369(6510):eaay4497.
36. Locke DH, Hall B, Grove JM, Pickett STA, Ogden LA, Aoki C, et al. Residential housing segregation and

urban tree canopy in 37 US cities. npj Urban Sustainability. 2021;1(1):15.

37. Grove M, Ogden L, Pickett S, Boone C, Buckley G, Locke DH, et al. The legacy effect: understanding how segregation and environmental injustice unfold over time in Baltimore. Annals of the American Association of Geographers. 2018;108(2):524–37.
38. Nardone AL, Casey JA, Rudolph KE, Karasek D, Mujahid M, Morello-Frosch R. Associations between historical redlining and birth outcomes from 2006 through 2015 in California. PLoS One. 2020;15(8):e0237241.
39. Lane HM, Morello-Frosch R, Marshall JD, Apte JS. Historical redlining is associated with present-day air pollution disparities in U.S. cities. Environmental Science & Technology Letters. 2022;9(4): 345–50.
40. Nardone A, Casey JA, Morello-Frosch R, Mujahid M, Balmes JR, Thakur N. Associations between historical residential redlining and current age-adjusted rates of emergency department visits due to asthma across eight cities in California: an ecological study. Lancet Planetary Health. 2020;4(1):e24–31.
41. Childers DL, Cadenasso ML, Morgan Grove J, Marshall V, McGrath B, Pickett STA. An ecology for cities: a transformational nexus of design and ecology to advance climate change resilience and urban sustainability. Sustainability. 2015;7(4): 3774–91.
42. Pickett STA, McGrath B, Cadenasso ML, Felson AJ. Ecological resilience and resilient cities. Building Research & Information. 2014;42(2):143–57.
43. Meerow S, Newell JP, Stults M. Defining urban resilience: a review. Landscape and Urban Planning. 2016;147:38–49.
44. Vale LJ. The politics of resilient cities: whose resilience and whose city? Building Research and Information. 2014;42(2):191–201.
45. Seddon N, Chausson A, Berry P, Girardin CAJ, Smith A, Turner B. Understanding the value and limits of nature-based solutions to climate change and other global challenges. Philosphical Transactions of Royal Society B Biol. Sci. 2020;375(1794):20190120.
46. Hobbie SE, Grimm NB. Nature-based approaches to managing climate change impacts in cities. Philosphical Transactions of Royal Society B Biol. Sci. 2020;375(1794):20190124.
47. Anguelovski I. From toxic sites to parks as (green) LULUs? New challenges of inequity, privilege, gentrification, and exclusion for urban environmental justice. Journal of Planning Literature. 2016;31(1):23–36.
48. Wolch JR, Byrne J, Newell JP. Urban green space, public health, and environmental justice: the challenge of making cities "just green enough." Landscape and Urban Planning. 2014;125:234–44.
49. Anguelovski I, Corbera E. Integrating justice in Nature-Based Solutions to avoid nature-enabled dispossession. Ambio. 2022;52(1):45–53.

Index

Note: Tables and figures are indicated by an italic *t* and *f* following the page number.